Maine Feldspar, Families, and Feuds

A History of Mining and the Feldspar Industry, Oxford County, Maine

Chapters by

Laura Wiley Ashton
Bruce Barrett
Jan Brownstein
John Davis
Barry Heath
Vandall T. King
Lorraine Parsons
Frank C. Perham
Addison W. Saunders

Edited by

Vandall T. King

2009

The editor and senior author acknowledges generous financial support from Larry Stifler and Mary McFadden and help from New England Historical Publications.

Additional financial support has come from a variety of sources including:
Bob Brown, Russell Clement, Carole Clement, Gary Freeman, Louise Jonaitis, Richard"Yukie" King, David McDermott, Pam McDermott, Janet Nemetz, and Peter Stowell

Copyright @ 2009 New England Historical Publications

Library of Congress Cataloging-in-publication Data

King, Vandall T., 1948-

    Maine Feldspar, Families, and Feuds. Oxford County Mining History / Vandall T. King.

    Includes bibliographical references and index.

    ISBN 978-0-615-29183-3 (hard cover)

1. History – Regional, Maine 2. History – Industrial, mining, minerals, feldspar, mica, beryl, gems
3. History of Science – mining, exploration 4. Technology – processing technology
5. Porcelain and pottery – manufacturing.

Printed and bound in China by Lithopak Ltd

ISBN 978-0-615-29183-3

$49.95
ISBN-10:   0-615-29183-X
ISBN-13: 978-0-615-29183-3

54495>

9 780615 291833

Dedication

Good friends are always important to us, but sometimes good friends
have also made history and such is the case with

Jane Charlotte Perham and Frank Croydon Perham.

Jane and Frank are well-known in Maine for their many contributions.

Jane is a Gemological Institute of America trained gemologist and
owner of Perham's of West Paris. She has written books and articles on
Maine gem and mining history which are fresh and vibrant covering
many details never before published. Her writing will always be as close to primary
sources as possible. As the "record keeper", she is always the authority
people turn to when they want to know "The Rest of the Story".

Frank is a Bates College-trained geologist. He is well known for his mining successes
and his community service. His advice is invaluable for the many naturalists who wish to
explore Maine's treasures. Through his experiences at many Oxford County localities,
Maine has continued to receive international attention for mineral wealth.
Perhamite, first found at Newry and Greenwood, is named in his honor.

It is with great pleasure that the editor acknowledges their friendship and their
contributions to history of gems and mining in Oxford County, Maine.

Chapters

Introduction

There are many philosophies on writing history. A book about a large densely populated place is frequently dominated by events and politics, while that of a very small place is primarily dominated by genealogy. A large town or city just has too much to research and write about and most of the "little" people are lost. For these reasons, the histories of Bangor, Portland, and Augusta have not been frequently updated. The structure of a good local history book generally includes writing about: the founding fathers, development of infrastructure (schools, churches, factories, railroads, clubs and associations), list of prominent people and fami-

is the perception that specialized professions have dull or routine development. Industries which feature labor strikes or armed conflicts, such as the shoe manufacturing industries, sometimes have caught the attention of authors.

Another problem for historians is that some personalities avoid public scrutiny. When the famous *Youth's Companion* author from Norway, Dr. Charles Asbury "C. A." Stephens, wanted to do geriatric research in his "laboratory" near the south end of Lake Pennesseewassee, he wanted to insure his building, but did not want to reveal the nature of his laboratory work. He asked his insurance agent, Freeland

C. A. Stephens' Laboratory and Residence, Norway. c1910 postcard.

lies with short biographies, list of veterans of wars, particularly the Revolutionary and Civil Wars, and unusual events such as the burning or flooding of the town, peculiarities of geography, etc. The unusual businesses, particularly those that did not have a large labor force may get only a few lines. There should be an index, but there almost never is one. Pictures are all important, but mostly the authors like to read the text, while readers like to look at pictures. Maps are important, but rare.

The problem of histories of large, densely populated places is that individual lives are seldom recorded unless they have a position of power and influence in town politics. Genealogists and family historians have always been frustrated by large cities, or even small cities, for that matter. Historical research and writing is ordinarily the work of one or a few people and the information included is limited by the individuals' inquisitiveness, patience, bias, and energy. Biographies must include the richness of a person's associates, otherwise we might conclude that the biographee had few friends and one is left to wonder how the person could have become so influential.

With the exception of lumbering, ship building, and lobstering, many of the individualized Maine professions receive little attention from historians, perhaps because there

Howe, Sr., "What is the highest insurance rate?" Howe replied that a blacksmith shop had that rate. Stevens then said: "I'll pay that rate."(Emily Foster, personal communication, 2000). Similarly, when most people think of the history of insurance, one imagines a dreary lifestyle. When George S. Jackson wrote his history of Maine's Union Mutual Life Insurance Company of Portland, he managed to include enough life in his story to insure the reader's attention. Many motivations of Maine's past leaders can be obtained from a detailed reading of *A Financial History of Maine* by Fred Jewett. Similarly, Dean Lunt's *Here for Generations*, the history of the Bangor Savings Bank, is essential reading for those interested in Bangor, Maine's mid-nineteenth century history.

Miners, spruce gum harvesters, trappers and hunters, bands and orchestras, lightning rod manufacturers, etc. have left a different footprint than the large manufacturing businesses and because of their specialized nature, have been misunderstood or ignored by historians writing about a town. Such benevolent neglect is better than the frequent newspaper articles that try to paint a picture of an activity for which the reporter has insufficient background in, in order to get the story correct. For example, too many writers have assumed that just because someone had been in a particular business for a long time, even for a lifetime, that the active person made a "comfortable living" or that business was "good". Almost always, life was a struggle.

Mining reporting, in particular, is tainted by local folklore, Hollywood portrayals, contrived glimpses from literature, and imaginative and inexperienced writers frequently fill in the blanks. One reporter, born in Oxford County, worked for the Associated Press, a news distributing company, and was proficient at writing a good story when there were missing details or those that were known were too boring. In the 1960s, this writer was a sports reporter for the *Bangor Daily News*. Basketball managers would call in sports results to the office. With the bare facts of total score and that by quarters, who scored more than five points in a quarter and whether any athletes scored more than 10 points in the game, and the vital answer to "Did anything unusual happen?", it was possible to write an exciting two paragraph summary of the game – "The fans cheered!" Many historical writings may be found that take fragments of information and boost them to unsupportable heights. It is still frequently repeated that Oxford County has one of the largest number or highest concentration of mineral species in the world. In fact, one mine in New Jersey exceeds Maine's entire total of mineral diversity, not to mention concentration. The number and size of quarries in Oxford County are few and small when compared with those of other states. They are worthy to write about because they are "our" localities.

The writing of the history of a place such as the Bumpus Quarry in Albany is really no different from the above concerns of writing about a town. A simple "natural history" of the locality's minerals would tell us very little. In fact, most *natural history,* or for that matter genealogical tabulations, are little more than cataloging, unless there is a discussion of how the subject matter changed through time. For example, Oxford County's forests were primeval until the end of the eighteenth century. By the end of the nineteenth century, many towns had relatively few trees. With the abandonment of subsistence farming and the migration to town centers, almost all of the farm area once clear cut is now in the process of reforestation. Many Maine places have a hundred times more trees today than in the late nineteenth century to mid-twentieth century, when a farmer formerly might be able to see the houses of all of his neighbors from his own house to the horizon. Oxford County has a different landscape today than it did a hundred years ago. Through time, cleared land has become covered with a succession of bushes and trees. To paraphrase Maine's poet, Edna St. Vincent Millay, the fields "went in little ways".

The aspects of the Bumpus Quarry include: the people who lived near it, the natural history of the rocks that were of interest and why they were of interest, the efforts of various people to harvest minerals for use in manufacturing, and how did the actions of the people actively involved in

mining relate to other Oxford County's residents? As it turns out, there were significant connections between the Bumpus Quarry's minerals and international events, although the thread connecting them is very fine at times. The understanding of why people did what they did needs a summary of the outside influences. The Bumpus Quarry can't be made into an international spectacle, although, at one time, it did hold the world's record for the size of one of its beryl crystals, although the first reports which actually made the claim in the 1920s were wrong. The Bumpus Quarry did not have the world's largest beryl crystal until 1950. The Bumpus Quarry

Upper view: Lake Pennesseewassee, Norway showing open fields now long overgrown. Noyes Mountain rear right, Patch Mountain center left. (1910s postcard).
Lower view: with artistic license emphasizing open fields and painted-in rock walls, hills, etc. from what appears to be the exact image of the earlier postcard. (1940s postcard)

was an early leader in the local feldspar mining industry and for almost all of the time that feldspar mining was important in Oxford County, the Bumpus Quarry was active.

Unfortunately, history frequently is researched ten to twenty years too late. The author remembers talking with many of the now historical Maine mining personages. They were full of stories and they entertained friends and visitors, but the stories were not written down and only impressions remain. There are now no opportunities to ask questions that were never thought of a the time.

First-hand experience is the best for historical reporting and, when recorded while "fresh", is reasonably close to what happened. However, three different personal accounts the author has heard, concerning a particular day at the Dunton Quarry in Newry, separately recorded when "fresh" by the three principals, do not particularly resemble each other except with an agreement of location and that tourmaline was found.

Memories fade and change and the passage of time frequently results in new memories replacing old. Twice-told stories, likewise, color our thinking about history and while children sometimes have a fair idea about their parent's history, grandchildren rarely have the true perspective. A few historians express the opinion that we can never know what really happened, but that opinion doesn't mean that we shouldn't try to capture the threads of what happened. Unfortunately, many historical accounts contain words rich with opinion. Words such as "obviously", "certainly", "eagerly", etc. are frequently seen in print, but those words must be red flags to any faithful historian and reader. This work also has some of these opinion revealing words and the writer has tried to be careful in choosing them.

The author has long been interested in mining and minerals of Oxford County. As mining history became a more important topic, books, articles, documents, correspondence, court records, and folklore became more valuable. During the late 1950s, his friends, Ben and Mary Shaub,

Albany portion of Songo Pond, 1910s postcard.

Elliott and Bartlett Spool Mill, Lynchville, 1910s postcard.

began researching a history of Maine gem mining and gem crafting, but their plans never reached fruition. In 1991, Ben and Mary gave the author their library of tape recordings of old timer's interviews and unpublished photographs of Oxford County miners and mines to supplement his own research which had a broader focus. While the Shaubs took a folklore approach, it has been found that the factual historical evidence remaining is very rich, although equally as widely dispersed as it is extensive. The author has already provided summaries of Maine's mineralogical natural history (1994, 2000), as well as its mining history, including

much that relates to Oxford County. Nonetheless, Oxford County's mining heritage is very rich and the Bumpus Quarry is only one star in the County's constellation of quarries.

Albany is a small town. Today, except for one small store, there isn't a genuine place where one can spend money. Albany's post office opened January 9, 1815 and closed October 31, 1903. There was also a Lynchville post office [August 14, 1882-July 24, 1895], but it was discontinued as it was so close to the East Stoneham and North Waterford post offices. In 1930, everyone's mailing address in the town was either R.F.D. (Rural Free Delivery) Bethel, North Waterford, or Lovell. There was one factory, the Elliott and Bartlett Spool Mill, beginning in 1886, but the mill was no longer listed in the *Maine Register* in 1919. If one looks at the snapshot afforded by the *Maine Register* for 1900, one would have thought the town was thriving. Famed Civil War veteran, Amos G. Bean, was still the postmaster of the Albany post office (located on Picnic Hill Road just off the Hunts Corner Road and south of the Congregational Church. (The "Congo" Church was frequently without a pastor and usually was supplied with one in a time-sharing agreement with Waterford' church.) Bean was also Town Clerk, Town Justice, on the School Committee, and owned the General Store. Charles P. Pingree was also on the School Committee. Wallace B. Cummings, who was Wallace E. Cummings' cousin and uncle to Laura Josephine Cummings, was Town Treasurer. Some "manufacturers" included: Stephen W. Libby (lumber, cider, and grist mill), H. I. Bean and L. J. Andrews (carpenters), E. T. Judkins and John Flint (blacksmiths), L. H. Burnham (lumber), Elliott and Bartlett (spools for spinning mills), McAllister & Sons (spool stock and dowels), Flint & Fernald (lumber and spool stock), Hilborn and Herrick (lumber and spool stock), and Thomas G. Kimball (mason). The town library had 200 books and the Round Mountain Grange met on the first and second Saturdays of the month.

One of the prominent geographical features of Albany is Songo Pond. The very northernmost shore area is in

Bumpus Quarry, Albany. c1975.
Courtesy John Kimball.

Bethel, but almost all of the Pond is in Albany. The Crooked River extends through the western side of Albany and has a famous natural phenomenon which was formerly a tourist interest, the Kettle Holes. There are erosional features carved in rock due to the swirling action of stones caught in eddies of glacial melt water.

Several tiny places, mere ledges nestled on knolls or hills, some say mountains, influenced Oxford County's people, since Maine became a State. The ledges contained minerals, not enough to make one wealthy, but good enough to help make a living. One mineral, feldspar, enjoyed a high reputation in the nation's porcelain industry. When Maine was the nation's leading feldspar producer in the early 1900s, this mineral was hand-selected from the Keith Quarry, Auburn, to make the dinnerware for the White House during president Woodrow Wilson's term (unpublished notes, 1938, Maine State Library).

The rare minerals found in Maine's ledges also contributed to scientific studies as early as 1823. There were frequent pilgrimages by internationally known professors and scientists even from the earliest days of Maine mining. By the mid-1920s, a new industry emerged in Oxford County, feldspar mining for use in ceramics and innumerable other uses. Initially, dozens of local people were employed. By decade's end, perhaps hundreds had participated in some aspect of the effort to mine raw rock, refine it, and sell processed feldspar. In the course of half a century, thousands of Oxford County residents had experienced working in some aspect of the industry, even if peripherally, in the construction, vehicle repair, gemstone, or tourist businesses. While no one of these tiny mineral deposits can be said to take precedence, the subject of

Rose Quartz, Bumpus Quarry, Albany.
Addison Saunders Collection.

this history will concentrate as much as possible on one of these localities: the Bumpus Quarry in the small township of Albany. Many will note the absence of details among others of the famous mineral deposits in the County: Mount Mica, Paris; Dunton Gem Quarry, Newry; Lord Hill, Stoneham; Deer Hill, Stow, Noyes Mountain Quarry, Greenwood, Bennett Quarry, Buckfield, and numerous other "famous" places. These locations will have their own story told elsewhere, although none of them will have been excluded here, in the telling of what impact mining had on central Oxford County. For exactly half a century, the West Paris mill was supported by the local quarries and the legacy of local mining continues to this day. There were successes as well as tragedies. Occasionally, concerns over miniscule amounts of money tore apart families.

The development of mineral-producing quarries, along with a manufacturing business, a feldspar grinding mill, located in West Paris, attracted many of Oxford County's families to become miners. Some families, such as Bumpus, Douglass, Kimball, Perham, Stearns, Tamminen, Wardwell, and others were either quarry owners or leased mineral deposits. Some such as Nevel, Pechnik, Penley, Perham, Wiley, and others started small businesses mining and selling their mineral products to the mill and in turn countless individuals from families such as Immonen, Inman, Koskala, Oja, Ross, and others labored in the mines and/or for the mill. Family ties across the County meant that although there were a variety of surnames represented, many miners could trace their local ancestry back four or five generations to common ancestors as migration into the region was slow after initial settlement, until the latter part of the nineteenth century.

The Bumpus Quarry is one of Oxford County's most famous mineral localities and this book contains its story. This locality is located south of Bethel, along Valley Road (State Route #5), approximately 1.5 km south of the Town House in Albany township and is mapped on the USGS 7.5' East Stoneham Quadrangle at about 44.312° N, 70.781° W. [7] It has been visited by countless naturalists, including many of the world's famous mineralogists. It has a unique and complicated twentieth century history with participation by many of the County's principal miners. The County's quarries left very small "footprints", many no greater than typical house or farm building foundation, but were influential in the local economy and history. Relatively few quarries were over 30 meters long. Counted as a whole, mining helped in a small way to keep Oxford County's economy from col-

Beryl, Bumpus Quarry, Albany.
Ann and Bill Cook Collection.

lapsing even lower than it did, during the Great Depression of the 1930s.

Mining is a specialized profession and, because of that fact, many details are included to introduce that profession to the reader. A quarry's life also includes the people interested in it, not just the minerals it contained. For this reason, biographical information of significant personalities

Oxford County Ores: M Mica, Albany; F Feldspar, Buckfield;
P Pollucite, Newry; T Tourmaline Specimen, Newry; S Spodumene,
Newry; G Gem Tourmalines, Paris; B & Q Beryl in Quartz, Albany.

is included, whenever possible. Additionally, a single mineral locality has a relationship to the other mineral localities around it and there is included information about the various quarries surrounding the Bumpus Quarry.

The Bumpus District extends into the towns and townships of Albany, Batchelders Grant, Gilead, Mason, and Waterford with very similar mineral occurrences as well as cultural roots of its miners. [63] The Rumford Mining District to the north of the Bumpus District includes Newry and Rumford, famous for tourmaline and rare minerals. [90,93] Geographically and geologically, the Bumpus Mining District deposits are distinctly different from the mineral deposits immediately to the east in the adjacent Paris Mining District, which includes Greenwood, Woodstock, West Paris, Sumner, Peru, Hartford, Paris, Norway, Buckfield, and Hebron, having localities rich in gem tourmaline and rare minerals, as well as feldspar and mica.

To the west of the Bumpus District, the towns and

Albany Town House, 2007.

Map of the Bumpus Mining District

1 Bumpus Quarry
2 Douglass Quarry
3 Farwell Mt Locality
4 Pingree Ledge Quarry
5 Bennett Prospect
6 Donahue Prospect
7 Songo Pond Quarry
8 Rattlesnake Mt Locality
9 Square Dock Mt Locality
10 Fleck Quarry
11 Wardwell Quarries
12 Holt Prospect
13 French Hill Locality
14 Stifler-McFadden Quarry
15 Wentworth Quarry
16 Scribner Quarries
17 Stearns Quarry
18 Johnson Quarry
19 General Electric Quarry
20 Flint Prospect
21 Round Mt Area
22 Thunder Mt Prospect
23 Knight Quarries
24 Burnell Hill Quarry
25 Amethyst Locality
26 Beech Hill Quarry
27 Saunders Quarry
28 True Quarry
29 Stearns Hill Quarry
30 Joseph Sanderson
               Locality
31 Bear Mt Quarry
32 Oliver Stone Locality
33 Peabody Mt Quarry
34 Brown Quarry
35 Wheeler Quarries
36 Peaked Hill Quarries

1 cm ≈ 3 km.

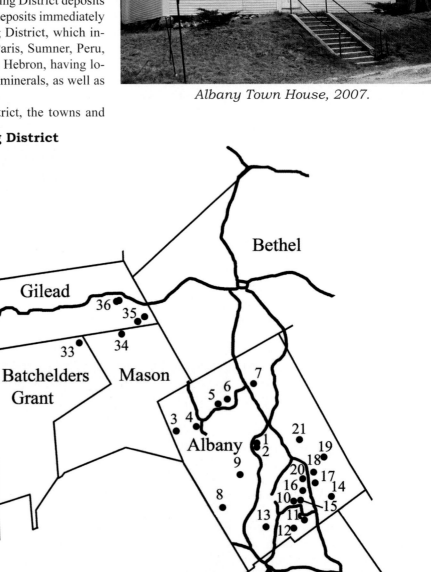

townships of Stoneham, Lovell, Stow, Sweden, Denmark, and Fryeburg, mostly well-known for amethyst occurrences, constitute an identifiably interesting mineral area: the Stoneham District. The somewhat similar mineral occurrences ranging southeast to Auburn and Minot, and to the extreme in Topsham, Georgetown, and Warren, have unique geological origins, as well and also have uniquely important cultural and economic differences. While it would be possible to include Stoneham in the Bumpus District or Minot in the Paris District, etc., the author has chosen to make these districts as "uncomplicated" as possible.

Round Mountain Grange Hall (left) and Albany Congregational Church (right), 2008.

Oxford County's Ore Minerals

The primary commercial "industrial minerals" now utilized in Oxford County are sand and gravel and far more money has been received from this commodity than any other in the County's history. Gravel is not mined, however, and, as an industrial mineral, it is normally used as found, although it may be sifted and sorted. In some places, particularly California, gravel pits are important sources of gold. The true "hard rock" mineral quarries in Oxford County have produced: feldspar, mica, quartz, rare ores, mineral specimens, and gems.

Feldspar has many uses and the purest and highest grades have been used in making ceramics or "non-abrasive" cleaners. Since the 1930s, feldspar has been an increasingly important constituent in glass. Mica was once used in fireproof windows in lanterns and stoves, but that usage has become obsolete and mica is now mostly used as an electrical insulator or weatherproof filler in paints and construction materials, but decorative uses are known in wallpaper and simulated snow on paper products, etc. Rare ores that have been recovered in Oxford County include lepidolite and spodumene as sources of lithium, etc. Columbite minerals have been important in providing niobium and tantalum used in specialty steels, particularly for surgical instruments, nuclear power plants, etc. Pollucite and lepidolite have been sources of cesium and rubidium used in chemistry, electric eyes, radio tubes, etc. Quartz has been used as an ore in making specialty flint glass, high-temperature glass products, or telescope mirrors, as well as in products as mundane as sandpaper. (River sand is unsuitable for sandpaper as the grains are rounded and quartz sandpaper requires sharp broken edges.) Gems, of course, are used in jewelry, while mineral specimens are found in private and public museum collections. Other uses of the County's minerals will be discussed in the following sections.

The story of the Bumpus Quarry, in particular, is centered on the mineral beryl, but the everyday "bread and butter" was paid by its feldspar output. Beryl was the ore of beryllium and is also represented in various Oxford County locations by several gems: aquamarine, golden beryl, and morganite. Beryllium metal was used at first as a component in specialty non-sparking and light weight alloys, but it later found a use in light weight alloys and atomic bombs. Beryl is no longer an important ore of beryllium, but the switch away from beryl is relatively recent and occurred after the Bumpus Quarry ceased to be operated. (See Appendix D).

The reader interested in Oxford County will find a twentieth-century slice of local history in the mining industry. The miners were generally not leaders of people, although several miners eventually held political office and/or were leaders in community organizations. Harold Perham was one of the notable exceptions, being elected to the Maine House of Representatives for Paris in 1929 and 1931. Many miners were volunteers in local services, fire departments, etc. Superimposed on the pure cultural history is pure natural history. Oxford County has enjoyed a concentrated attention by both academic and amateur mineralogists since the discovery of Mount Mica just after Maine achieved Statehood. [124] By the turn of the nineteenth century to the twentieth, the rise in amateur mineralogists resulted in increased tourism as well as an increased interest in nature in Maine. The Bumpus Quarry has remained an attraction because of its minerals and its history is naturally of interest.

Notes Concerning the Text

The transition from using English measurement units to metric units is incomplete at this writing. While the mixing of different system units in the text is "inelegant", it was decided to keep older units whenever quoted, because the text would be constantly interrupted by conversions or insertions. The reader can make "soft" approximate conversions if necessary: 1 mile = 1.6 km; 1 yard = about 1 meter; 1 foot = about $^1/_3$ meter; 1 inch = 2 ½ cm; 1 gallon = almost 4 liters; 1 quart ≈ 1 liter; 1 pound ≈ ½ kilogram; 1 avoirdupois ton = 10% less than 1 metric ton; one long ton ≈ about 1 metric ton.

The text contains closely spaced citations to substantiate specific facts. The citations vary from personal communications, to records in the Oxford County Registry of Deeds, scientific journals, popular magazines, manuscripts, correspondence, etc. Newspaper clippings are cited only in the text. Short appendixes contain detailed information that would otherwise interrupt the flow of the main text. Source data are included with the illustrations, where known. Uncredited illustrations or specimens are from the senior author's collection.

The end notes contain bits of history or expanded information about personalities. The appendixes contain summaries of the business histories of Maine feldspar, mica, and beryl. Depending on the year, an Oxford County mica miner might have been better off financially than a feldspar miner or *vice versa*. Developments in other mining states affected Oxford County's fortunes, but the Great Depression of 1929-1937 was one of the greatest factors in the County's mining industry. Descriptions and illustrations of minerals have been published separately. Two earlier volumes containing descriptions and illustrations of minerals as well as historical sections for areas outside of the Bumpus Mining District have been published: *Mineralogy of Maine, volume 1* by V. King and E. Foord (1994) and *Mineralogy of Maine, volume 2* by V. King (2000). Additional historical works currently in progress include a full-length biography of George Howe and George "Shavey" Noyes, for which manuscripts have been circulating among knowledgeable local historians since

1998; a history of the Paris-Greenwood-Buckfield-Hebron mining district with emphasis on gem mining and historical influence of these localities on the general history of science, and a history of mining in Newry and Rumford as well as the gold fields in Byron – a project which started everything off.

Acknowledgements

The author is grateful to many people, past and present, for sharing information about the history and mineralogy of the Bumpus Quarry and adjacent areas. Particular thanks and appreciation go to: Laura Wiley Ashton, Don Bennett, Randy Bennett, John Bradshaw, Bob Brown, Jan Brownstein, Ava M. Hutchinson Bumpus, Irene Ross Card, Levi Chavarie, Priscilla Stearns Bryant Chavarie, Roger Clapp, Carole Clement, Russell Clement, Ben Conant, Cathy Kimball Cox, Dennis Creaser, John Davis, Dana C. Douglass, Jr., Barbara Douglass, Roberta Gordon, Sidney Gordon, Robyn Greene, Richard Hauck, Dennis Holden, Ron Holden, Gary Howard, Lorna Cummings Howard, Stanley Howe, Barbara Stearns Inman, Eleanor Inman, Milt Inman, Dan Jonaitis, Louise Jonaitis, John Kimball, Nathaniel King, Richard "Yukie" King, Tom Klinepeter, Mickey Liimata, Seabury Lyon, Jim Mann, Stuart Martin, Phil McCrillis, David McDermott, Pamela McDermott, Mary McFadden, Janet Nemetz, Lorraine Parsons, Frank C. Perham, Jane C. Perham, Joseph Perham, Mary Perham, Stanley I. Perham, Bobbi Peters, Francis Pike, Bruce Richards, Linda Ashton Richards, Dean Rippon, Winsor Rippon, Wayne Ross, Ruth Brown Salo, Addison Saunders, Ben Shaub, Mary Shaub, Sally Irish Spencer, Larry Stifler, Katie Tamminen, Nestor Tamminen, Woody Thompson, Neil Wintringham, and Ray Woodman. In 1990, Benjamin Martin Shaub [January 12, 1893-March 23, 1993] and Mary Sumner Church Shaub [March 30, 1913-March 18, 2004] gave the author a portion of their Maine mining and mineralogy photograph collection and this gift has been of enormous benefit. Similarly, Sid Gordon's photographic archives have been invaluable in telling the story of Oxford County's mining and milling history. Raymond Woodman, Robyn Greene, Janet Nemetz, and Lorraine Parsons, traveled with the author collecting information and, oftentimes, "opened doors" or discovered important historical items that the author might have otherwise missed. Janet Nemetz and Woodrow Thompson acted as early outside readers and editors. The author is particularly grateful for grants from Larry Stifler and Mary McFadden, Gary Freeman, Bob Brown, Louise Jonaitis, Richard "Yukie" King, Russell and Carole Clement, David and Pam McDermott, and Peter Stowell which helped make this book a reality.

Bumpus Quarry (water-filled = black), showing Route 5 and access roads (left), Maine Information Survey, 2007.

Some Oxford County Mining Scenes - Perham Quarry 1925-1926

Top: Perham Quarry, January 2, 1925
Second: Perham Quarry Spring 1925
Third: Perham Quarry 1926
Bottom: Ore cart with straight track
Photos by Vivian Akers Courtesy Sid Gordon.

Top: Perham Quarry, January 2, 1925
Second: Perham Quarry Spring 1925
Third: Mill from Perham Quarry ~1925
Bottom: Ores carts and frog junction track
Photos by Vivian Akers Courtesy Sid Gordon.

Chapter One: A Day in the Life of an Oxford County Mineral Quarry

The richness of Oxford County's mining history includes the details of what the work was like. Every day of any miner's employment is really different. The weather is always a concern. As few Oxford County mineral quarries had tunnels, on particularly rainy, snowy, cold, perhaps even strongly blowing days, bad weather might mean that no work would be done. The crew would be dismissed and no wages would be paid. Fringe benefits were non-existent, as most Oxford County contract miners were "self-employed". The same was true for any of the outdoor trades: logging, house painting, road construction, farming, etc. Quarrying days were also governed by the calendar. Summer work might continue for 12 hours or more, usually five days, but also six days, a week. Sunday work was almost unknown. In the 1920s and 1930s, wages were about 35-45 cents per hour. A miner might be paid nearly twice as much in the summer as in the winter just because of the increased number of hours that could be worked. Many quarries shut down for four months in a normal winter, but some struggled on during inclement days. Miners, as well as farmers, frequently found work lumbering during winter.

Above: Richard Edwards (in yellow shirt) at Mount Mica Quarry, June 2004 adjusting water and air hoses at the start of the day.

These few lines hardly explain the duties of the various personnel, some of which were specialized. The "day" described here would vary from quarry to quarry and is not the story of any one locality. The foreman in most Oxford County quarries was also a worker who saw to details of nearly every job. The shift would start with readying the equipment, assuming there were no pieces of broken equipment needing to be fixed, before normal operations could continue. If there were drilling to be done, the air compressor would be filled with fuel and started. Someone had to remove the jackhammer from the equipment shed and bring out the air pressure hoses and hook them to their couplings. Of course, there were many untidy mines where all equipment remained exposed to the weather overnight.

A small to large amount of water might accumulate on the floor of the work area each idle weekend, if not each night. The water pump would be started, although a few quarries employed long hoses as siphons. If a quarry were being worked during months with snow, the low temperature combined with a temperamental piece of equipment might mean that a miner had to spend considerable time just starting compressors, excavators, trucks, etc. before actual mining could begin.

In earlier days, before carbide tipped drills and removable drill bits, a quarry had to have a blacksmith among its crew members, who would sharpen the steel drills when they became dull. Blacksmiths would not only shoe work

Double-jacking demonstration at the Noyes Mountain Quarry, Greenwood c 1900 (Note Sunday-best clothes.) Tim Heath with hammer, George "Shavey" Noyes with drill.

Hanness Hakala drilling at Mount Marie Quarry in Paris, c1949. Courtesy Wayne Ross.

horses or mules, he was also the quarry's principal mechanic. When there were no smithing to be done, he might also work as a driller, mucker, etc. Some days, the drilling would occur in the morning or drilling might be postponed until the next day.

Hand-drilling, generally considered obsolete in Maine's commercial quarries even by the turn of the nineteenth century, consisted of one miner hitting the drill with a sledge hammer. The drill holder, frequently the youngest member of a crew, sometimes a teenager or younger in a family operation, had the easiest, but most dangerous job. After the drill was hit by a hammer blow, the holder would rotate the bit a quarter turn or so and the drill would be hit again. After five or so turns, some of the dust would be blown out of the hole, because the dust would absorb the energy of the hammer blows. The two miner combination, one striking-one holding, was called double-jacking, after an old generic nickname for a Cornwall, England miner – "Cousin Jack". Mines frequently had two miners alternatingly hitting the drill steel with a third holding it, a method also called double-jacking. (Single-jacking was an individual miner holding a short drill and who used his other hand to hit the drill using a short-handled sledge hammer.) Double-jacking obviously resulted in faster progress. Manually drilled holes were frequently shallow because deep holes, greater than four feet, might have too many accumulated rock chips and dust in them to allow rotation of the drill or to allow any forward advance in the drill. Despite the obsolescence of hand-drilling, part-time miners might do it to make money on their farms when there was no other choice for economic survival.

Bastin (1910) wrote about styles of quarrying at Maine pegmatites and noted of the G. D. Willes Quarry (now referred to as the Trenton Quarry), Topsham:

"The methods of operation at this quarry are somewhat antiquated for a working of this size, the drilling all being done by hand and the blasting by black powder. A tramway carries the waste to dump piles and the good rock to stock sheds, from which it is loaded into wagons and hauled 1 ¾ miles to the mill … near Cathance [railroad] station."

Harold "Red" Perham was able to utilize his mining experience when he "changed" careers,

Drill bit artifact used at Newry 1920s. Note broken points.

to work on the Portland Pipe Line. He was initially hired as a ditch digger, in the Summer of 1941:

"The writer's first nickname on the Portland-Montreal Pipe Line Company in 1941, was 'The Lucky Bastard'. This came about because the Ditch Boss could not use me until the new batch of shovels came in the following day.

Immediately, I hiked up the Right of Way, and run into a red-headed Boss swearing at the Jackhammer Man who had failed to appear on the job that morning.

The first question he fired at me was this – 'Are you the man they sent up to run that Jackhammer?' I said – 'No, sir, but I have had 5 years experience at such work, and can hold down the job.' 'We will soon find out! came back the answer. 'Grab that Jackhammer and put in a Four Foot Hole in that ledge, right where I am pointing this stick.'

I grabbed the Jackhammer, but before putting it in operation, requested that he designate 12 Holes ahead by placing pieces of rocks as markers. I then drove down one two foot hole,

Contract miner, Wendell Pike of Waterford, standing in front of a "boxy" diesel-powered air compressor at Shawnee Peak Ski Area, c1985. Courtesy Francis Pike.

withdrew the steel and began on Hole No. 2. He yelled, 'STOP, and finish the Four foot hole.' So, I threw in a four foot steel into the No. 1 hole and finished it. Then immediately was drilling on No. 2 hole. He recognized the value of the system, and left me alone.

Soon the Ditch Boss walked by, saw me running the Jackhammer, and stopped in his tracks, and said – 'Damn, who hired you as a Jackhammer Man?' I told him – 'A red headed Boss.' 'Well, all I can say is you must be a Lucky Bastard!' came back the answer as he continued on his way. Thus, I became known as 'Perham, The Lucky Bastard.'

Knowing that we were to be working Eleven Hour Days, I judged my work accordingly in order to conserve my strength. At the end of two-weeks I was way-wise to the job and doing well." [187]

On the first day, Harold drilled a record number of holes for the Pipe Line construction project. The management was impressed, but soon objected to the fact that he kept sending his drills to the blacksmith for resharpening before they had dulled. Harold said that he was an experienced hard-rock miner and that the only way to drill holes quickly was to have sharp drills. This wisdom changed the philosophy and rate of progress of the pipeline's construction. Perham continued:

"Soon my brother, Roy, and I were working as a team, and trouble struck one morning. The man was delivering 'sharp bits' from the Blacksmith Shop, started to give me Hell for using too many Drill Bits.

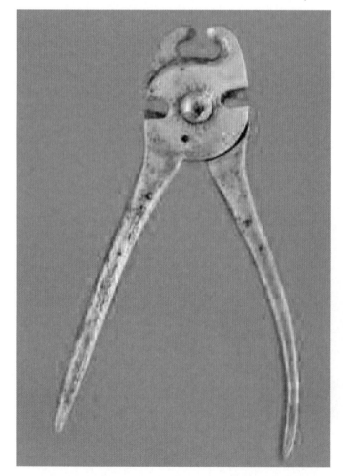

Dynamite pliers with wire crimping and wire stripping design. Made by Hercules Dynamite Company and marked with company name.

Frank Perham using jackleg mount for horizontal drilling at Dunton Quarry, Newry. July 1974.
Courtesy John Marshall

I did not take kindly to the approach, and told him that the Perham Brothers were drilling double the amount of holes that the other Jackhammer Crews were producing, merely because we were 'Hard Rock Drillers' and we were bright enough to send the 'Dull Bits' to the Blacksmith Shop.

At that, he 'blew his top' and said – 'You will either be Fired or Hired by tomorrow morning!' My answer – "Be sure you send some one up with the authority to do it, and with more brains than you are showing.'

Sure enough, the 'big trouble shooter' arrived the next morning, and his first question was 'What is the story about the Perham's drilling double the footage of other Jackhammer Crews?'

I answered – 'Just simple fact, and nothing else. The Perham Brothers are Hard Rock Jackhammer Men and trained never to use Dull Bits. That is the main reason why we get the extra footage. Of course, the man delivering 'Sharp Bits' don't like it, and the Blacksmith don't like it. But you can not get increased footage out of men who have become Rum Dum from being 'battered by using Dull Bits', and from Jackhammers that have

Mid-twentieth century safety poster.
Courtesy Dick Hauck

Chapter One: A Day in the Life of an Oxford County Mineral Quarry   13 become worn out from the same internal battering.'

Furthermore, if you require that the Perham's use Dull Bits, the Perham's will soon be in as bad shape as the other crews, But when the other Jackhammer Crews know that Drilling with Dull Bits is not required, you will soon see an increase of 50 per cent in footage drilled, and a much happier crew.

Also, your Oklahoma men are not entirely to blame, for on their last job they were drilling 'soft rock', and not the hard rock of Maine, New Hampshire, and Vermont [the USA route of the Pipe Line]. The Perham brothers simply had the advantage of Hard Rock Training.

This is the morning that Red Perham was to be either Fired, or Hired. Either way, I prefer to be known as an intelligent and truthful man'

The final outcome! The next New Air Compressor and Jackhammer unit was awarded to the Perham Brothers, and the Drilling with Dull Bits was 'frowned on', as a company policy."

At the quarries, as well as construction sites, motive power for drilling the dynamite holes varied from manual labor, to steam-driven air compressors, and, by WWII, most compressors ran on gasoline or diesel fuel.

The action of compressed air solved many problems. The compressed air pushed a piston to provide the force necessary for the mechanical hammer to hit and rotate the drill, as well as cleaning the hole of dust and chips. The name "jackhammer" was an easy adoption from "jacking", as a drilling term. The drills were usually part of a set of steel rods of varying lengths, usually in two foot increments. Mines that had air compressors were able to drill deeper holes and frequently had drill steels ranging from 2 to 8 feet, rarely longer. By the early 1940s, hardened steel drill bits were available and required fewer sharpenings, By the late 1940s, carbide-tipped drill bits were also available and could be screwed onto threaded drill steel as replacements for dulled bits. George Crooker, superintendent at the Whitehall Quarry (Nevel Quarry), Newry in the early 1950s had an experience when a traveling drill bit salesman visited him on the mountain:

"[George Crooker] remembered that the drilling through spodumene, lepidolite, and other minerals was particularly difficult. Once a carbide drill bit salesman tried to convince the miners to use his product saying that they would perform in a superior fashion to the iron drills that required constant resharpening. Since the disposable carbide drills were $15 each [then more than several day's pay], only one was provided as an experiment, but it was only good for six inches going straight through spodumene. George noted: 'Ordinarily we'd try to drill between the spodumene blades and set short charges as the going was so tough.' The salesman may not have known he'd been tricked." [125]

In the 1930s, drills could be mounted on portable stands called "jacklegs". The mounting was a relatively simple pole, usually with a telescoping interior pole for length adjustment, and frequently had a foot-rest the miner could push against in order to exert more force against the drill. This mounting technique made it easier to drill horizontal holes (Twitty, 2001) Larger quarries might have owned a carriage assembly to hold the drill in place, relieving the personal force a miner had to exert.

In small quarries, the driller would usually hold the jackhammer vertically on the rock above a working face and begin to drill into rock. The first moments of drilling were particularly difficult and required strength to keep the drill in one place until a hole had started to advance into the rock. Drilling was usually dry, without water lubrication, and before the days of OSHA (Occupational Safety and Health Administration) regulations, few drillers used a filter mask to reduce dust inhalation. Many drillers, in particular, developed silicosis from inhaling quartz dust. The safety helmet was also a long time coming and probably no Maine miners used helmets until the 1950s and even by then, the safety helmet probably spent more time in a truck than on the miner's head. All of the drilling mentioned was accompanied by a constant clanking noise of metal against rock, as well as the noise of air releases after each stroke, the noise of the compressor, and, of course, dynamite explosions. Because most miners just put their hands over their ears, ear plugs were similarly unknown until the 1960s. The general absence of safety equipment was commonplace in Maine quarries. Miners wore old clothes, but tried to have good gloves and rugged boots. It is unknown when steel toed footwear came into use, but it is certain there were many Oxford County miners who would arrive at work with little more than what would be generically called "work clothes". Neither the mill nor the small contract mining companies provided employees with safety apparel.

Drillers would start drill holes with short drill steel lengths and replace short lengths with longer ones until the

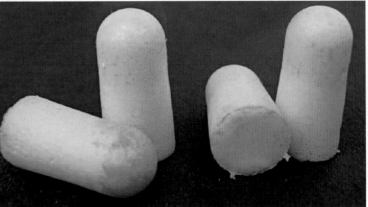

Soft rubber ear plugs, 2008.

desired depth was reached, usually 8 feet. As revealed in the Perham anecdote, the driller would usually not change drill steels until the entire set of holes were completed to the same depth. "Air releases" helped miners by letting some of the compressed air through a central hole in the drill steel which would blow the rock dust out of the hole. An accumulation of dust might clog the hole or even seize the drill steel so that it would not turn or could not be removed from the hole. A single 8-foot drill hole might require between fifteen minutes to a half an hour to drill if done by compressed air power. Very few quarries continued to drill holes by hand to the end of the Depression, and an 8 foot hole in unfavorable conditions might

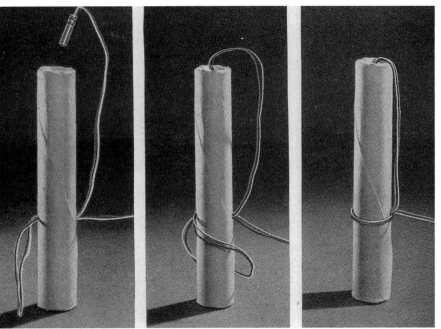

Dynamite, Blasting Cap, and Ignition Wire.
Blasters' Handbook

require an entire day's shift to complete by hand. In Oxford County, small quarries rarely had more than one drill team active at a time, but, again, there were exceptions. If a quarry did not have a compressor, it was a must to have extra drill teams to assure there were enough drill holes to keep production up. Depending on the plans for the day, a number of drill holes would be made, frequently five to ten and occasionally more for "development work" when a lot of rock was to be moved. The drill holes might be placed several feet apart. (By the 1960s, very small quarries, usually operated for gems or mineral specimens, might use relatively small "Cobra", "Pionjär", and other models of portable, gasoline powered drilling machines. These devices had the advantage of relatively low cost, low weight (25-30 kg) and were self-powered, not requiring an air compressor; resulting in another savings. They had the disadvantage of having a relatively short life-span, short drilling path length (~1.5 m), and would never last very long in a quarry producing tons of feldspar, etc.)

Of course, the reason the drill holes were bored was for the use of dynamite. Since the 1950s, the number of safety regulations regarding dynamite have become strict and complicated. By the time the Bumpus Quarry was being worked by Northern Mining Company, dynamite sticks were frequently kept in a separate shed from the equipment and wise miners would store their blasting caps in still a third separate shed, although this was not always the case. Modern Maine quarries now have to have approved and regularly inspected safety-locked magazines, one each for caps and dynamite. Magazines must be separated from each other by

Jim Mann, miner and gemologist with his original
Pionjär portable drill which he used in
Oxford County gem mining, 2008.

*Perham Quarry February 1926
Drillers and blasters with
their equipment and lunch pails
Vi Akers photo. Courtesy Sid Gordon.*

prescribed distances as well as away from work areas. The amount of dynamite is now planned and the exact amount of dynamite, for the day's work, is brought to the working face.

Many Oxford County farmers were already familiar with dynamite and many a local hardware store sold dynamite for agricultural use, especially for tree stump or rock removal. The *Blasters' Handbook* included a chapter on placement of explosives for different kinds of roots, as well as rocks, ledges, soil blasting, and ditches. Common dynamite came in one foot lengths, but large mines and construction sites could take advantage of a variety of sized sticks, cartridges, and canisters. In Oxford County quarries during the 1920s to 1950s, common stick dynamite was universal. The blaster would insert a blasting cap, complete with wire, into the firm but gelatinous interior of the dynamite stick. While the cinema still enjoys the use of lighting a fuse on theatrical dynamite, so that tension and suspense may be increased from the anticipation of an explosion, by the 1920s, most real dynamite was ignited by use of electrical devices connected by wires to a blasting cap inserted into the charge.

The blasting cap's wire would usually serve as the method of lowering the dynamite into the drill hole. After the dynamite charge had been placed, fine sand from the previous blast would be poured into the drill hole to prevent the expanding gases of the explosion from rapidly escaping from the hole instead of pushing against the rock. The fine sand that filled the "stem" of the hole was called stemming. A wooden pole slightly less than the diameter of the hole would be used to tamp down the stemming in

*Proper Method of Placing
and Loading a Charge
Near a Boulder for
Breaking or Moving It with
a Snakehole Shot.
(Blasters' Handbook).*

Blasting Machine (detonator), c1940s.

case a few large particles clogged the hole and there was a cavity in the stemming. (Wood did not spark or prematurely ignite caps or dynamite.) Blasters might use one or several sticks of dynamite per hole, depending if they wanted big pieces of rock or wanted to have extra shattering power to move tons of waste rock a considerable distance to avoid excessive handling.

The potential rock that could be moved from 5 blast holes spaced 3 feet apart and 8 feet deep set back three feet on an ideal exposed face would amount to about 288 cubic feet of rock equivalent to about 25 tons and frequently larger blasts would occur.

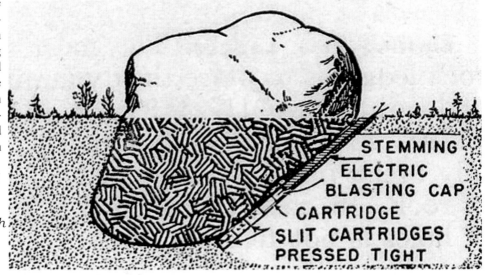

STEMMING
ELECTRIC
BLASTING CAP
CARTRIDGE
SLIT CARTRIDGES
PRESSED TIGHT

Drill hole in feldspar at the Bumpus Quarry.

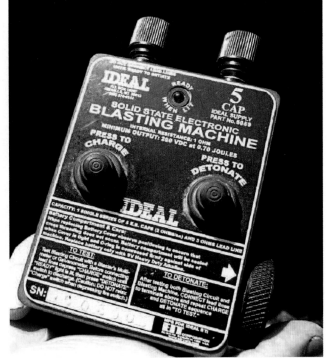

Blasting Machine (detonator), 2008.
Photo courtesy Nathan King.

The blasting cap wires would be unrolled from a reel and lead to a safe place, far from the working face, and all workers would be directed to designated safe locations. When everyone was accounted for, the wires would be attached to a blasting machine (detonator), but poor mines, in the 1920s to the 1950s, might have used a 12-volt automobile battery. The use of an automobile battery was very dangerous as the power drop off between blasting caps might result in some caps not igniting, leaving unexploded dynamite as a hazard in the blast debris. With the workers assembled in their proper locations, a series of warnings had to be issued lest an unexpected visitor or trespasser had entered the premises, a not uncommon problem. The warnings had to be called out loudly, clearly, and separated by a number of seconds: "Dynamite!" … "Dynamite!" … "Dynamite!" Some miners preferred shouting: "Fire in the Hole!", etc.

Some quarries, such as the Bumpus Quarry in the 1950s, also had a steam whistle to alert people of an impending blast.

The Blast

A development blast would shoot rocks weighing over 50 kg a considerable distance depending on the placement of the charge and the topography. The above blast shows rocks over 1 meter long traveling off a cliff face. The blast was designed to remove worthless rock leaving a bench over "productive ground". Smaller controlled blasts would expose rock so that it could be carefully examined. The force was only enough to "move" the rock. Closer inspection using various kinds of power equipment would reveal the fruits of the effort.

One anecdote involved a mining crew in the 1950s at the Perham Quarry that wanted to leave "early". The blaster agreed to stay behind to ignite the dynamite at the actual end of the shift so that the absent miners could collect full pay. The quarry was in earshot of the mill and an early blast would have been noticed by the mill manager who also would have noticed the time. The blaster waited for the normal quitting time, called out his warnings, but heard the sound of children frolicking. He investigated and realized that there were echoes from far away. Returning to his post, he called out the warnings and ignited

Mine Model "similar to Newry" made
by Stan Perham and D. I. Shaw.
Perham's of West Paris Collection.

the dynamite. Unfortunately, a car had driven up the mine road and parked at the entrance of the quarry near the blast site. The rocks crashed through the auto's side window, caved in the roof and hood, and sprayed the driver with sand and small rocks. The visitor, who didn't know which miners were still at the quarry, had walked around some trees to climb up the ledge looking for their relative whom they were supposed to pick up at the end of the shift. The passenger was more than startled as he was an experienced quarry supervisor. Both unexpected visitors were unhurt, but the driver had some explaining to do to the cousin he'd borrowed the car from. The passenger apologized for entering the quarry during the usual blasting time and the repentant blaster spent the remaining decade of his career making sure

Two styles of dynamite blasts. Upper view is a development blast seen from the Scotty Quarry toward the Crooker Quarry at Newry. Width of the bench equaled about 20 meters. Right view shows a controlled blast at the Crooker Quarry intended to fracture rock, but not enough to destroy potentially interesting minerals. Note that rocks were still flying vertically up to 35 meters (see top of photo). Photograph by Nathan King. June 24 and 26, 2008.

Above: Shovel rack on right and rails at Perham Quarry, 1920s. Courtesy West Paris Historical Society.

he was not the only crewman working at shift's end.

Modern blasting innovations include: testing the dynamite circuit for continuity, newer stable, safer explosives, more powerful explosives, delayed firing sequences, wireless detonation, etc. (The use of wireless detonators means that charges must not be connected during electrical storms and radios must be turned off lest accidental power surges ignite the dynamite caps.) Dynamiters now require training, apprenticing, and licensing; in previous times "apprentices" might be loading dynamite without supervision on their first day of work. Fortunately, only two Oxford County fatal mining accidents are known: one from dynamite, sometimes sus-

Mucking crew with full wheelbarrows. Perham Quarry, February 1926. Photo Vi Akers. Courtesy Sid Gordon.

Board ramp to allow access to ore cart. Perham Quarry. Vi Akers Photo. Courtesy Sid Gordon.

pected as a suicide, and one from a rock falling.

When dynamited rock explodes away from the working face, very small rocks might travel 75 meters. The majority of the rocks would move less than 7-9 meters, some only one meter or less. Blasters were frugal and tried not to overload a hole for a particular blast lest the valuable mineral be all reduced to worthless "fines" or shot into the woods. Gem miners frequently use fractions of a dynamite stick when near a crystal pocket, although there are legends of entire gem pockets taking a rapid trip into the woods, when miners were not anticipating finding a pocket. The miners still had to wait for the fumes to clear after a blast and many a hasty miner suffered from painful headaches, particularly in the early days when nitroglycerin was a component of the explosives. Eventually, it became standard practice to ignite dynamite at the end of a shift and let the fumes dissipate overnight. Today's choice of explosives is varied and, while dynamite is still used, there are many replacements and enhancers available.

Removing the Rock – Good and Bad

After the blast fumes cleared, the laborers or "muckers" would inspect the loose rock. The useful minerals had to be shipped to market and the waste minerals had to be moved to a dump so that the quarry area was as clean as possible. Mucking was really a very important position and did require some critical thinking. The mucker was the quintessential miner. He was positioned near the working face of the quarry and was expected to be able to recognize the minerals being sought and when to reject a piece of rock when it contained detrimental impurities. A careless, untrained, or uncritical mucker would soon lose his job when it became apparent that shipments to the mill were being rejected because the load had a bad mix. The mucker needed to understand what was acceptable in terms of percentages of the target minerals being loaded onto the truck. The foreman

was expected to inspect loads before they left the quarry, but when the foreman was away, standards might sometime slip. Some miners were quick to spot the difference between the various common minerals. while others never acquired the knack. Much of the spotting was based on color, but when the target mineral was not very distinctive, such as some beryl, all pollucite, etc., the quarry foreman had to constantly reinforce accurate mineral recognition among crew members. There were undoubtedly many quarries that were worked only when easy to identify high-grade minerals were obvious and when work slowed because the minerals began to require too much time to separate, another prospect was tried in hopes of making easier money.

By using sledge hammers with varying handle lengths, the feldspar, mica, beryl, etc. would be separated from the waste rock. Picks might be used to loosen the rock pile. In its earliest days, the Perham Quarry had an ore car siding so that ore cars could be loaded for the trip to the waiting wagon. Feldspar could be put into one car while waste could be put in another, although it was more likely that only feldspar would be taken in several cars in order to get the feldspar started on its journey to the mill, while waste rock would be hauled to the dump when there was no one waiting for the feldspar load.

The Perham Quarry had a miniature rail system for hand-pushed ore carts, one of the few such systems ever used in Oxford County. Some muckers would select the feldspar, beryl, mica, etc. and load it by hand or shovel onto a metal skiff. The skiff might be perforated to allow "fines" to sift back into the quarry. The loaded skiff could be lifted by a crane or winch and be dumped into a truck. Alternatively, a front end loader truck could receive large chunks and that equipment could dump the rock into the truck. In the 1920s, shovels and wheelbarrows were the primary tools used to remove rock from the working face.

Frequently, there would be a place where a loader could drive up a dirt slope and be able to dump its bucket directly into the truck. Rocks, crashing into the truck bed or sides, damaged the truck and some operators put wood planks on the bed and/or interior sides of their truck to reduce damage. The loaded truck would then begin its journey to the mill or railroad station. The waste rock and fines from the blast would be shoveled into the loader and placed on the dump. As this operation produced no money, the dumping spot was as close to the quarry as possible. If the working face were progressing uphill, the waste might be used as fill to raise the quarry floor so the equipment could conveniently reach the new work area. Quarries had to have an attention to neatness to be efficient. While mucking was occurring or truck drivers were hauling minerals, drillers would drill new set of holes. If the working face had reached the end of the prepared area, someone would have to shovel overburden, usually soil, brush, etc. covering the ledge and move them to the dump. The mineral leases issued by land owners frequently required a large enough crew, usually at least six, so that effective mining could occur with large enough production to make the mining worthwhile. There were technical people to drill and blast, and relatively unskilled people who moved the minerals onto trucks, to the dump, or to a delivery point.

Dump trucks were relatively small compared with the huge specialized earth and rock moving vehicles that began to appear after WWII. Few small quarries invested in big trucks, but used vehicles commonly available. Many trucks carried only a ton, but the hefty vehicles might carry a load of five tons. Eventually so-called construction vehicles were commonly available in large sizes and when a second-hand dump truck came on the market it offered both convenience as well as lowered price for small quarry operators. New vehicles were always rare.

The Derrick

The removal of rock from the strewn debris of a dynamite blast was not always a wheelbarrow operation. Quarries, especially those that had to go deep into the ledge for useful minerals might literally dig themselves into a hole. Before the days of excavators and front-end loaders, the only way to get rock out of the deep part of a quarry would be a ramp or a derrick. The ramp was a time consuming and body breaking solution. Taking the time honored experience of deep quarries, such as granite quarries on the coast of Maine, Oxford County had derricks in use by 1899 at Mount Mica. The commonly seen postcards produced from c1905-c1920 of Loren Merrill's operation there show very neatly placed blocks of rock, forming a wall immediately next to the working face. The derrick required a system of strategically anchored stressed cables to keep the central pole stable and prevent the load on the boom from collapsing the entire apparatus. The operator would winch the cable or rope supporting the boom to remove buckets of rock. The boom would be pulled to the place where the load was to be dumped, either the receiving wagon or truck or a dump pile. The derrick had to be tall enough to support the boom so that dumping would be far enough out of the way. This allowed work to be done neatly and at a distance which would keep the dumped material from preventing expansion of the pit. Otherwise, waste rock would have to be moved again. Derricks were almost universal at working quarries in Oxford County, even into the 1950s. After that time, efficient earth moving equipment became more affordable and available. The last derrick system used in Oxford County was used at Mount Mica Quarry by Frank Perham in 1964 and 1965.

Other Concerns of the Mining Day

Breakdowns of equipment, whether the compressor

Derrick at Mount Mica with Loren Merrill at center. Postcard postmarked 1905.

Loading skiff at Mount Mica with Loren Merrill at center. Note shallow working distance between waste rock wall and exposed ledge. Black powder had little shattering power favoring the production of large blocks. Worker with horse and wagon at upper left. Postcard postmarked 1915.

or the truck wouldn't start, etc., meant that all efforts had to be directed at fixing what was inoperative, in order to resume the routine. When a drill steel was really stuck in a hole, it might have to be freed by dynamiting it out, especially when it was the only one of that length owned by the miner or company. Horses or mules were frequently used in Oxford County quarries until the end of the Depression and there might be a corral and barn at some locations where the animals spent the night. The Bumpus Quarry had two horses, Tiger and Veery, to haul feldspar out of the pit in the 1920s and 1930s.

The Bumpus Quarry activity by Northern Mining Company meant that the quarry operator's lengthy commute from Portland had to be reduced. In 1950, Dana Douglass, wife Barbara, and their four daughters, moved to Bethel and were thus conveniently located between Dana's two operating beryl quarries in Albany and Newry. During this period, Northern Mining Company's mining equipment and plan at the Bumpus Quarry were:

"Two roads and ramps run into the pit, where trucks are loaded by a ¾-cubic yard gasoline shovel. Steel barrels cut in half and flat steel trays are used to store hand-sorted material. When full, they are picked up by the shovel and loaded onto trucks.

Blast hole drilling is done with a heavy-duty, wagon-type drill operated by air from a Diesel compressor. A jack hammer is used for secondary drilling, compressed air being supplied by a small,

Plan View of Early Perham Quarry, Showing Tracks to Dump and Loading Platform. February 1926.
Photo by Vi Akers. Courtesy Sid Gordon.

gasoline-powered compressor.

The ore is blasted with 40-percent gelatin dynamite and instantaneous electric detonators.

Stripping with a bulldozer and the gasoline shovel has been started on the east end of the present pit." [176]

At the end of the day, prudent miners would put away their tools and unused explosives before leaving the quarry. Equipment was cleaned for use the next day. Broken equipment meant that no money was made while it was inoperative. Rocks caught in the cleats of loaders might merely separate the cleat or damage it, both resulting in "down time". Teeth might break on the bucket loaders, etc. Any piece of equipment was prone to disaster and poor mines, with old equipment, could have prolonged periods of inactivity due to breakdowns. Oxford County quarries rarely employed night watchmen before the 1960s.

Alcohol use, while not rampant, was an issue in Oxford County mines. When a worker showed up for work obviously intoxicated, he would frequently be put on mucking duty as that was the safest spot for him. While in this miserable condition, he was expected to work steadily and it was hoped that he would be reasonably sober by noontime. There are stories that several contract miners were plied with alcohol so that they might better endure the harsh weather, either winter or summer, and some contract miners said that they had received some wages in the form of alcohol. At least once, there was a need of miners to go to New Hampshire to work and they were treated to free alcohol after work and found that when they woke up, they were at the Ruggles mine in Grafton, New Hampshire and without

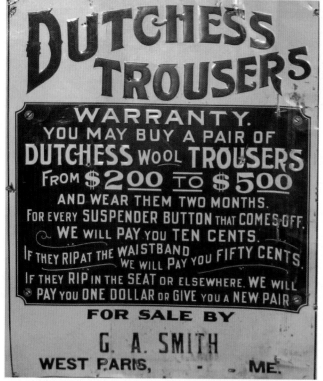

Good clothing was essential for miners.
Guy A. Smith, located two doors up on High Street from the cement bridge, sold clothing and groceries, during the mining period (1907-1951, afterwards owned by Reynold Chase). Gammon and Martin, at Main and Pioneer Streets, sold clothing (1923-1948), as did other village stores. Courtesy Eleanor Inman, West Paris Historical Society.

an immediate way to get back home, except at the discretion of the foreman.

Typical Quarry Plans

Quarries have a design dictated by the terrain they are located in and the physical orientation of the mineral deposit. Quarries would be excavated by digging "benches". A bench was the work level of the quarry, usually the depth of the longest drill holes. If the head wall, or cliff, were unattractive to mine, the miners might dig into their working bench and widen it to the head wall. The new limit of the excavation would be twice the depth it was originally. The one bench would have had two levels. In Maine, it was rare for a deposit to be neatly sliced deeper into the rock. The miners could mine *across* the rock face or *into* the rock face. Miners did not like to blast good minerals into a deep pit from a high headwall. The rock had to be hauled out of the pit resulting in extra costs. The action of falling the extra distance might pulverize good minerals and pulverized good mineral would be sent to the dump and lost.

The miners would remove the rock from the level of the bench and they would continue work until the interesting minerals "played out". If the mining reached the contact of the pegmatite, with barren "country rock" beyond, the choice was to dig deeper, widen the bench, or to make a new level elsewhere. The choice of digging further horizontally or deeper was an important consideration. Frequently, it would simply be inconvenient to continue widening the bench. The hillside slope might become unfavorably steep, etc. The drillers would start drilling at the former bench's working face level and the blasts would either take advantage of an exposed slope or they would, in essence, make a new pit. Digging deeper was not unlike making a vertical tunnel. Starting a new bench could involve a lot of hard work and for this reason, smart foremen had a plan in mind on which direction the quarry should progress. When good

Lower Level of Bumpus Quarry, looking toward lower tunnel. Giant beryl outcrop was on right. 2006.

minerals disappeared, the foreman would have to decide to take a chance going in the current direction or changing where they worked the ledge. Foreman who only chased high-grade minerals might discover they were mining themselves out of business. There are many quarries with good minerals remaining, but where there isn't enough room to move mining equipment around in. There are times when the pit had to be developed through barren rock, with all of the blasted mineral going to the dump. The investment in development was necessary to keep a quarry workable.

In the case of the Bumpus Quarry, the pegmatite had a definite vein-like structure and one end of the vein was exposed at a low flat point on the Cummings farm, inviting the miner to start at the conveniently placed low point and follow the slope up the hill. The huge beryls of 1928-1930 were found not very far into the pegmatite in this lowest bench. The Bumpus Pegmatite was quarried by long benches of about 3 meters deep at its low point and new attacks on the ledge could be made by cleaning off a thin layer of soil and glacial gravel. Deeper into the hill, the miners began lowering the floor of the pit so that the floor of the bench would remain roughly horizontal. As the head wall of the quarry continued to rise up the hill, an additional bench had to be carved. The Bumpus Quarry pit depths vary from 10 to 15 meters deep, but the floors are partly infilled by rock debris. In some cases, the bench floor is infilled by at least three meters of waste rock. Frequently, miners would decide that the quarry floor would not be lowered any further and a new ground-level bench was started uphill. There are at least three "ground-level" bench approaches into the Bumpus Quarry. Part of the reason to have a new approach into

Harlan Bumpus (right) and two of his children, Edwin (left) and Ruth Ann (center) on the truck used to transport feldspar, mid 1930s. Courtesy Ava Bumpus.

the quarry was to reduce the distance needed to haul minerals and waste rock out of the quarry. The muckers were well-aware of the amount of time involved to have to push a wheelbarrow through the quarry to a waiting vehicle. The Bumpus Quarry does not seem to have had a derrick in its earliest years. When Dana Douglass's Northern Mining Company had to lift minerals out of the pit, by 1949 or 1950 he seems to have had a P&H [Pawling and Harnischfeger] crane. Because of the lack of space, due to the confining trench of the quarry and its linear shape, the Bumpus Quarry was sometimes inconvenient to work.

In contrast, the A. C. Perham Quarry was not such a linear deposit. The first opening was made on the western slope of a hill and the pegmatite was not dike-like, but was more of a horizontal lens-shape. The Perham Quarry worked though all of the seasons, weather permitting. The crew included members of the Perham family plus many Albany, Waterford, Bethel, and/or West Paris residents. Many of the early Perham Quarry photographs reproduced here were taken by Vi Akers. Vivian "Vi" Akers was a well-known photographer and artist in Norway. He had been a member of George Howe's youth group, being a member of the "original nine" from 1898. [129, 130] The Howe groups studied nature with a specialization in mineralogy. Although Vi was not an active mineral collector in later life, he did keep a

Ore Cart, full and empty, at Loading Platform. Perham Quarry, February 1926. Photo by Vi Akers. Courtesy Sid Gordon.

few good minerals in his home. (At one time, Vi was married to Addison E. Verrill's daughter. Addison, born in Greenwood, was an early mineral collector in the 1840s to 1860s and eventually became a world famous biologist at Yale University. Among the one thousand creatures Verrill scientifically described for the first time, the Giant Squid is the most famous.)

The Perham Quarry foremen opted to deepen the pit rather than widen it. Part of the reason may have been the use of derricks. The derricks had a limited length of the booms that could be used, but it was only a matter of making sure that the cable or rope used at the quarry was long enough to reach the bottom. The use of motorized winches meant that the amount of time raising a bucket of feldspar

was not excessive. The re-location of large derricks would have been costly and time-consuming. By the end of its life, the Perham Quarry did not use derricks and the mouth of the quarry was returned to ground level. The days of expensive and slow manual labor were over. Earth moving equipment and trucks became the tools of mucking, loading, and transporting feldspar.

Blacksmiths and Equipment Repair Services

The West Paris mill had an active crew to handle its normal repairs. About 1950, mill supervisor, Harlan Childs was quoted: "Breakdowns of the big crushers that do the initial job of mauling the feldspar have caused some delays the past year. One grinder broke last Monday but Childs figured it could be fixed in a day if repair crews worked around the clock." There were also many other repair jobs at the mill: broken belts were always a problem and the Babbitt metal bearings of the ball mill had to be replaced every 12-24 months.

The mill repair crew would also be dispatched to a quarry when mining equipment repair jobs were too big or too specialized for the miners themselves. As the mill sometimes supported ten or more quarries, there were often times when outside help was contracted from local commercial garages. In West Paris, in 1930, Howard McKeen and R. L. Cummings had vehicle repair shops in West Paris – otherwise more than enough repair services for a small village. Interestingly, there were also four blacksmiths: Elroy R. Davis, B. M. Richardson, A. W. Walker & Son, and H. W. Chapman. The number of blacksmiths eventually saw an expected decline, to two, but they survived for a while: H. W. Chapman worked into 1940 and B. M. Richardson into 1948. Walter T. Dougherty, originally in South Paris, was in the blacksmith business in West Paris in 1948 and 1949. These were not the last blacksmiths in the village as Ellingwood Brothers and I. H. Mace, also formerly of South Paris, opened in 1949, but only Ellingwood's was listed as late as 1955. The blacksmith not only worked with horseshoeing, but he also manufactured low-tech replacement parts, keeping frequently obsolete equipment in service. Blacksmiths could repair braces on stiff-leg quarry derricks and could make specialized metal couplings. Truck and automobile repair shops may have also tried to manufacture special pieces. As equipment became more complex and replacement parts were stocked by local stores, not to mention the numerical decline in work animals, the blacksmith, as a named profession, was taken over by repair shops and more modern "machine shops".

By 1935, there were three vehicle repair shops in West Paris: Brock's Service Station, the Service Garage, and the Trap Corner Garage, at least as far as the *Maine Register* was concerned. In 1940, Brock's Service Station and Trap Corner Garage were joined by A. R. Cummings.

The number of trucks and automobiles increased through time, but by the decade's end, 1950, only Trap Corner Garage was advertising vehicle repairs in West Paris. In 1951, West Paris' repair services increased to also include Farm Equipment Service, Omer Colby, and McKeen's Garage. By both 1953 and 1954, West Paris had more repairing garages listed than were listed in South Paris: Trap Corner Garage, Mill House Garage, Hibler Brothers, Farm Equipment Service, Omer Colby, McKeen's Garage, and Bob's Auto Shop. (It is certain that the South Paris automobile dealerships had repairing garages as part of their businesses, but that number would only make the number of *advertised* South Paris and West Paris repair shops equal at six each. Additionally, new car dealerships were unlikely choices to repair mining vehicles, particularly because they would have charged a higher hourly rate than independent repair shops.) One mechanic, Peter Collette worked in West

Paris and was frequently called by the feldspar mill to travel to the Bumpus Quarry, the Dunn Quarry (in Norway), and other localities when the mill's repair crew was already out repairing. The mill did seem to spread the work around, more or less. No one repair shop seems to have had a decidedly greater share of the quarry equipment repair work, though certain mechanics may have been favored when a head gasket was blown in contrast to a mechanic who was better at repairing axles.

Miner's Competition

Typical of close knit communities, local miners frequently knew each other from childhood and many were relatives. Harold Perham remembered an incident from about 1940 when he and his brother were working on the Portland Pipeline, not long after Harold had recommended that sharp drills improved drilling efficiency and had mentioned the Perham efficiency being greater than that of an Oklahoma trained crew:

"Only once in the writer's life time did I ever see my brother, Roy, worried over a competitive event, and that was when we were working as a Jackhammer Team on the building of the first pipeline from Portland to Montreal.

He approached me one night with the news that the Barton brothers, Jackhammer Men, of Oklahoma had spread the word along the Pipe Line Right of Way, that on the following morning the Barton's of Oklhoma were

Blaster (Harold Perham) at the Perham Quarry, February, 1926. Photo by Vi Akers. Courtesy Sid Gordon.

going to 'Whup The pants off the Perham's at the Jackhammer Work, and do it up quick '.

To me, it was a piece of good news, for I had expected it ever since the Perham Brothers had been awarded the first New Air Compressor and Jackhammer Unit. The Barton boys were sold on the use of 'sharp drill Bits' and should be able to make the event interesting.

But, to Roy, the fact that all the Cat [Caterpillar brand tractor] Drivers were placing their bets on a Three to One basis on the

Left: Shovel men. Right: Drill holes plugged with birch stakes at Perham Quarry, 1925
Photos by Vi Akers. Courtesy Sid Gordon.

Barton boys, seemed to get on his nerve. In this state of mind he accepted my plan of attack, and promised to follow it to the letter.

I pointed out that the Barton brothers had chosen their own spot to compete, a smooth section of down grade ledge which must be perforated by 4 foot holes, zig zagging 3 foot apart. Also, that it was a different section of ledge than they had worked on the day before. This section had been planed smooth by the Ancient Glacier of Thousands of years ago, and presented problems.

Special care needed to be taken on the entering of the first 2 inches of each drill. So, Roy was to handle the Jackhammer when starting each hole, and brother Harold was to place his big No 12 boot on the lower side of the Drill Bit. This rather simple procedure was to save the day later on.

The morning of the Big Event arrived, and it looked as if every Cat Driver from Portland to Montreal was there with plenty of Betting Money.

Roy started up the Air Compressor, and attached the jackhammer Hose, while his brother, Red, set out the separate Pairs of Drill Bits and made the Drills in readiness for use.

At the starting signal, all worked smoothly for the Perham brothers, and every second the drill bits were digging their way into the hard ledge. 4 foot holes were being produced in consecutive order, and at maximum speed.

But grief and agony had accompanied the activities of the Barton brothers. At the starting signal, their Jackhammer and Drill Steel skidded down hill on the inclined ledge, and was captured by 2 Red Faced Bartons. On the second try, the Bartons changed places, and ended in another failure. By this time the "Barton Rout" was well under way.

The Perham brothers had foreseen this possible calamity, and had agreed that what-

ever happened to the Barton brothers, that one of the Perhams would always be driving steel into the ledge.

The Barton brothers were both wild by this time and getting worse. But they had adopted the Perham method of entering drill steel into the ledge.

Then the oldest Barton decided to "step over the line" and borrow some of the Perham Drill Bits that seemed to possess magic for the Perhams. But he never crossed the line, for brother Red met him with doubled fists and the promise that one move toward the valuable drill bits meant a broken face in the Barton family.

Barton retreated, in preference to forcing the issue, and during these tense moments, brother Roy stuck to his jackhammer with full speed ahead.

Soon the demoralized, and thoroughly beaten Bartons "threw in the sponge", the Cat Drivers left, and the Perham brothers simply went on with their day's work."

1938 Caterpillar RD-6. photograph courtesy "Willie" at www.westernfarmcollectibles.com

Chapter Two: Early Mining in and near the Bumpus District

The Bumpus Quarry was not the first mining attempt in Oxford County nor even the first in Albany, but it was among the first which produced feldspar as an ore. The Bumpus Quarry, although primarily a feldspar producer, became widely known for its huge beryl crystals soon after the quarry began operation in 1927. The mineral location was originally on the Cummings family farm [16] and the quarry was initially and informally known as the Cummings Quarry, but that name appears on only a few documents and it has been subsequently and universally known as the Bumpus Quarry.

Mount Mica actually led the way in the County's pegmatite mining in 1821, but that mining mostly sought gems and rare specimens. Augustus Hamlin finally purchased the land Mount Mica was located on, in 1880, and organized a company to recover gems. There were also other mining adventures by Hamlin, who then resided in Bangor. He was active in developing the Douglass Copper Mine in Blue Hill, Hancock County, in 1880-1884, the Winslow Tin mine in Winslow, Kennebec County, and the Blanchard Slate Quarry in Blanchard, Piscataquis County. Hamlin also invested in the aquamarine locality, Beryl Ledge quarry, at South Royalston, Massachusetts as well as other gem occurrences.

There were also attempts at silver mining in Oxford County during Maine's unrestrained, if not truly "Wildcat", Silver Boom in the 1880s and that activity and other County metal mining interests are summarized by John Davis (this volume) and Appendix A.

Regarding the earliest mineral reports in Albany,

Nathaniel True c. 1865

Jackson (1838) knew of unspecified mineral localities there. Bethel's well-known naturalist and former Gould Academy principal, Nathan True, announced his discovery of Albany beryl at the Portland Society of Natural History [42] in 1862, although it was published years later in 1869: "Small, but very fine specimens, have been recently discovered on Farwell Mt., in Albany. A twin [sic] crystal, found there by the writer, is of a bluish green color, transparent, and sufficiently free from fractures to be cut into jewelry." [35] Kunz (1885) reported: "At Albany, Maine, Mr. N. H. Perry has recently found beautiful transparent golden yellow beryls that would cut into perfect gems of over 2 carats each." later (1887) adding the locality: "A few small, rich yellow [golden beryl] stones were also found at Round Mountain, Albany, Maine." Another occurrence, the French Hill Beryl Locality, in southwestern Albany, was known many years before 1906, when U. S. geologist, Edson Bastin, was guided there. The French Hill locality, as well as the other early Albany mineral locations, were mineral collectors' localities only. The reader will note the innumerable repetition of old finds in various

West Paris, 1892, from a stereoview.

mineral books leading the novice to imagine a veritable gem rush when the old stories are repeated in fresh wording. It is equally astonishing that it was reported by Williams (1888): "Feldspar… extensively mined at Topsham, Sagadahoc county; Canton, Oxford county; *and* [emphasis added] Brownville, Piscataquis county" and "Quartz rock and glass sand… Mined extensively in connection with feldspar in Sagadahoc county; Paris, Woodstock, Hebron, Canton, Greenwood, and Albany, Oxford county; Auburn, Androscoggin county." Again, the novice would imagine that Canton was the first feldspar quarry in Oxford County, rather than a discovered ledge containing this mineral. The Reynolds Quarry in Canton was worked for mica. Of course, it is much more astonishing that Brownville, in Maine's famous slate district, was "extensively mined" for feldspar. Except for Brownville, granite pegmatite veins were known, all over the towns mentioned, by the mineral collectors combing the hills: Nathaniel True, Addison Verrill, Ezekiel Holmes, Elijah Hamlin, Nathan Perry, Edgar Andrews, Edmund Bailey, Augustus Hamlin, Samuel Carter, and their like. The interest in Oxford County gems is such a rich story that it has already been the subject of several histories and the reader may want to consult Perham (1987) and King (2000) concerning some of the interesting events of that activity.

There was a small interest in metal mines in Oxford County after 1900. Mount Glines Gold and Silver Mining Company in Milton Plantation was organized in 1902 by William N. McCrillis, Dr. Joseph Abbott Nile, Stanley Bisbee, Ralph T. Parker, and, perhaps not too surprisingly, George Howe's brother-in-law, Judge Arthur Eben Morrison. McCrillis, Nile, and John Johnson also opened the U. S. Nickel mine in Rumford about 1909. (Mineralogists might smile at the thought of a nickel mine in Rumford. Unfortunately, the local metamorphic formations contain the iron sulfide mineral, pyrrhotite. Minute amounts of nickel may be associated with this mineral and, the unfortunate part, the chemical field test for nickel is extremely sensitive and a positive test may be had on only trace amounts of nickel.) McCrillis and Nile also promoted the Rumford Paint mine, an ocherous deposit being deposited by a spring, and, just before WWI, they attempted to start a mineral fertilizer company on the site of the failed nickel mine.

In view of the increased mining activity in the State, the Maine Mining Bureau was established by law. BILS (1908) reported: "Notwithstanding the fact that this law as been on the statutes for several years, no organization of the bureau has ever been effected. The reason for this is that the statutes prescribes duties which necessitate the expenditure of money, but as no appropriation has ever been made to carry out the provisions of the law it is for all practical purposes dead." The proposed mining bureau does not seem to

Downtown Hartford, c1900. Scott Robinson house is second house from left, small bright white. Courtesy Lorraine Parsons.

have had a practical existence until the early 1960s.

The Lust for Mica Led the Way to Exploiting Feldspar

With regard to mica, commercial quality sheet mica is primarily found in granite pegmatites and the only common kind of desirable sheet mica in Maine is the white-mica called muscovite. The granite pegmatites of Maine are distributed in a discontinuous band principally ranging in an arc from Georgetown to Fryeburg. The principal centers eventually included: Topsham, Auburn, Paris, Greenwood, Albany, Waterford, and Stoneham, with important quarries as far north as Newry and Rumford.

Maine was never an important state for mica and New Hampshire, North Carolina, and South Dakota can justly boast of their enormous productions. Mica mining began in the U.S.A. at the Ruggles mine, Grafton, Grafton County, New Hampshire in 1803 and New Hampshire was the only U.S.A. producer until 1867 when North Carolina became a mica producing state, as well. Sheet mica was produced in Amelia, Amelia County, Virginia, "before 1870". Nonetheless, mica provided some interest in Maine. BILS (1902) stated: "As far as we have any records, Mount Mica in Paris, furnished the first mica for commercial purposes in Maine in 1871." There is a myth that the Mount Mica mica miners were inexperienced and that they discarded the contents of gem pockets onto the dump never to be seen again, but the land owner, Odessa Bowker, was a part-time mineral dealer and certainly would have taken advantage of this short-term ignorance. Mount Mica was not a good producer of sheet mica and the miners were only attracted to the locality because of the whimsical name it had been given.

There were two small early mica boom periods in

Maine, the first in the latest 1870s and the second in the last few years of the nineteenth century and the early twentieth. Market and production summaries are not available for the first period and the economic stimuli leading up to the second are fragmentary. Before mica from India began depressing the international price of mica, this mineral was costly enough to induce some exploration for it in Maine. Pingree Ledge Quarry, in the northwest corner of Albany, had been prospected briefly in 1878-1879 during a short mica boom period in Oxford County: "About 2,000 pounds of sheet mica, valued at $2,000, was produced in Maine in 1880, according to the Tenth Census." [21]

India mica mining started in 1885 and that country was soon able to produce nearly all of the mica needed in the world, due to its extensive deposits and low paid work

Board of Big Gem Mica Company
L->R: J. M. Gooding; George Storer; Philip
Larrabee, H. M. Yorke, Charles Bisbee,
A. S. Bisbee. Rumford Times, August 3, 1901.

force, and consequently their mica almost instantly out-competed the world for price. Bowles (1947) indicated that mica mining changed its "philosophy" in 1894 when an industry innovation resulted in the manufacture of "built up" mica, where fine splittings and thin mica sheets could be shellacked or glued together. The U.S.A. producers did not take advantage of the huge market for built-up mica as did India's young mica industry and built-up mica could be used as an inexpensive substitute for some sheet mica. Although rare minerals and gems were the products most sought in Oxford County in the 1880s and 1890s, there was another attempt at mica mining in the late nineteenth and early twentieth centuries.

The Dingley Tariff of 1897, named for its author, Maine Congressional representative, Nelson Dingley, Jr., "protected" a wide variety of USA industries with an average tariff of 49% on competing imports, and was a stimulus which spurred the search for domestic mica, although the

tariff rate for mica was only about 20%. North Carolina producers still complained that the cheapness of Indian mica was a problem. Domestic mica prices rose, briefly, and became relatively attractive by about 1898, but by 1905, Oxford County mica production was essentially finished. The tariff accomplished one of its initial goals, protection of domestic mica mining. It was so successful that the nation's new mica producers cut each other's financial "throats" in order to sell their own mineral. The year 1900 broke previous national records of production, but mostly because of increased domestic competition, the mica was valued, by the pound, below almost every year since records had been kept. The national output of sheet mica in 1899 was 108,570 pounds valued at $70,587, while the national sheet mica output for 1900 had an amazing increase to 456,283 pounds valued at only $92,758 (USGS, 1901). The price of $0.65 per pound in 1899 was attractive, but in 1900 the price was only about $0.20 per pound. The next year, 1901, saw very little improvement with an average price for sheet mica of $0.27 per pound. While it is possible that the lower average price was influenced by smaller, and therefore cheaper, mica pieces entering the sheet mica market, the result was that increased production was still extremely depressing to the average. The effect was that the second Maine mica boom was bust after only about two years. There was no Maine mica production reported for 1899, but Maine probably had the lion's share of 9,200 pounds of unprocessed sheet mica in 1900. The Consolidated Mica Company may have been the State's only producer of note and according to industry standards, the report of Maine's output was combined with that of Idaho, Nevada, and Rhode Island so that no one could precisely figure out what a specific company's production was. Maine's share was probably just over 1% of the national total. [48] In 1901, the U.S. Geological Survey reported that no mica production was recorded for Maine, despite the fact that scrap mica was still being produced at Black Mountain.

The Maine Mica Excitement, as it would have been called in the time period, did do one positive thing. It resulted in the exploration of the Oxford County's pegmatite deposits and revealed that feldspar was present in significant amounts. The feldspar industry already existed in Topsham, Sagadahoc County, from the 1850s, but by the beginning of the twentieth century, the Auburn District, Androscoggin County, initially a gem and specimen producing district, began to be well-known for its feldspar. Maine soon rose to be the nation's leading producer of feldspar, a crown that was struggled over for more than twenty years, before diminished State production and increased out-of-state competition denied Maine from ever holding the title again.

Ezra Warner Truesdell, 1900.
Courtesy Paul Truesdell, Jr.

Oxford County became a feldspar-producing region with feldspar being mined from the Twitchell Quarry in Paris, about 1901, when the Portland Mineral Company organized in the spring of 1901 and began receiving that quarry's production. The Portland Mineral Company had the second feldspar mill built in Maine and it shipped much of its ground feldspar to Trenton, New Jersey.

Details of the Search for Sheet Mica in Albany and Elsewhere in Oxford County

There was one individual who particularly searched for Maine mica and feldspar: Winfield Scott Robinson (February 15, 1840 Hartford, ME – September 6, 1927 Togus, ME). Robinson opened the George Elliot Mica Quarry, Rumford in 1898, and also prospected the F. H. Bennett quarry "5 miles west of Hunts Corners" in 1899 and, in 1900, re-prospected the C. P. Pingree quarry, the latter two quarries both in Albany. An undated (early 1900s), unidentified newspaper wrote of Mount Mica Quarry: "At one time, Scott Robinson, who has been closely connected with mica mining in Maine, wished to gain control of this mine."

"Scott" Robinson prospected at least one mica quarry that yielded muscovite commercially and that was on John Elliott's property west of Route 5 near the Newry and Hanover town borders. Robinson signed a deal with Mr. Elliott on December 27, 1898 for a ten-year lease. There was no annual rental payment, but there had to be development work of at least 40 cubic yards or 100 tons of rock moved annually. Robinson was to pay for sheet mica "squares" of any size or shape at $0.25/pound and scrap mica at $0.25/ton. The next year on August 12, Robinson registered 14 leases he had that were apparently sub-leased to a "Northern Mica Company". The agreement included the Elliott Quarry. The interest in the Elliott Quarry attracted the attention of the USA's largest mica mining concern: the Consolidated Mica Company of West Virginia and on February 2, 1901 this company obtained a quit claim deed for all "mined

and unmined mica" on the Elliott property from John E. Elliott [b.~1838] and John W. and George W. Elliott. The resolution of the deal with Northern Mica Company is unknown. It was the Consolidated Mica Company that did most of the mica mining at the Elliott Quarry, also known as the North Rumford Mica Quarry, but the quarry ceased operations by 1902. Probably, in an effort to distance itself from litigation, the Consolidated Mica Company within a short time sold a quit claim deed for $1 to its Saco Office. The "Saco Office" was apparently far from official. The "office" seems to have been the office of a lawyer, Hampden Fairfield, son of former Maine Governor John Fairfield, of Bloodless Aroostook War fame. Fairfield was probably only legal counsel for the Consolidated Mica Company's activities in Maine. A sheriff's Deed was issued October 24, 1902 to seize all of Consolidated Mica Company's "rights, titles, and interests" in the Elliott Mica Quarry and presented to Consolidated Mica Company of Saco. It would appear that the Consolidated Mica Company had not paid for its acquisition of the Elliott's "mined and unmined" mica. The local superintendent for the Consolidated Mica Company was Clarence Leslie Potter [1870-1906] of South Waterford. (Potter was with the newly organized Oxford Mica Mining Company of Waterford, which company would sell its services to interested mining ventures, and was soon to be associated with Dr. Hiram Francis Abbott of Rumford Point and Hollis Dunton, of the Virginia Section of Rumford, at the nearby gem tourmaline locality on Halls Ridge, Plumbago Mountain, Newry, according to the now lost Hiram Francis Abbott diary.)

Scott Robinson also seems to have had a personal investor, an Ezra Truesdell of Boston. Although Truesdell genealogists are uncertain who this particular Ezra Truesdell was, there was only one known from the time period who seems to be a provocative fit. Ezra Warner Truesdell [August 16, 1854 Liberty, PA – July 16, 1903 Minneapolis, MN] moved to Newtonville, MA, 6 km west of Boston, in the mid-1890s, before Robinson was arranging mica leases (Paul E. Truesdell, Jr. and Carol Wagner Truesdell, personal communication, 2007). Truesdell was a grain exporter in Boston. Unfortunately, he developed Brights Disease and his

Main Street, West Peru. From a 1907 postcard.

West Peru railroad station.
From a 1910 postcard.

doctor advised him to find a new climate in the West and, about 1899, also advised him to invest in land development in Port Angeles in Washington State.

Given Truesdell's entrepreneurial nature, including investments, it is reasonable that he also had other interests than land. The timing of Truesdell's life also coincides with Robinson's ceasing his mica searches, and just about the time of the collapse in mica prices. Ezra moved to Washington in July 1902. Ezra's health worsened and the next year he died in Minneapolis. The Robinson-Truesdell leases were made in late 1899 through late 1900, when prices were still relatively attractive and when Ezra was not yet in poor health. One of the last of the 14 Truesdell-Robinson leases, with Zenas Kneeland and concerning his pasture on the west side of Waterford Road, Waterford, lists Truesdell as treasurer of the Consolidated Mica Company. As this is the only mention of any person specifically associated with the Consolidated Mica Company's efforts in Maine, the mention of Truesdell's role with that company seems either a mistaken record or a misrepresentation. It would also seem that Robinson sold his interest in the Elliott Quarry so that Consolidated Mica Company could gain control of the location, but it is, admittedly, uncertain what relationship Robinson had with the Oxford Mica Mining Company or the other companies he dealt with. Most of the business deals seem to have been informal at best, despite the recorded leases.

The 14 leases held by Robinson were generally for a 20 year period with a $6 per year rent. The royalties offered and agreed to, as mentioned above, were also paltry, even by the time period's value of money. Among the mining agreements recorded (John Davis, personal communication, 2008) they were from: Charles Bisbee with a ledge on his farm in Woodstock; George V. Childs for a ledge in Franklin Plantation; Jonathan Bartlett, L. H. Burnham, and David D. and Emma Brown for mining rights on their farms in Albany and Stoneham; Elden B. McAllister and John L. and Calvin H. Stearns for mining rights to their land in Stoneham; Amos Barker for mining rights to his farm; G. E. and J. A.

Leighton of Gilead for mining rights on the "York Lot", Albany Mountain, Albany; with an agreement with Alonzo Felt, Ronello Davis, and Abby Curtis for mining rights to their land on Perham Mountain, in Woodstock. The Burnham lease was for 50 years at $1 per year, a bargain if you were only interested in the first year's prospecting. Except for Childs and Bisbee who would get only $0.25 per pound for their "squares", the other lessees got a munificent 5 cents per square. Although there is a long paper trail of interest, most of the agreements soon lost their economic luster and most of the contracts simply lapsed.

During the mica "boom", a feldspar prospect in Peru on the Charles E. Knox farm on Hedgehog Hill became known as the Big Gem Mica Quarry. It was worked in 1901 (*Dailey Kennebec Journal,* February 10, 1902): "It is said by those competent to judge that the mica found at Hedgehog mountain is superior to that found in India." – the kind of comment always seen in the press regarding new Maine mining ventures. It was also projected that feldspar was going to be recovered as a by-product and a grinding mill was to be erected. Both statements were conjectures. The Big Gem Mica Company was organized with J. M. Gooding (Portland) as president; George A. Storer (Brunswick) vice-president; Phillip J. Larrabee (lawyer in Portland) secretary-treasurer. The *Rumford* Times (August 3, 1901) revealed that Dr. Charles M. Bisbee of Dickvale, Dr. H. M. Yorke (dentist) of Kennebunk, A. S. Bisbee, Brunswick, and H. Pierce of Foxcroft were board members and it was revealed that mica mining began in June of that year: "Mr. Loren B. Merrill of Paris… was at the Peru mica mine Saturday and looked over the property. He expressed confidence in the prospects of an inexhaustible supply of the minerals found in this mountain. At present about a dozen men are employed in the mining and packing operations. Several tons of mica are almost ready for shipment. Bastin (1911) erroneously said that the Big Gem Mica Quarry was worked "only in the summer of 1902". Nonetheless, the locality was short-lived, but many of the late nineteenth century-early twentieth century mica prospects were eventually re-prospected as Federal subsidies again made mica mining artificially profitable during and after WWII and the Korean War. In the succeeding years, after the mica boom, there was some hope of rekindling pegmatite mining in Peru. The *Daily Kennebec Journal,* ever interested in Maine mining revealed (June 14, 1916): "Writes the Dickvale [Peru] correspondent of the Oxford Democrat: 'A Mr. Stackpole has been in town buying feldspar mines.'"

The Black Mountain Quarry, which will have its history discussed elsewhere, was also operated during the mica boom by Oliver Gildersleeve [March 6, 1844-July 27, 1912] of Portland, Connecticut. Gildersleeve was from a famous family of shipbuilders and he was a leader in that company, but as that Portland was in the feldspar and mica mining district of Middlesex County, Connecticut, he was familiar with market interests and also may have formed a company working in Portland, Connecticut as a primary investor among a group of interested businessmen. (A feldspar grinding mill had existed in Portland, Connecticut since 1879.) His mining operations, under the name Maine Products Company, leased the Black Mountain Quarry in Rumford, Maine from 1901 through 1905, but only scrap muscovite was obtained, although reported production was about 500-750 tons per year, with only 250 tons during the company's last year. Maine Products, although a "Maine Corporation" had its offices on 270 Madison Avenue in New York City, had control of Black Mountain Quarry after they purchased it on August 3, 1905 (OCRD v. 409, p. 596), and later leased that quarry for ten years to United Feldspar and Minerals Corporation on August 17, 1943.

In 1906, Robinson, along with James A. Gerry of Mechanic Falls, opened the Hibbs Quarry in Hebron. Although there was probably not much work done: "All [feldspar] output must be hauled by teams 3 miles to Hebron station, on the Rangeley division of the Maine Central Railroad." It was also revealed: "The development work was suspended for reasons wholly aside from the quality of the deposit." [26] The reason was probably the dismally low price for mica then existing. The term "development work" was well-chosen as Sterrett (1907) reported that no mica was produced in Maine in 1906. A little later, it was indicated that a feldspar grinding company was organized at Buckfield by Robinson's daughter, Winifred. The Maine and New Hampshire Feldspar Company Incorporation papers were filed with the State, but "with $200,000 capital stock, of which nothing is paid in." (*Daily Kennebec Journal,* February 28, 1908). Winifred Mary Robinson, of Buckfield, was president and Joseph D. Fisher of Boston was treasurer. Despite the filing, the company does not appear to have sold much stock, did not build a mill, nor have done any work.

The Robinsons certainly realized that a successful, economically competitive, feldspar industry required a mill nearby to the pegmatites as well as a railroad connection to the feldspar-consuming regions. The parallel with the future West Paris feldspar grinding mill is interesting. The pioneering feldspar mill efforts of the Robinsons failed, but they must have had some influence on the Perham family concerning the idea of having a mill in West Paris sixteen years later. Robinson was a well-known Oxford County personality and lived primarily in Hartford and Sumner, near West Paris. However, on May 1, 1869 his marriage intentions toward Hattie A. Fogg of Hartford were recorded with certificate issued five days later. In this year, Robinson was a resident of Canton. In 1870, Robinson was a justice of the peace in Hartford (Foster, 2004), but is not listed in this office in 1871 by the *Maine Register*. The Robinsons had two children, Charles W., and, seven years later, Winifred, but Charles did not reach adulthood. In the 1880 census, Robinson was listed as a stone mason.

He was active in the 1880s in the Republican Party and gave speeches promoting candidates. He was sometimes moderator of the Hartford Town Meetings and was on the school board. He was a speaker at Memorial Day ceremonies in Stoneham in 1899 (*Daily Kennebec Journal,* May 26, 1899) and gave a lecture on the Battle of Gettysburg at the Turner Grange (May 22, 1914, *Daily Kennebec Journal*). For a while, in the 1880s, he was an agent for the Keeley Institute of Portland and gave speeches in Oxford County concerning the evils of alcohol abuse and how one might receive rehabilitation at the Institute. In the 1910s, Robinson turned his efforts towards raising chickens commercially. In 1913, for example, he won prizes for his Orpington chickens (cockerels and pen chicks), and a Best Exhibit at the poultry show in Madison (*Daily Kennebec Journal,* December 22, 1913). In the 1920s, Robinson, one of Oxford County's oldest Civil War veterans, was an inmate at the Togus Veterans Hospital, near Augusta. Robinson had served in the 5th Maine Infantry, also called the Forest City Regiment, first in Company E from Lewiston and was mustered out as a corporal.

After the turn of the Nineteenth Century, the Bureau of Industrial and Labor Statistics (BILS, 1902) reported that there were nearly 100 pegmatite miners in all of Maine. There were "about 15 hands" working in South Paris. Another fifteen pegmatite miners were at work in Hebron. Presumably, Mount Rubellite may have accounted for some of these Hebron miners: Merrill, Stone, Hamlin, and several assistants, while there was a crew at the Mount Marie quarry. (Mount Marie Quarry was located in Hebron until the early 1900s, when the town line shifted, yielding land to Paris.) The remaining pegmatite miners were from the Auburn and

Winifred M. Robinson, c1910.
Courtesy Lorraine Parsons

Peru mica mining crew demonstrating double jacking. Rumford Times, August 3, 1901.

Topsham districts. Given the collapse of the mica boom in Oxford County, we can reasonably assume that many of the County's miners had to seek new jobs in late 1901.

The *Lewiston Journal* (1902) reported of Mount Marie: "A steam drill is on the way for the feldspar quarry on No. 4 Hill. Mr. W. E. Mills arrived in town on Monday. It is understood that the Mount Marie Mining Company, of which Mr. Mills is president, is to begin operations at once, with a crew of workmen. Some talk has been heard concerning his intention to erect a mill at the mine. Prospects now are that the work of mining for quartz and feldspar will be pushed rapidly during the coming year."

Bastin (1911) visited the Mount Marie Quarry, "near the Paris-Hebron townline", in August, 1906 and revealed that the so-called Mills Quarry had been worked in 1901, but that it had been abandoned since that year: "No mining machinery was installed at this locality." The feldspar was hauled 5 miles, mostly down grade, to South Paris on the Grand Trunk Railway. Only a few tons of it was shipped, and much spar now lies in stock piles at the quarry." The earlier newspaper announcement of 1902 may have been part of a publicity event, perhaps to entice investors to the company, but the report of hearsay may have been unfortunate. The "quarry" actually consisted of two main excavations. One was 75 x 30 feet and was 10 feet deep, while a second pit was about 30 x 30 feet, also 10 feet deep. The value of the feldspar in the excavations was assessed to be very low and the value of the quartz was not even mentioned. It was revealed: "Muscovite has been saved during the mining, but most of it is what is known as wedge mica and would be valueless except as a source of ground mica." The specimen value of the locality was mentioned, but not evaluated: "The coarsest and most highly feldspathic portion of the deposit as exposed in the larger pit contains some cleavelandite and granular lepidolite and a few colored tourmalines of pink and green tints, which are translucent to opaque. A few small pockets occur and several less than a foot in diameter were exposed at the time of the writer's visit. In some of the pockets a few translucent tourmalines

of gem quality were found during the mining operations. ... Even if the mica and tourmalines were marketed as accessories it is probable the deposit could not be made to pay."

Oxford County pegmatite mining saw some activity following the mica boom, but most of the active properties were being worked for gems or mineral specimens. Loren Merrill seems to have been interested in feldspar at the Bennett Quarry before 1920. The interest that resulted in a half-century feldspar mining run in Oxford County was the discovery of commercial feldspar on the Perham farm, West Paris, the erection of a feldspar grinding mill in that town, and the subsequent discovery of feldspar on the Cummings Farm in Albany. With two strong sources of supply, the mill had enough stability to exist in its earliest years.

Top: Togus Veteran's Hospital, c1905 postcard. Center: Garnet 3x3 cm, Hedgehog Hill Quarry. Bottom: Hydroxylherderite, beryllium mineral, 2.8 cm tall crystal, Keith Quarry, Auburn

Chapter Three: The Birth of the Feldspar Mill at West Paris

Typical corn shop/cannery buildings, upper = Buckfield; middle =New Sharon; lower = West Minot. From old postcards

In order to understand Oxford County's feldspar mining and the stimulus to have a mill in West Paris, it is necessary to know what was happening regarding other feldspar mills in the State. Feldspar isn't used in its raw state, fresh out of the quarry. However, many industries face the "Which came first – the chicken or the egg?" problem. The cost of transporting raw materials rises with distance from a mill. Even large gravel pits, with sorting and crushing facilities on the property, have found that they saved money by moving their huge sieving equipment closer to the working face of the gravel pit as the excavation expanded. The savings in fuel, time traveled, etc., eventually more than repaid the cost of moving equipment and machinery.

Economical shipping rates were also important to the feldspar industry. Feldspar must go through several stages of crushing and, often, purification, before it can be used by pottery and porcelain manufacturers or any of the other businesses which require feldspar as a starting material. There is a 10-15% waste loss between the crude ore and the finished product, but as the mill buys long tons of crude feldspar and sells short tons of ground feldspar, the number of "tons" works out to be practically the same. However, if the ground feldspar is destined for more than one end-user (customer) the extra handling costs, etc. begin to cut into any profit. The mill ended up in the hands of a single ground feldspar distributing company which in turn sold directly to end-users. Nonetheless, a mill (the egg) encouraged mining (the chicken). West Paris was a small village within the relatively small town of Paris [population 1930 was 3761].

Other Oxford County industries, such as agriculture, saw an increase in production as service mills or "shops" were established in the area. As economics became unfavorable to operate local food processing plants, the local shops went out of business, commercial agriculture declined in a spiral related to lower production and there were even fewer shops. When corn shop after corn shop closed, Oxford County's commercial corn farming greatly diminished. Processing mills, of all kinds, are frequently located next to railroad tracks, or high quality roads in order to take advantage of inexpensive transportation. (Before the 1930s, many of Oxford County's important roads were still unpaved and the time during a winter thaw or early spring, "mud season", saw many vehicles on these public roads mired in ruts and mud.)

Railroad transportation was important for almost all producers and it was usually cheaper to ship entire carloads than to ship by truck (John Davis, personal communication, 2007). Small shipments by rail were not always competitively priced and trucking would have been cheaper for small shipments (e. g. less than 30 tons). Of course, trucks had to go through an evolution of increased capacity and 10 ton trucks were a luxury item even in the 1950s.

The railroad line servicing Bethel to West Paris was the Grand Trunk Railroad. The line servicing East Hebron, Buckfield, and West Minot was the Rangeley Branch of the Maine Central Railroad, formerly the tracks of the Portland and Rumford Falls Railway. These two railroads connected at Mechanic Falls and crude feldspar could easily go to any of the Maine feldspar grinding mills or to those out-of-state. However, the Littlefield Corner feldspar grinding mill closed on October 1, 1929. Any feldspar going to Topsham, either to the Trenton Flint and Spar Company or to the Consolidated Feldspar Company, probably went through Lewiston Junction and would have been transferred there to the Maine Central Railroad (John Davis, personal communication, 2007).

Early Maine Feldspar Grinding Mills

Topsham First Mill. Maine's feldspar industry was active very early in coastal Maine: "The first feldspar mining in Maine was done between 1852 and 1860 by a Mr. Jordan, who worked the property of John W. Fisher in Topsham. In the late 1860s a Mr. Carr of Grafton, West Virginia, worked one season at the quarry, Jordan had begun, and then sold the quarry to Golding Brothers of Trenton, New Jersey. In the meantime, W. D. Willes was opening additional quarries on the Fisher property, and about 20 years later he purchased the original Jordan quarry (Mitchell et al., 1904)." At this time, there was no grinding mill in Maine and the crude feldspar had to be shipped out of State. Golding Brothers had, at various times, grinding mills in East Liverpool, Ohio, Trenton, New Jersey, and Wilmington, Delaware.

After the Civil War, American tastes changed dramatically. Previously, homes were generally stoically decorated. Photographs were still an innovation and wall pictures were just becoming available after printing advances were made in the 1850s. Jewelry, especially, was rare before the War, but increased in popularity afterwards. The excesses of the Victorian period rapidly gained influence in architecture and the contents of homes. Chinaware was thick and heavy before mid-century, but by the last third of the nineteenth century, potteries in Czechoslovakia, Germany, France, and England began to produce more delicately designed and glazed products than previously and no well-appointed Victorian residence could be without these new luxuries. The U. S. feldspar-glazed pottery producers began to catch up with the foreign imports.

The birth of Maine's first feldspar processing mill occurred after the feldspar resource was well proven:

"In 1869 The Trenton Flint and Spar Company, of Trenton, New Jersey, purchased the property where their mill now stands, and at the same time leased their quarries, then unopened. Work was soon after commenced and the quarries were operated. In 1872 a mill was built [on the edge of the Cathance River] for grinding the feldspar. The ground feldspar is

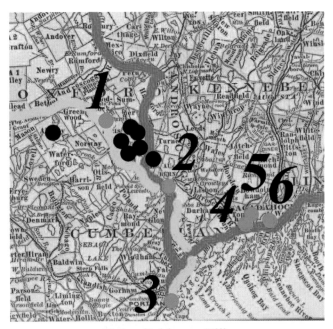

Maine Feldspar Mills, Blue (dots) = 1 West Paris. 2 Littlefield Corner. 3 Portland. 4 & 5 Topsham. 6 Bath. Yellow (line) = Grand Trunk Railroad. Red (line) = Maine Central Railroad. Black (dots) = Harold Perham Quarries. Red joined yellow from the north at Mechanic Falls. Maine Register base map.

used in the manufacture of crockery by the various potteries at Trenton, New Jersey, to which place it is shipped. About a dozen men are employed in operating the quarries and grinding the spar. Mr. George D. Willes [d. March 1918], of Bath, has been superintendent from the commencement until the present time (Wheeler and Wheeler, 1878)."

It is unknown whether the earliest mentioned quarries remained in business, remained independent, or were purchased by the Trenton Flint and Spar Company. The only definite statement concerns the Trenton Company's "unopened" quarries, implying a time when that Company was prospecting for feldspar. The early mining results were certainly shipped to New Jersey for experimentation and with success, the Company built a mill in Topsham.

The feldspar mill, at the head of tide on the Cathance River, originally shipped its ground product about 2 km downstream to Merrymeeting Bay, where the Androscoggin River joined with the Kennebec River, and on to Bath. From Bath the feldspar was loaded on ocean-going ships to New Jersey, but by the mid-1880s, with railroad tracks passing near it, the Cathance mill's feldspar was shipped to New Jersey by railroad cars. [44] Perham (1955) indicated: "Part of the Cathance River was diverted for a settling area to separate the fine from the coarser grindings. This method was employed for twenty-six years, until 1898." The settling pond had to be protected from contamination from the river, especially from suspended silt and mud that would have contaminated the settled fines which formed a "mud" of its own.

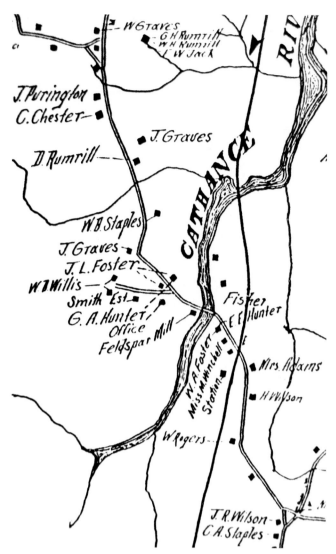

Falls of the Cathance River immediately next to Trenton Flint and Spar Mill, 2008. There was probably a loading dock below the falls in the earliest years of the mill's operation.

Unidentified 1880 river survey map showing Feldspar Mill on west side of Cathance River, Topsham. Map edge orientation is north - south. 1 cm ≈ 0.5 km.

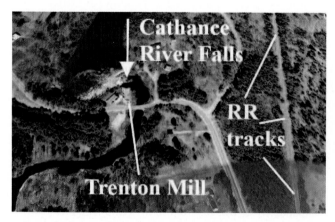

Aerial view of Trenton Flint and Spar Company Mill on Cathance Road, ~2007, Railroad tracks cross Cathance Road almost immediately south of map boundary. Courtesy Google Earth.

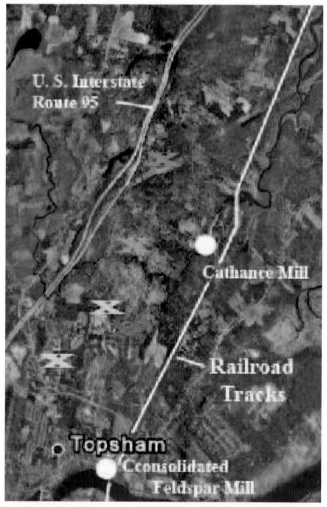

Topsham Feldspar Mills and Some of the Quarries. Yellow = Mt. Ararat Quarry; Red = Consolidated Feldspar Quarry Group; Gold = Swamp #1 Quarry; Blue = Fisher Quarry; Green = Trenton Quarry. Redrawn from Google Earth.

Upper view: Dual ball mills in ruins of Trenton Flint and Spar Mill, 2008. Note different colors of ground feldspar emptied below the mills suggesting they were of differing grades. Lower view: Flint pebbles in ground feldspar below ball mills 2008.

Cathance Railroad Station, Facing northerly to intersection of Cathance Road, ~1940s.

The Topsham feldspar grinding mills, of course, did much to promote pegmatite quarrying along the central coast region from Freeport to nearly Rockland. *Dailey Kennebec Journal* (July 14, 1899) noted: "Feldspar mining has given Cushing an exciting land boom." One might imagine the "boom" was short-lived.

The presence of the Cathance mill encouraged the opening of more feldspar quarries in Sagadahoc County and by 1904, "Ten men, two teamsters and a foreman are employed at the mill, and around 40 men at the quarries. Seven or eight cars are shipped per month. G. D. Willis [sic] is general manager and agent, Elisha Potter foreman at the mill and James Prest at the quarries. Most of the men employed are Italian laborers (Mitchell et al., 1904)." The Cathance Feldspar Grinding Mill closed about 1936. In the 1980s, there were a few residential apartments in the building, but by the 2000s, the mill was abandoned and the Topsham Fire Department burned the mill as a controlled exercise in the Spring of 2008 (Gary Howard, personal communication, 2008).

Portland. The second feldspar grinding mill in Maine was organized in the spring of 1901 and built by the Portland Mineral Company on Commercial Street. Their first quarries included the Twitchell Quarry, South Paris and the A. R. Berry Quarry in East Poland, although they may have received feldspar in small quantities from the Big Gem Mica Quarry in Peru or Mount Marie in Paris/Hebron: "At Hedgehog mountain in the town of Peru, where the Big Gem Mica Company has been operating during the past year, feldspar in large quantities and of excellent quality is found, and the company proposes to erect a mill for grinding feldspar near the mountain in the near future. (*Dailey Kennebec Journal,* February 10, 1902)." The mica boom collapsed in this year and along with it, any hope of using by-product feldspar to make the quarry profitable.

The Portland Mineral Company was never large.. It never appeared in the *Maine Register,* for some unknown reason and may have been a subsidiary of another company. Portland Mineral Company was listed in BILS (1902) and Dunnack (1920) and those references seem to span the company's business life. Watts (1916) did not mention the company's existence in a detailed list of Maine's, and the nation's, feldspar grinding mills. BILS (1902) reported: "There are at present two mills for the grinding of feldspar in the State, one in Portland and one in Topsham. ... At the Portland mill the feldspar is ground finer than flour. Ground feldspar is used extensively in the manufacture of stoneware, and the feldspar of Maine is said to be the finest in the country." The Portland Mineral Company seems to have bought flint grade quartz from various independent Oxford and Androscoggin county pegmatites and it may have evolved, principally, into a quartz grinding mill. The *Dailey Kennebec Journal* (February 10, 1902) wrote of early mining products sent to the Portland Mineral Company: "There are also quarries of feldspar in South Paris which are being worked at the present time. Feldspar from these quarries is sent to Portland, and shipped from that city to New Jersey and other points. About 15 hands are employed in the quarries at South Paris. ... Quartz is also ground, some of which goes to glass works and some to sand paper works. The demand is greater than the supply, so there is no difficulty in finding a market."

It is interesting that the Swasey Pottery Company was located across the street from the Portland Mineral Company. There was an abortive attempt to organize another feldspar grinding mill in Portland. *Daily Kennebec Journal* (December 19, 1913) reported: "Eastern Feldspar Co. organized at Portland, for the purpose of carrying on the business of mining and quarrying, milling, treating and preparing for market, ores, with $180,000 capital stock, of which nothing is paid in. Officers: President, David E. Moulton of Portland; treasurer, David E. Moulton of Portland." Although there is a mention of a Moulton Quarry in Auburn, no mill was erected.

Auburn. In 1902, the Maine Feldspar Company was organized and started quarrying operations in the Mount Apatite District, Auburn, and by 1903 the Maine Feldspar Company's feldspar grinding mill was operational near Littlefield Corner, north of Danville, a neighborhood in south Auburn, Androscoggin County although the company did not make its first appearance until 1906 in the *Maine Register*. [250] It was about that the Mount Apatite Mining District, about 3 km to the north, began to flourish as a feldspar producer and for a short period was the largest feldspar producing district in the State. (Gems were produced on Mount Apatite as early as 1862, however.) It is somewhat provocative that the second and third feldspar mills were organized so close in time. Certainly, the price of feldspar must have been attractive. The fact that the Maine Feldspar Company controlled Mount Apatite must imply that there was a period of exploration and leasing before there was mining. Addi-

Commercial Street, Portland. Harbor and wharves to the right. From Elwell's Portland and Vicinity, 1881.

several men from Boston who hope to open up some pockets of valuable tourmaline. Work was commenced, Tuesday, and a crew of men were busy all day starting the first vein. It is expected that machinery will arrive soon and then work will commence in earnest. A one year's lease with an option for buying at the end of the time has been taken on twenty-five acres of land owned by C. J. McAllister. The rich ledges will be opened with the aid of dynamite and the feldspar will be mined as well as the tourmaline." No Summit Quarry is known on Mount Apatite, and while there may have been some hand tools used, it is doubtful if the news release was anything more than a scam. It was not very sensible to have an unequipped crew working and waiting for "equipment" as the land was cleared at the time and a month's worth of hand tool attacking a ledge would not have accomplished that of several hour's worth of machinery work. Paraphrasing the *Maine Mining Journal* from over 30 years before: "If this was not a fraud, it bears a striking resemblance to one." Although mineralogists were enthusiastic about mineral specimens in Auburn, that product seldom made the news, so when specimens did become publicly announced, they had to have been good (*Daily Kennebec Journal,* May 29, 1914): "Workmen at the mine of the Maine Feldspar Co., on Mount Apatite, Auburn, opened a large pocket of magnificent rock crystals Saturday. A large number of cabinet specimens were taken out and the indications are that other pockets are near by."

The Maine Feldspar Company expanded its feldspar grinding capacity and had a mill in Topsham that was operational by December, 1912. In the post-milling history (q.v.) of that mill building, the building has been greatly altered and renovated for office space. The Auburn mill closed on October 1, 1929. The Maine Feldspar Company sold its products to potteries in both Trenton, New Jersey and East Liverpool, Ohio. The Charles Franzheim Company of Wheeling, West Virginia was a principal agent for selling the ground feldspar from the Maine Feldspar Company and the end-users were among the most famous of the brand-named potters.

Littlefield Corner, Auburn, had more than one company that bought local feldspar:

"I have a copy of the 1922 **Grand Trunk**

Swasey Pottery Company, Commercial Street, Portland. From an old postcard.

tionally, it would seem that the Portland Mineral Company had been effectively shut out of the Mount Apatite District and the only attractive pegmatite remaining for the Portland Company was across the Androscoggin River in Poland.

In the early days of gem quarrying at Mount Apatite, there was little hope of by-products "paying the freight" so that gems and specimen recovery could be subsidized. The mill may have offered hope to some (*Daily Kennebec Journal,* July 1912): "The Lewiston Sun says that a new mine, which is to be known as the Summit Mine, has been opened upon the top of Mt. Apatite at Haskell's Corner, Auburn, by

Maine Feldspar Grinding Mill, Auburn, 1906.
Courtesy U. S. Geological Survey.

Littlefield Corners mill ruins after much expansion from original building, Danville, ~1958.
Unidentified Lewiston newspaper

Shippers Guide for Carload Lots listing those who shipped and received and whether their sidings were private owned or using the team track.

The only station between Portland and Island Pond showing mineral shipments is the Littlefield agency which was located 1.79 miles in on the Lewiston Branch from Lewiston Junction. Actually there was no station building or agent there except a flagstop shelter for any passengers, so the paperwork was probably handled by the Lewiston Jct agent. Most interesting is that the only shippers even listed for Littlefield are both mineral firms, the Eureka Spar Company and Maine Feldspar Company. Unfortunately there is no indica-

tion whether raw spar was received or it was processed spar being shipped out, but both firms are listed as having private sidings. (Randall Bennett, *The Railroad Journal,* www. thebetheljournals. info/ Railroad%20Journal. htm; accessed 2007)"

The *Maine Register* never listed two feldspar companies in Auburn and the Eureka Spar Company did not have a grinding mill in Maine, but apparently bought feldspar from independent miners. Watts (1922) indicated that Eureka Flint and Spar Company had its grinding mill in Trenton, New Jersey. Eureka processed feldspar from Maine, Connecticut, New York, North Carolina, and Canada.

Yet another mill was to open in the Auburn area, later (*Daily Kennebec Journal,* June 12, 1918): "Lewiston Feldspar and Mineral Mining Company., organized June 7 at Portland. Capital stock. 100,00: all common; nothing paid in; par value, $10; shares subscribed; 11. President, R. L. Johnson, Portland; treasurer, E. Connor, Portland; clerk, James J. Manter, Portland; directors, R. L. Johnson, E. Connor, James E. Manter, David F. Drew and Charles W. Hamilton, all of Portland. Purposes, a general business in the mining and manufacture of feldspar, mica, quartz, etc." The follow-up was encouraging: (*Daily Kennebec Journal,* August 6, 1918): "A feldspar mill is soon to be erected on the Ferry Road near Pleasant Street, Lewiston, by the Lewiston Feldspar &Mining Co. The site has already been secured, and plans made for the mining of feldspar on an extensive scale. A building has been secured to accommodate the workmen." The mill was never constructed. It is possible the that mill was intended to process Oxford County feldspar as the Mount Apatite District in Auburn was already working to capacity.

Although mining accidents do not seem to have been a problem in Maine, any that come up seem to be in the Mount Apatite District (*Daily Kennebec Journal,* September 20, 1921): "Fulvio Gigli vs. the Maine Feldspar Co. and the Federal Mutual Liability Insurance Co., insurers; petition to fix the amount to be allowed for medical surgical and hospital services…" In the same issue, the same case had a hearing for a "lump sum payment".

The Auburn pegmatites were a proposed destination for a field trip (*Daily Kennebec Journal,* August 5, 1922): "Plans for the first meeting of the Association of Maine ge-

Consolidated Feldspar Grinding Mill (abandoned), Topsham, c1958, Courtesy Ben Shaub.

Consolidated Feldspar Grinding Mill (renovated into offices, etc.), from railroad tracks, Topsham, 2008.

Feldspar in train cars near Trenton, New Jersey potteries, destination of some Maine feldspar, 1930s postcard.

Goldings Sons, Feldspar Grinding Mill, East Liverpool, Ohio. Destination of some raw feldspar from Sagadahoc County. 1910s postcard.

ologists, to be held in the region around Lewiston and Auburn on Friday, August 11, have progressed rapidly and arrangements have now been made which will insure an interesting and profitable day and evening for all who participate."

Topsham's Second Mill. The Maine Feldspar Company expanded its feldspar grinding capabilities and opened a second grinding mill, in Topsham, in 1912, although it isn't clear in what years, subsequently, the new Topsham facility was in operation. The mill was located on Elm Street along the railroad tracks opposite the former Topsham town hall. Many of the Company's quarries were located at the north end of Tedford Road Topsham.

In 1928, Herbert P. Margerum of the Goldings Sons Company of Trenton, NJ purchased a number of companies to form the Consolidated Feldspar Company. Goldings Sons merged with Maine Feldspar Company, Brunswick, Maine (including feldspar mills at Auburn and Topsham), Bedford Mining Company, Bedford, NY, Dominion Feldspar Company and New York Feldspar Company, both of Rochester, NY, Isco-Bautz Company, Murphrysboro, IL, and the Erwin Feldspar Company of Erwin, TN, as well as properties near Keystone, SD and Kingman, AZ. [351] (See further historical details in Feldspar, Mica, and Beryl Appendix.)

Bath. The J. W. Cummings feldspar grinding mill was built in Bath, on the west shore of the Kennebec River about 5 km south of Merrymeeting Bay. John W. Cummings [b. August 12, 1859] was the son of Thomas Cummings [d. 1897], both of whom were from Pottsville, Pennsylvania. Thomas Cummings was a mining engineer and his son pursued a career in feldspar processing, beginning in Connecticut and later in Trenton, New Jersey, before moving to Maine in 1915. The *Daily Kennebec Journal* reported the expansion of Cummings' mining ((September 13, 1916): "J. W. Cummings shipped his first shipment of feldspar from his new quarries in Georgetown this week consisting of two car loads. Work is progressing on the wharf at the quarry and a crew of 20 men are employed. The first shipment was to Trenton, N.J. Mr. Cummings has several quarries on his recent purchase in Georgetown." A follow-up article revealed *Daily Kennebec Journal* (October 20, 1916): "John W. Cummings has a crew of men employed in his feldspar quarries in Georgetown and is shipping about 120 tons a week by rail to Trenton, N.J. The spar is shipped to this city by lighter [freighter?]. The work will be continued as long as real weather conditions are favorable, which usually lasts until the last of December." The presence of the feldspar mill in Bath encouraged many prospectors (*Daily Kennebec Journal,* May 29, 1916): "Another of those deposits of feldspar for which Maine is famous has been uncovered, this time at Sebasco {Phippsburg] and it is said to be a great vein of clear stuff. A crew of men will at once be put to work getting out the valuable feldspar." The *Portland Herald,* October 3, 1928 indicated that the Cummings grinding mill was incorporated August 15, 1915 and the mill was built for $84,000 and was operational in 1919. Its location was the south end of Reed's Wharf on the Kennebec River. The mill also in-

Thomas Cummings Feldspar Grinding Mill, Bath, Feldspar stockpile between railroad tracks and Kennebec River estuary. Portland Herald, October 3, 1928.

creased its capital stock in 1927, suggesting it was also expanding capacity. In 1928, Cummings mill had quarries shipping to it from Bowdoinham, Georgetown, Phippsburg, and even South Paris. Although located on the banks of the Kennebec River, the Cummings Mill was apparently connected to its South Paris quarry via the Lewiston, Augusta & Waterville Street Railroad by which South Paris feldspar was also shipped. ["A spur track from the then existing street railway ran directly into the [Cummings' Mill] yard and for a number of years the industry thrived (*Bath Daily Times,* September 23, 1944)." However, this reference only indicated that ground feldspar was shipped from the mill to the Maine Central Railroad yard by street trolley. When the street trolley was discontinued, the cost of transporting product to the station apparently increased. In the 1920s, Cummings' best customer was the Champion Spark Plug Company and during the company's run, Georgetown feldspar was all that Champion used in its spark plugs, but the mill also shipped ground feldspar to Trenton, New Jersey and to East Liverpool, Ohio. The Cummings mill was shut down in 1930, presumably for economic reasons, but the death of John Cummings not long before, as the mill's founder, probably had much to do with the cessation of the business (*Bath Daily Times,* September 23, 1944). There were up to 30 employees in the mill. (At one time, Golding and Sons controlled some quarries in Georgetown and shipped some of their crude feldspar to New Jersey for processing.)

In 1937, the *Maine Register* indicated that the Cummings Company had been re-organized as the Ceramic Feldspar Company, but there was not a subsequent listing of that company in 1938. The new officers of the mill were: Thomas J. Cummings (John W. Cummings's son: treasurer and general manager), Richard Wainford of Trenton, NJ, president), Charles Wainford (of Bath, plant superintendent), and Lauren N. Sanborn (of Portland, clerk). In 1936, the new company had exercised their option to buy the idle milling company (*Portland Herald,* October 12, 1936). Output of ground feldspar from the Bath mill was stated to be about 600 tons per month with a mill crew of 11 men and the sup-

John W. Cummings, c1920. Courtesy of Lorna Cummings Howard.

porting feldspar production from 24 Sagadahoc County quarries, with each quarry having a crew of 3-6 miners. Their ground feldspar was mostly exported to England, Germany, and Canada. Thomas J. Cummings, president of the J. W. Cummings Company, surrendered his mining lease on the Lester P. and Cora M. Twitchell mining property in Paris on June 4, 1937. [350] The fact that the Cummings mill held a mining lease on the Twitchell Quarry, as had the previous Portland Mineral Company, makes a provocative link between the two companies.

After the closing of the feldspar mill, it was rented to river dredgers mostly for offices and access to the Kennebec. The land saw a few new owners, unrelated to feldspar and in 1944, the property was purchased by Bath Iron Works and

J. W. Cummings Feldspar Grinding Mill,
c1936, Reeds Wharf, Bath.
Courtesy Lorna Cummings Howard.

the feldspar mill buildings were torn down in September of 1944.

The Organization of the Oxford Mining and Milling Company and the West Paris Feldspar Mill

With five feldspar grinding mills in Maine, one might wonder about organizing a sixth. Maine was the leading producer of ground feldspar in the Nation in the early years of the twentieth century and the market for that product continued to expand. Before the mid nineteenth century, the USA was dependent on European china and porcelain manufacturers. In succeeding times, domestic potteries filled the market needs for fine stoneware,

For a mill, or any business, to be organized, there are logical steps: 1. Recognize an opportunity, 2. Examine the economics, 3. Raise money to get the business going, 4. Run the business. The first two steps have been the hardest to document. The third follows in the next paragraph, while the fourth stage occupies a large portion of this book. The course of human events never seems to follow straight line logic, at least for extended lengths of time. The **Third Stage**, raising money to get the business going, is a matter of public

record, although the details are widely scattered.

The Oxford Mining and Milling Company [OMMC] was organized on April 23, 1924, with offices originally at 85 Exchange Street, later at 102 Exchange Street, Portland, Cumberland County. William L. Adams, Sr. was president and treasurer. Directors included: William L. Adams, Sr. (of Portland), William L. Adams, Jr. (of Portland), Harry M. Packard (of Bethel), Arad T. Barrows (of Brunswick), Arthur R. Stowell (of Greenwood), and Charles H. Tolman (clerk, of Portland). [359] By its incorporation, OMMC became the sixth feldspar grinding mill in Maine.

The *Maine Register* listed the company with offices

West Paris village looking to SW, c 1925 postcard
Arrow above knoll indicates Perham Quarry location.
Mill off scene far right.

in both (102 Exchange Street) Portland and (RFD #1) West Paris only for its first two years, and then only at West Paris until 1943. In the 1915 Portland City Directory, both William L. Adams, Sr. and Jr. were listed as superintendent and foreman, respectively, at 33 Pearl Street, the location of Megquier & Jones Company, builders of steel frame buildings and bridges. Arad Thompson Barrows [1881-1928] was a Civil Engineer who graduated from the University of Maine in 1907 and wrote a thesis on railroad bridge construction, (*Brunswick Record*, v. 27 no. 7, January 3, 1929, obituary and picture). Charles H. Tolman was a lawyer in Portland and Justice of the Peace. Arthur R. Stowell was the postmaster of Locke Mills, Greenwood (1915-1934).

West Paris "new cement bridge" over Little Androscoggin River. Route from Perham Quarry proceeds left after crossing railroad tracks immediately before business district. Mill off scene to left. c1930 postcard.

Each OMMC corporation member had one share of stock valued at $100, but the entire amount of stock available would have given the corporation a potential of $10,000 operating capital. [360] The single share may have been "voting stock". (Harry M. Packard was not Harry A. Packard [November 1, 1886 Norway-July 31, 1965 Paris], one of George Howe's youth group members of twenty years before in Norway.)

The new corporation found itself under-capitalized, however. Later in the year, on December 30, the board of OMMC voted to amend the bylaws allowing an increase of the capitalization of the corporation to $100,000. OMMC obviously raised its needed investment money and soon started work building its mill.

It was later revealed that (*Portland Herald*, May 3, 1929): "The principal [new] financial backers of the Oxford Mining and Milling Company have been C. H. Farley, William W. Mitchell and William Adams, all of Portland; A. C. Perham is a stockholder and director of the company." Perham (1987) credits the efforts of C. Alton Bacon, brother-in-law to Alfred Perham, and a J. A. Cheney with attracting some financial participants to the formation of the company. Bacon was a barber in West Paris and had a built-in connection with the area's social network. [47]

The effort to get the enterprise going required some stimulus. Why would one want to start a new business? The feldspar mining business was barely known in Oxford County when Alfred began to think about the possibility of mining his property in 1923. He had been in the farming and lumbering businesses, but those activities were becoming marginally profitable (*Lewiston Journal Magazine*, April 3, 1948). In order to unify the story of the mill, it is necessary to understand that the prepared mind recognizes opportunity.

The history of Oxford County always had flirtations with mining. Gems were found at Mount Mica in Paris as early as 1821 and had been mined frequently for over a century. Geological surveys were made in the late 1830s and again in the early 1860s. Silver mining was a dream in the 1880s. However, the Perham Farm had been closer to the fray than most pieces of land. Additionally, there was at least one mineral enthusiast in the family, a young Stan Perham.

An unidentified 1970 era newspaper clipping, featuring an interview with Rupert Aldrich, indicated: "The first feldspar rock ever brought from a mine in the West Paris area was by Charles Aldrich, the grandfather of Rupert Aldrich, back in 1915. [The outcrop was apparently the same as the future Perham Quarry.] He loaded the feldspar into the horse-drawn wagon and then drove to Auburn for the processing operation." Aldrich did not mine feldspar for very long, however.

About 1918, the Maine Feldspar Company, of Auburn, had "cruised" the West Paris side of Curtis Hill in search of commercial deposits of feldspar. (Cruising is a form of searching for minerals. A "cruiser" will inspect ledges for interesting minerals and when a promising spot is located, might drill a small test hole for blasting. Cruisers might, or might not, have obtained permission from landowners, in advance of their search.)

There isn't a contemporary genesis story of discovering feldspar on the Perham Farm and consequently, there are a variety of later stories to choose from. There are also several stories regarding what happened to convince people that they should invest their hard earned money in a factory. What is known is that, by 1924, the price of ground feldspar was at an all time high and the economics of starting a feldspar grinding mill in Oxford County seemed like an opportunity: the necessary **First Stage**. It's still a problem of

West Paris looking westerly on Main Street toward railroad tracks. Maple Street near tall Elm tree on right. Mill located off scene far right at end of Main Street. From a 1927 postcard.

West Paris looking easterly toward Trap Corner area. Wooden sidewalk on left. 1920s postcard.

West Paris looking southwesterly from railroad tracks across Little Androscoggin River up High Street toward the Perham Quarry Area. 1920s postcard.

the chicken and the egg. Do you see that the prices of a product are high and go out to satisfy the demand or do you make a discovery and see how the good fortune can be used to best advantage?

Unbeknownst to Alfred Perham, and other members of the Oxford Milling and Mining Company's board, the reason the price of ground feldspar was artificially high in 1924 was because of a long labor strike in the porcelain industry. Stoneware companies stopped buying raw materials. Reserve stocks of ground feldspar were very low as distributors of the raw materials to the factories did not buy ground feldspar and, consequently, grinding mills around the nation greatly decreased production and almost stopped buying crude feldspar for a while. Affected miners stopped mining. There were, of course, customers for ground feldspar not in the ceramic industry. The settlement of the strike and resumption of production in the ceramics industry quickly depleted available ground feldspar stocks in distributors' hands. Local feldspar grinding mills had to respond by resuming production levels and, therefore, previously inactive feldspar quarries across the nation had to scale up production for their associated customers. Ground feldspar temporarily became more valuable due to the demand and it was about a year before prices began to decline through re-supplying. Unfortunately, ground feldspar production capacity also increased across the USA, not just in West Paris. By decade's end, compounded by the advent of the stock market crash and its consequences to the housing industries, the glut of ground feldspar supplies caused prices for crude and ground feldspar to rapidly decline, eventually by more than half.

Despite the fact that prices were attractive in 1924, the establishment of a new feldspar mill in a new district, having no real historical feldspar production, was a courageous move. To be sure, Alfred and Harold Perham, and other family members, worked hard to develop a large feldspar resource, but their quarry was not the only feldspar producer in the immediate area. The Twitchell Quarry had been operating sporadically in South Paris, by about 1901, and the Bennett Quarry in Buckfield, about 12 miles away was also working, perhaps in the late 1910s.

Stage Two - Proving the Economics

Although feldspar is easy to identify,

the probable extent of a feldspar deposit needed to be verified. OMMC contracted a site appraisal from the leading local expert on minerals: "In 1923 or 1924, George Howe was asked by the organizers of the feldspar mill at West Paris to examine the property on which the [future] A. C. Perham quarry was located. George was paid for several days to examine the pegmatite and estimate its extent for commercial feldspar production (Richard Box, personal communication, 2000)." [130] There was also a coincidence in that Harvard student, Kenneth K. Landes, was investigating the Noyes Mountain Quarry in Greenwood and the Bennett Quarry, Buckfield and he may have searched the area for additional mineral localities. It's even possible that Kenneth K. Landes was being referred to in a quote from Stan recorded by Woodbury (1957): "Alton Bicknell, a man who knew something about mines, told Stan's father, 'I'd put more stock in what that boy [Stan] says than in many of the city experts [Landes].'"

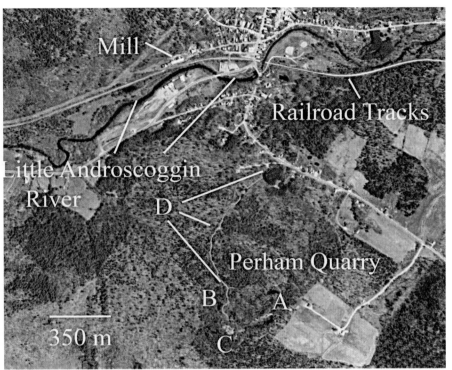

West Paris and Perham Quarry Area. A Original headwall of Perham Quarry B Current Quarry Entrance C Dump Area D Quarry Access Road. A. C. Perham farm formerly on Mine Road "D". Courtesy Google Earth.

After Howe's investigation, actual prospecting had to begin. "A first blast proved worthless, but then young Perham recalled a spot where there was a breakdown in the rock structure and where evidences of feldspar were to be seen. A blast at this point ripped out between 75 and 100 tons of ledge and the Perhams found themselves at the heart of the feldspar vein (*Lewiston Journal Magazine,* April 3, 1948)." The two decade old story doesn't mean that Howe chose poorly or that the later blast yielded all of the approximate 100 tons of feldspar, especially as a single explosion capable of "ripping" that much feldspar out of the ground would have been alarming. The separate stories are probably told from a point of view and probably are best combined into a single episode. George Howe, a very popular man, with considerable mineral knowledge would probably have had an entourage of onlookers during the site evaluation, not just Perham family members. At any rate, the evaluation and blasting were necessary to go on to the next stage, organization of a company and the raising of capital.

An additional variation of the feldspar discovery story involves Alfred's son, Harold Perham, who became an important miner. The Perham Farm feldspar outcrop was investigated by Alfred and Harold, in 1923: "... a new phase of life developed in conjunction with the opening of the 'Paris Crystal Spar' Feldspar Quarries by his father... They broke into the Feldspar game together." (Stan, who was then 17 and had his own store from which to sell mineral specimens, also certainly took an active part in exploring for feldspar.) [A newspaper clipping, 1970 era, suggested that brothers, Ronald and Stan Perham made the

first who did the drilling and blasting at the Perham Quarry,] Woodbury (1957) provided additional detail:

"'I don't suppose Dad had a single penny left to buy dynamite when we quit [the first time we prospected] ... But I knew there must be feldspar down there somewhere. Every sign was right.'

A few days later, the boy [Stan] was driving the cows in from the pasture beyond the hill when a moss-covered ledge caught his eye and he scraped it bare with his knife. Here was

*Original West Paris Feldspar Mill, c 1926
View from near Perham Quarry
Photo Vi Akers. Courtesy Sid Gordon*

Perham Quarry, full crew, c1929: L-R, Alfred Perham, Leon Proctor, Valentine Oja, Eugene Wilson, Gus Roberts, Roy Perham, Earle Bacon, and Charlie Bane. Note railroad tracks.
Courtesy of West Paris Historical Society. Robyn Greene photograph.

Bates railroad station, West Paris, nearly adjacent to Feldspar Mill which is off to the left, 1920s postcard.

the work myself,' Stan says. 'I was built for it – close to the ground – and I was tough. Sometimes we would blast rock all day at 20 below zero, done up in reefers and mittens and lamb's-wool boots. …

A short time after the first blast, General Electric offered a substantial sum for the huge Perham quarry. The feldspar there would produce high-quality electrical insulators. But the Perhams refused to sell. They had built a processing mill in West Paris, employing village people. If the big company took over the mine it would ship the rock away and the mill would close down. Remembering that fateful decision, Stan comments, 'We're idealists, I guess. A thing must benefit everybody or it won't help us.'"

Woodbury (1957) seems to contain some shuffled details. For example, the approximate $300.00 received from the sales from the initial blast, divided by $7.50, means that about 40 tons of feldspar were initially sold. It is unlikely that the 40 tons of feldspar came from a single blast, although the initial blast may have been in a ledge of solid feldspar without impurities. If there were no West Paris mill, where was the feldspar sold? Was it sent to Auburn or Topsham as a market experiment? It is interesting that General Electric Company wanted West Paris feldspar. General Elec-

the feldspar, 500 feet from where he had set off the unsuccessful blast. Next day the boy and his Dad dug an extra deep hole for the last charge of dynamite their slender budget could afford. When it blew it laid open the ledge and mined up a whole carload of the finest feldspar they had ever seen. That rock sold for more than $300.

The Perhams drove their new mine into what proved to be a huge vein of prime rock. In the next few years they took out 10,000 tons of spar that brought a premium price of $7.50 a ton. But every cent went back into machinery and workmen's wages and the building of a heavy road to cart the stuff out. 'I did a lot of

West Paris looking westerly with Church and Maple Streets on right.
Probably 4th of July Crowd.
1920s postcard.

West Paris Feldspar Mill, February 1926. Steam from locomotive engine. Snow covering feldspar stockpile in front of mill building. Photo by Vi Akers. Courtesy Sid Gordon.

Feldspar Mill, February 1926. Trap door leading to jaw crusher in lower level. Note oversize rocks in pile, also bagged ground feldspar. Photo by Vi Akers. Courtesy Sid Gordon.

tric may already have been prospecting for flint-grade quartz for high-temperature glass in Maine, possibly Oxford County. General Electric also was having Dick Nevel mine for pollucite in Buckfield in 1924-1925 and later Newry in 1926 and he was having prospecting done in a wide variety of nearby locations. The General Electric Quartz quarry on Round Mountain, Albany was possibly in the exploration stages by the early 1930s.

Stage Four - Oxford Mining and Milling Company's Plans

When the OMMC investors and board members began to schedule operations of the projected feldspar mill on December 1, 1924, the Oxford Mining and Milling Company leased, for ten years, the "total output of feldspar of all mines now owned and controlled" by Alfred C. Perham and "Deliveries shall begin as soon as the mill is ready to take care of the materials and shall continue without interruption… [A. C. Perham] shall so deliver not less than four thousand (4000) tons during the first year from date of this agreement and not less than five thousand (5000) tons during each year thereafter; and such further quantity, not exceeding ten thousand (10,000) tons per year, as [OMMC] may require." [303] The accompanying aerial image shows the current Perham Quarry location obscured by vegetation. The original excavation of the Quarry progressed down hill and changed direction from westerly to northwesterly. The plan view of the Quarry is an open "J". Although the "proper" name of the West Paris mill was the Oxford Mining and Milling Company Feldspar Mill, it was usually known as the West Paris Feldspar Mill, or informally as the "Perham Mill". Alfred Perham was a member of the American Ceramic Society from 1924 to 1929.

OMMC had originally purchased a large parcel of land on Bird Hill Road, adjoining the Birds Hill "burying ground", Bethel, very near the village of Locke Mills in Greenwood on April 13, 1924, possibly in anticipation of building a mill there. The nearest land to the railroad tracks was about 1 km via a dirt road, however.

In March of each year, the mill management and Alfred Perham were to agree to an amount of feldspar to be mined during the up-coming season, but if the mill required more feldspar than Perham could supply, beyond the specified amount, the mill could purchase feldspar from other suppliers (miners). The delivered trimmed feldspar had to be of pottery grade and

Overhead loading bins and wooden boxed delivery chutes in tower section of mill, 1926. Vi Akers photograph courtesy Sid Gordon.

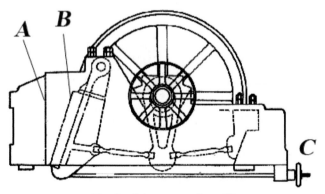

Typical jaw crusher. [243]

of $10.00 then this contract shall either cease or [Alfred C. Perham] shall allow [OMMC] to take over the mines owned by [Alfred C. Perham] and mine the feldspar themselves, paying ... fifty cents per long ton royalty..." [303] Unfortunately, the price of feldspar was to fall drastically in the near future.

The negotiations for the land on which the feldspar mill was built, undoubtedly started before OMMC had completing negotiations with Alfred Perham for feldspar. On February 26, 1925, OMMC formally purchased a lot of land, with a warranty deed, adjacent to Alfred C. Perham's land, from the Burnham and Morrill Company. [343, 45] The land was on the north side of the railroad tracks to the west of West Paris village. Despite the winter time purchase, the agreements in place meant that building had to begin immediately. Perham (1987) noted: "The steel for the mill was put in place in ten-degree-below-zero weather." The U. S. Bureau of Mines noted the presence of new mills across the country: "New [feldspar grinding] mills were established during the year in California, Colorado, Maine, and New York. ... In 1925 Maine ranked third among the States in the production of crude feldspar but showed a decrease of 5 percent in quantity and value compared with 1924. Androscoggin, Oxford, and Sagadahoc were the producing counties. The Maine Feldspar Co. is the largest producer of both crude and ground feldspar. During the year the Oxford Mining & Milling Co. established a mill at Bates, Oxford County." [161] (Before 1918, the railroad station was simply called "West Paris".)

The figures for Maine ground feldspar production between 1925 and 1926 reveal that there was a significant change. The State's entire output increased from 26,314 tons with a value of $497,348 in 1925 to 32,676 tons with a value of $587,406 in 1926. Maine replaced New Jersey as the second most productive state for ground feldspar. (Note: New Jersey did not produce its own crude feldspar, but imported all of its raw materials, primarily from Maine and North Carolina.) There were several factors at work affecting the Maine totals. The Mount Apatite District of Auburn and

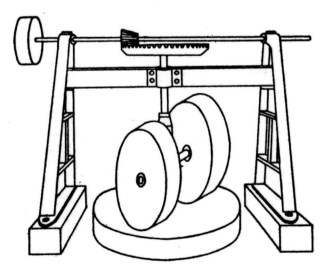

Typical buhrstone crusher. [243]

"fairly free of black tourmaline, black mica and garnets". (All three of the listed minerals contain iron which would badly stain any pottery made from ground feldspar which contained them.) Pottery grade feldspar was also called #1 Grade: "No. 1 is carefully selected, free from iron-bearing minerals, largely free from muscovite, and contains little or no quartz, usually less than 5 per cent." Feldspar would be purchased at $10 per ton with weekly accounting and payments. The $10/ton price was a munificent amount in the time period, but the lease also specified: "... if at any time the actual expense or cost of mining and delivering the feldspar herein contracted for shall equal the purchase price

Buhrstone used at the Cathance River mill.

Poland was rapidly declining and the Danville feldspar grinding mill closed in less than five years. It was probably this District's decline that resulted in a decrease in Maine production between 1925 and 1926, despite the West Paris mill's going online. The West Paris mill probably went through growing pains and adjustments and the first year probably saw only a partial year's production. The increase of Maine production in 1926 was the result of a full year's work in West Paris, in spite of Mount Apatite's shrinking market share of ground feldspar. (See Appendix B)

Inside Details of the West Paris Feldspar Grinding Mill

The interior of the Feldspar Mill was rarely pho-

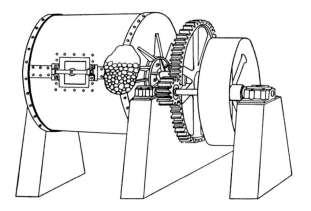

Typical ball mill. [243]

tographed. When the mill was still the property of the Oxford Mining and Milling Company, in 1926, local photographer, artist, and naturalist, Vivian "Vi (Vee)" Akers visited the mill and captured a few images. The process of milling and grinding feldspar has been well-summarized by Perham (1987) and the reader is directed to that excellent source for additional details.

The accompanying photograph shows the West Paris feldspar mill in winter. Note large outdoor feldspar dump

(probably over 500 tons or 1,000,000 pounds+; 5 feet high, probably 25 feet wide, 50+ feet long) covered with snow. Also note railroad car on siding on the left side of the building and passing steam engine with steam plume, on main track, heading northwesterly toward Bryant Pond. Feldspar from the Bumpus Quarry would be off loaded from the siding, and was initially stored outside, along with feldspar trucked directly from the Perham Quarry. The outside storage was continued for about 20 years and was a problem. Wet rock could not be ground as it would "cake" together. A paste of the milled feldspar "fines" would clog the process entirely. The feldspar had to pass through a drying oven before it could be ground. In winter, there was the added problem of removing feldspar from the stockpile as it became lodged in ice. During those times, the feldspar had to be "mined" a second time. There was a set of scales to weigh feldspar trucks in front of the office building. (The feldspar from the train, principally from the Bumpus Quarry, had already been weighed and the bill of lading was a record of the weight and fees charged.) At least during the 1960s, late arriving trucks would have a sign-in sheet available when office staff had left for the day. Truckers who reported higher

Ball mill, West Paris, 1926. Photo by Vi Akers. Courtesy Sid Gordon.

loads than their truck could hold were reprimanded if they tried to do it twice in a row (Frank Perham, personal communication, 2007). Contract miners were required to break up their "over-sized" rocks, rather than leave that job to the mill crew. The person receiving the load had to inspect the feldspar to verify that it was high-enough grade mineral, with the proscribed acceptable amounts of impurities. Watts (1916) wrote about a "secret" every Oxford County mineral collector already appreciated: "The feldspars of Maine, for example, are not naturally very pure feldspars. In fact, no feldspar producing section in the United States has so many injurious minerals associated by nature with its feldspars. Nevertheless, the Maine feldspars are furnished to the potter in a state of purity second to none, if we except the quartz content, which can not be eliminated and which must of necessity increase with continued mining operations in any district. This high state of purity is obtained by untiring vigilance on the part of the miner and at the mill to insure against even the smallest introduction of iron bearing impurity."

The Bumpus feldspar enjoyed a high reputation in this regard. Any feldspar loads with excess impurities were rejected at mills and there was the quandary whether to truck it back to the quarry for hand sorting, mixing with higher grade rock, etc. or to take it to the town dump or elsewhere. In fear of having rejected loads, miners were particularly careful about purity and many mineral collectors wonder about otherwise handsome blocks of feldspar remaining in the quarry dumps where they are collecting.

The optimum size of the raw rock coming into the mill was under four inches in diameter and which could be fed into a jaw crusher. Suitably sized rocks would be fed in the hopper on the left and the rotary action of the crusher would push a moving steel plate, "B", against a stationary steel plate, "A", and the rocks would break until they would drop through the maximum opening of the plates at the bottom, usually one inch or less. A screw adjustment wheel, "C", could change the width of the opening. A bin would catch the fragments to be moved to another crusher such as a buhrstone mill.

The buhrstone mill was merely a set of stone wheels, up to five feet in diameter, made of granite or quartzite, which would crush mineral fragments simply because of the weight of the wheels. The wheels would run over the fragments until they had a uniform fine sandy size. In both stages of crushing, mica, iron minerals, and excess quartz and other minerals had to be removed before they were crushed. The people tending the machines, therefore, had to be trained and vigilant. Buhrstones were probably not used in West Paris in its latest years of operation. The final grinding stage was accomplished by using a ball mill or a tube mill; the principal difference between the two is, as the name implies, that the tube mill is longer. In effect, the ball mill was a rotating drum that was lined with flint blocks. The flint prevented wear to the ball mill walls and helped in the grinding process. The ball mill would be partially loaded with flint pebbles up to 3 inches in diameter.

Ball mill, West Paris, 2007.

The ball mill was rotated at a speed, about 10 rpms, so that the feldspar and flint pebbles would cascade over each other and against the hard rock lining of the mill. The smashing of the pebbles against the feldspar would eventually reduce it to powder finer than flour. If the mill were rotated too fast, there would be no cascading of the pebbles and they would ride around with little or no movement. If the mill were filled more than half full, the grinding action would be much reduced and the operator of the ball mill had to be particularly trained. The ball mill was emptied through a coarse screen so that the flour-like feldspar could quickly sift out, but the flint pebbles were retained on the screen. The procedure was excessively dusty and, as the dust always contained quartz, mill workers frequently developed lung conditions, many having their lives cut short from silicosis. Ed Herrick, Slim Bubier, and Reg Ross were among the mill workers who died because of silicosis (Frank Perham, personal communication, 2007). Cigarette smokers were particularly hard hit by the doubled damage of the cigarettes themselves, and the frequent deep inhalation smokers have.

The ground feldspar was transported through the mill by conveyors and rotary elevators, which reduced the amount of handling. The ground feldspar was conveyed into a hopper and an 80-pound bag would be filled and the bagman would be the dustiest person on the crew. Eventually, hopper train cars would be loaded just by using a boom feeding tube to let the ground feldspar flow into the car.

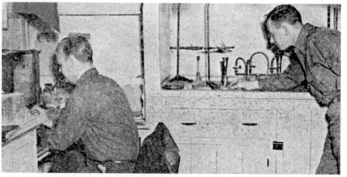

L-R. Orin O. Cole and Durward O. "Bunny" Ring in West Paris feldspar testing lab. c1948.

The small residence style house next to the sheet metal sheathed mill was not just the receiving and weighing station, as well as the shipping and payroll office, it also housed the feldspar testing laboratory. The ground feldspar would be chemically analyzed and a certificate of analysis produced for each lot. The laboratory would also make a standard sized cone of the ground feldspar and would fuse it recording the temperature at which it did fuse. If the fusing temperature was off from specifications, there might have to be a reformulation of the particular ground feldspar lot by blending it with pure feldspar.

In some of the last years of the mill, the quality control people would keep analyzed samples of the mill's ground feldspar and would provide those samples and their analyses to customers, as the analyses usually did not vary much. Many of these samples would be months old and did not come from any particular shipment. One out-of-state customer tested a 100 ton shipment using their own samples and rejected the full shipment as it did not meet the claims of the mill's analysis. The hopper car was expensive to rent, and union railroad labor was also too expensive to hire to clean out the car. The mill sent two of its staff, Wayne Lawrence and Leo Morrison, to unload the ground feldspar. At first, the mill men tried to convince the railroad to let the ground feldspar slowly filter out along the railroad track bed as the car was being hauled to a new station, but the railroad refused. The men worked two weeks using five gallon buckets to hand bail the ground feldspar out of the railroad car and take it to a dump. The "fine as flour" ground feldspar was like quicksand to stand in and footing was treacherous.

The Discovery of Minerals on the Cummings Farm

With the West Paris mill assured a constant minimum amount of feldspar, exploration continued to locate new sources of feldspar. Harold Perham, one of Alfred's sons, began leasing properties for the purpose of mining and he eventually worked at seven feldspar quarries. One of his best leases was at the Bumpus Quarry on the Cummings Farm. For the next half century, this quarry was involved in a saga of discoveries beyond feldspar, particularly beryl and rose quartz, resulting in family feuds, predation by businesses, and government intervention.

The Cummings family originally purchased their farm on Valley Road in

1870 by deed to Abigail Cummings and Joseph Cummings. (Joseph and Abigail were married in 1865.) Two additional parcels of land were also acquired in 1870 and in 1899. [300] Abigail and Joseph had six children: Wallace Edwin [April 8, 1869 Albany – March 17, 1951], Cora Emma [July 3, 1870 Albany – April 22, 1909], Sybil Elizabeth [October 5, 1872 Albany- June 28, 1927], Laura Josephine [September 10, 1874 Albany – August 6, 1956 Auburn], Allen Everett [November 24, 1876 Albany-1943], and Viola Eleanor [November 7, 1878 Albany – January 15, 1951 Albany] and they all grew up on this farm. Laura graduated from Edward Little High School, Auburn, in 1892 and taught in the Auburn schools before getting married to Harry E. Bumpus on June 30, 1899 in Auburn. [15] Harry Bumpus [November 23, 1875 Auburn – April 10, 1953] was age 24, having just graduated from the College of Pharmacy in Chicago, when he married Laura. She and Harry were members of the Elm Street Universalist Church in Auburn.

The genesis story of the feldspar discovery on the Cummings Farm has become a twice-told tale, as have been

Side view of ball mill, etc., 1926.
Vi Akers photograph courtesy Sid Gordon.

Cummings Family, Summer 1906. Front. Edith Cummings [Stearns], Laura Cummings [Pinkham].
Second Row Annie Cummings Hazelton [Bumpus], Joseph Wiley Cummings, Abigail Ward Jackson
Cummings holding Harlan Maurice Bumpus. Third Row Etta Briggs Cummings, Cora Cummings
[Cummings] holding Adelia Cummings [Waterhouse], Sibyl Cummings, Viola Cummings.
Back Row Wallace E. Cummings, George Cummings, Allen Cummings,
Laura Cummings Bumpus holding Sibyl Bumpus [Staula],
Charles Dunham [husband of Viola Cummings]
holding Everett Dunham. Courtesy Barbara Inman.

Allen E. Cummings, 1906.
Courtesy Barbara Inman.

Wallace E. Cummings, c 1935
Courtesy Barbara Inman.

Harry Bumpus c. 1904.
Courtesy Ava Bumpus.

Laura and Harlan Bumpus, ~1904.
Courtesy Ava Bumpus.

many of Maine's mineral discovery stories. The earliest known <u>published</u> story relating to the discovery of feldspar on the Cummings Farm, comes from about 1952 (unidentified newspaper clipping dated due to reference to then current owners, see remainder of article in mention of the sale of United Feldspar and Minerals Corporation). In this newspaper article, it may be noted that Allen Cummings's name was variously mis-spelled in the mineral story and reminds us of comedic situations where fictional brothers might have the same names, poking fun at rural families which might have siblings named, "Larry, Darryl, and Darryl":

"*Brothers Hoed Garden Over Rich Beryl Deposit.* The Cummings brothers Allen and Alan, hoeing in their Albany farm garden back in the early 20s, might have tossed the rocks away a little less vehemently had they known their garden was the roof over something so valuable. It was these rocks hampering the Cummings boys [30] in their farming that led to the discovery of the beryl crystals.

Because some of the stones looked peculiar, the Cummings brothers eventually collected a small bag and took them to a young college student, Stanley Perham, at West Paris. They understood Perham was studying geology at Bates College.

After studying the stones Perham told the Messrs. Cummings that they had unearthed an especially fine grade of feldspar. He said their garden probably covered a valuable deposit of the mineral used in the manufacture of crockery, tiling and dentures.

All of this didn't excite the Cummings boys. They kept right on farming and contin-

Bennett Quarry operated by Maine Feldspar Company crew, Summer 1924. Note board walkway on narrow ledge allowing exit from upper level. Hole on lower level is the famous 1924 crystal pocket.
From Landes (1933) also in Landes (1925a).

Inside the giant Bennett Quarry crystal pocket of 1924. The pocket was eventually 25 feet long with a 9 foot wide x 5-8 foot tall cross-section. From Landes (1933).

ued to kick rocks out of the garden spot."

As with many newspaper accounts, the reality is hard to fathom. Was the account intending to give credit to Allen's brother, Wallace, as a co-discoverer, or did a newspaper compositor get carried away?

This short account, if accurate, gives some inkling why Harry and Laura Bumpus became involved with a lease. Whomever made the discovery, it was a good one, but the discoverer was not a miner or a businessman. Because of the presence of the new feldspar mill in West Paris, established in 1924 and open for business in 1925, and reports of feldspar being found in a number of Oxford County locations, as far away as Mount Rubellite Quarry, Hebron, once owned by Harry's great great uncle, Daniel Bumpus, the thought of making money from feldspar mining interested Harry (Harlan Bumpus, interview, 1971, BHS; [400]). There also had been a feldspar grinding mill in Auburn since 1903 +/- and because of Harry Bumpus's active position in Auburn City government, he was certainly aware of feldspar's value from many points of view. Harry was already a businessman and was the force that got the Bumpus feldspar quarry started. The royalty for the Cummings family from the feldspar removed was "found money" and Harry had the organizational skills to get an agreement from Harold "Red" Perham to market the

Josephine and Alfred Perham, 1938

Perham Brothers Left to Right: Roy Perham, Harold Perham, Ronald Perham, Stanley Perham. c. 1930

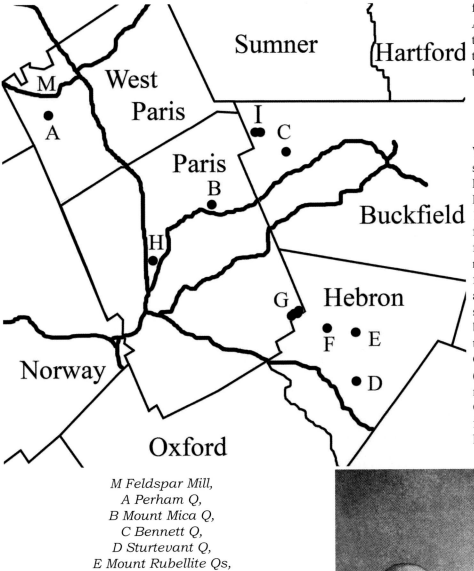

feldspar. It is unrecorded if either Allen or Wallace ever worked in the quarry as employees, although there is no family tradition that they did. [31]

Bennett Quarry, Buckfield

Feldspar soon came to the West Paris Mill from other sources. Both the Hibbs Quarry in Hebron and the Bennett Quarry in Buckfield were worked before 1925. The 1906 activity searching for mica already has been noted for the Hibbs Quarry. Paul Bennett operated the Bennett Quarry for feldspar in 1920 until 1923, although it had been prospected several years before, about 1917, by Loren Merrill. Bennett leased the quarry to the Maine Feldspar Company from 1924 until 1926 (Piret Inzarello, personal communication, 1999). The Bennett Quarry feldspar went to Auburn for processing until Harold Perham acquired the lease and began

M Feldspar Mill,
A Perham Q,
B Mount Mica Q,
C Bennett Q,
D Sturtevant Q,
E Mount Rubellite Qs,
F Hibbs Q,
G Mount Marie Qs,
H Corbett Ledge Q,
I Bessey Qs.
1 cm = 1.69 km.

Perham Quarry crew, Roy Perham in center with arm akimbo, Note absence of railway, etc. c1929. Courtesy West Paris Historical Society.

Harold holding his son Alfred Columbus Perham, 1928.

miner and mineral dealer interested in rare minerals and Nevel sold many of the Bennett Quarry rarities to Charles Palache of Harvard University, presumably as study material for Landes' research (Palache, 1924). The Bennett Quarry crystal pocket soon reached 25 feet by 9 x 5-8 feet and was present in the quarry for several years before it was completely excavated (Stan Perham, personal communication, 1964; [249]; Landes, 1933). The pocket eventually reached over forty feet and "was what killed the feldspar mining at the time" [because it was so labor costly to remove its contents to the dump]." [401]

Some of the understanding of which localities were worked in Oxford County depends on knowing which families were participants. Not only were the Bumpus's involved, but so were the Hibbs and Perham families. There has not been a continuing awareness among Maine miners and historians that these families were intertwined, but there certainly was some influence of these family relationships in the time period. One of the central locations in the web of feldspar locations was the Hibbs Quarry.

Perham Quarry crew, L-R, Chester Morey, Roy Perham and Gus Roberts. c1929. Courtesy West Paris Historical Society.

sending the feldspar to West Paris. In some ways, the Bennett Quarry was a pivotal and inspirational locality to encourage the development of a true feldspar industry in Oxford County.

The Bennett Quarry received international significance, in late 1925, as it was the site where much of Kenneth K. Landes' Ph. D. dissertation research was based, although he also leased the Noyes Mountain Quarry in 1923 and 1924, as part of his research on mineral replacement as a process in granite pegmatite evolution: *The Paragenesis of the Granite Pegmatites of Central Maine*. Although Landes' research is the foundation of many current ideas about pegmatites, Landes did not have the honor of publishing the very first paper on the subject of replacement processes that year as his college advisors had been talking about Landes' findings to other researchers in the field and the scientists who were "tipped off" managed to get several articles published just before Landes' article appeared and there were other papers which came out almost immediately after Landes'. When asked about his advisors' "spilling the beans", he replied that there wasn't much that he could do as he was then a "young and callow fellow" (Kenneth Landes, personal communication, 1978). The scientific excitement of Landes' theories did heighten local awareness. The huge crystal pocket described by Landes attracted the attention of Dick Nevel, who was a

The Bumpus Quarry and the Long-Term Association with the Perham Family

The West Paris feldspar mill started to process the feldspar of the Alfred C. Perham quarry, in 1925, and the Cummings Farm would never have been developed if it had not been for the Perhams' pioneering influence to encourage the organization of the Oxford Mining and Milling Company. The Cummings Farm feldspar was shown, either in the autumn of 1926 or early 1927, to Stanley Perham, who was at that time a Freshman at Bates College in Lewiston, ME and a part-time mineral dealer. Since graduation from high school, Stan had worked at the family's West Paris feldspar quarry, summers and earned enough money to go to college. [401]

Alfred Perham had married Josephine Inez Bacon on November 16, 1895. In early life, Alfred worked for the railroad and later farmed in the summers, keeping cows and selling milk on a route in West Paris village, and in both businesses the children participated. In the winter, he and his children lumbered.

The Perham sons, [33] Roy Francis Perham, Harold ("Red") Chase Perham, Ronald Alfred Perham, and Stanley Irving Perham were all also eventually employed by their

Penley Wood Products Mill immediately across Little Androscoggin River from the Feldspar Mill, c1940s.

father to mine feldspar for the feldspar grinding mill and frequently worked in the mill.

Earlier, Harold joined one of Maine's first Boy Scout troops and had become an Eagle Scout. He was always athletic and when he attended the University of Maine for a year, 1919-1920, he achieved "top honors" in boxing and wrestling. He was married to Phyllis Diantha Williams of Portland in 1920 and they had three children, but she died of pneumonia, September 4, 1925. The next year: "... he branched out as an individual. And was soon operating quarries at Hebron, Buckfield, and West Minot, and shipping feldspar out of state." [187] (It is well-documented that all of the Bumpus Quarry feldspar went to West Paris. The statement might actually refer to the time after his father was no longer and shareholder in the mill.)

On May 28, 1926, Harold registered a complicated "five-year irrevocable" renewable mining lease with Blanche Bennett owner of the Fred Bennett (Melzer Buck) farm of Buckfield. The working mining crew was to have

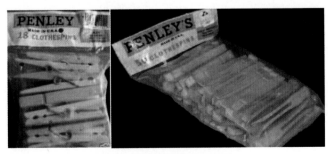

Penley brand clothespins.
Courtesy Nathaniel Edwards.

at least six men working a five day week, but failure to mine during three successive months would begin the opportunity for Bennett to cancel the lease. Perham could suspend his operations "when working conditions at the quarry or market conditions, render his business unprofitable". Feldspar royalty was the standard $0.50/ton, but there would be a adjustment in royalty if the price of crude feldspar increased over $10/ton or dropped below $8/ton. Flint-grade quartz royalty was 25¢/ton after the first year if the market value of the quartz remained over $3/ton. All gems and mineral specimens would be sold by Perham with a 50% payment of their market value, while mica's royalty was set at 50% of the net profit. Blanche's husband, Paul Mason Bennett, had the right to claim a laborer's job at the mine, at the highest laborer's rate, so long as he was physically able to work in a continuous way. It was revealed that workers received 38¢ to 45¢ per hour. The feldspar was hauled to the railroad depot in Buckfield and payments would be made on the actual tonnage shipped to West Paris. The minimum total royalties would be guaranteed to be $500/year with stipulations regarding a mortgage on the property. A codicil registered March 23, 1927 taking effect on the first day of 1928 specified that the mining season would average eight months long. In the midst of beginning his new career as a miner, Harold Perham married Mary Slattery of South Paris on September 6, 1926. His older brother Ronald had married Dorothy Wardwell of West Paris on June 24, 1925 and Ronald eventually became a secondary school teacher.

Other quarries leased by Harold included the Atlee B. Sturtevant Quarry in Hebron registered on February 26, 1927 also with provision allowing Atlee Sturtevant to work

Mann's on Pioneer Street, West Paris, c1943 postcard.

at the quarry as a miner. [301] (In the 1950s, Bell Minerals opened a Norman Sturtevant Quarry in Minot.) Harold also entered an agreement in early 1927 with Effie F. and Manley W. Bessey of Hebron on a property in Buckfield then known as the "old Andrews place", now known as the Bessey Quarry, with work to begin after January 1, 1928. [302] Although known for its attempted production of mica, the Reynolds Quarry in Canton was leased from Charles W. and Linda Walker on January 14, 1928. Harold also acquired a lease of mineral rights on Mount Marie on the Paris-Hebron border, May 21, 1928, from Kalli Piirainen and Matti H. Kahkonen.

An interesting addition to mineral leases was the provision for royalties on the rare mineral, pollucite. In the 1920s, Oxford County experienced a pollucite "rush", because of the radio industry's interest in developing an improved radio tube. The primary companies included the General Electric Company of Schenectady, NY and the Westinghouse Company with their research laboratories in Pittsburgh, PA. Pollucite contained essential cesium and the placement of cesium in a vacuum tube greatly increased its performance. Unfortunately, by 1930, it was realized that cesium was far too corrosive to be used in obtaining a stable radio tube and interest in the cesium-doped radio tube almost vanished, but there remained a minor interest in cesium for use in "electric eyes' and other sensing equipment. [232] By 1930, pollucite deposits in Newry and Buckfield had been depleted, but there was a later discovery of pollucite at the Emmons Quarry, in Greenwood near the Albany border. [104] In the mid to late 1920s, pollucite was worth $12.50 per pound. For this reason, pollucite was sometimes specifically mentioned in mineral leases. For example, a standard feldspar lease, to take effect on January 1, 1928, on the small pre-existing and active Corbett Ledge Quarry on the south slope of Paris Hill, agreed between Harold Perham and Howard E. Corbett [December 11, 1890 Paris – June 30, 1962 Paris] and Leah Day Corbett [August 16, 1898 Bridgton – August 6, 1984 Norway], registered May 11, 1927, specified: "On a material called Polusite [sic] said Corbett is to receive one-third of the net profit on such material mined." Corbett Ledge Quarry is about 4 km southwest of Mount Mica, along the Hoopers Ledge Road, Paris. Mount Mica was prospected for pollucite about this time by the General Electric Company and the talk of the pollucite quar-

Paris Manufacturing mill, West Paris, across from Bates railroad station. c1940s postcard.

ries in Buckfield and Newry must have been on every miner's lips as that mineral was worth about a week's wages per pound.

1929 - Oxford County Mining and Milling Company Sold to United Feldspar Corporation

The local ownership of the mill did not last very long. On May 2, 1929, it was announced that OMMC was purchased by the United Feldspar Corporation (OCRD v. 388, p. 497-501. United Feldspar Corporation (UFC) had been recently organized into one of the largest producers of ground feldspar in the nation, accounting for a quarter of the USA total. United Feldspar Corporation was merged from: United States Feldspar Corporation, Cranberry Creek, Fulton County, New York, Tennessee Mineral Products Company, Spruce Pine, Mitchell County, North Carolina, and the Oxford Mining and Milling Company (Bowles and Middleton, 1930).

The day before the announcement, on May 1, 1929, UFC purchased the mineral rights to the Perham Quarry from Alfred and Josephine Perham (OCRD v 388, p 317-318). It was announced that UFC would build "an aerial tramway" from the Perham Quarry to the mill and that "conveyors will replace trucks in getting the spar from the quarry to the mill. This deal in no way affects the Harold C. Perham feldspar quarries at Hebron, Paris, Canton, Buckfield, Peru, and Albany." (*Portland Herald,* May 3, 1929). Remember that Harold was an independent miner and that his business was not "under the control of A. C. Perham." At this time, it is not certain to whom Harold was selling his feldspar. He had a contract with the mill for the Bumpus Quarry, but he may have already begun to sell feldspar from his more southerly quarries to other buyers, including some "out of

Chair, Paris Manufacturing Company, c1920s.

State". OMMC may have felt that it couldn't compete with large feldspar grinding companies, as well as it had hoped. The pre-Depression sale may not have been a disaster or the mill may not have been as profitable as ground feldspar prices declined. Possibly the various board members just wanted to get out from under their obligations when approached by "out of State" feldspar companies? We do not know if Alfred Perham was in agreement with the change or if he felt manipulated and coerced. Harold's later actions seemed to indicate he was not an enthusiastic West Paris mill supporter.

UFC finalized the acquisition of OMMC rather painlessly by getting a ten-year mortgage at 6% from it on December 23, 1929 (OCRD v388 p499-501). In effect, UFC could improve the facilities and then pay off the mortgage with the income from the mill. No tramway was ever built, however.

OMMC, as UFC's agent, had taken over the operation of the Perham Quarry in the summer of 1929, seemingly taking advantage of the contract provision that they could do so if the price of crude feldspar dropped below a certain price.

Harold Perham's Feldspar Customers

Many have assumed that Harold always sold his feldspar to the Oxford Mining and Milling Company. Even when Alfred, his father, was still associated with OMMC, Harold made sure he had a market for his feldspar and got the best price he could. Behavior is always an important understanding of history. The mill may have had an adequate supply of feldspar from the Perham Quarry and the Bumpus Quarry, at least initially, and Harold may have had more production than OMMC wanted. Harold was also mining feldspar from the southern part of Oxford County as well as Minot in Androscoggin County.

By October 21, 1927, Harold had already contacted another feldspar consumer, Trenton Flint and Spar Company and signed an agreement October 21, 1927 (OCRD v. 376 p. 535-536): "This agreement concerns the future relations in the Feldspar business between" Harold Perham and the Trenton Flint and Spar Company (TFSC). Harold was to supply 4,000 long tons of crude feldspar at $8.00 per ton, distributed over the mining season. There were provisions for keeping accounts up to date, adjusting prices to market conditions, annually, and if the cost of mining exceeded market costs, Harold would permit TFSC to operate Harold's quarries and they would pay a 25 cents per ton royalty.

Predictions and intentions recorded in the Press and elsewhere frequently have an evanescent quality. Harold had no allegiance to an out-of-state company, as he might with a company run or influenced by family members. Additionally, there were many little agreements, special requests, etc. recorded in various mining-related documents which are not related here.

On June 24, 1930, Harold obtained a short-term

Miners' trulli at the Golding Brothers Quarry (Consolidated Quarry), Georgetown, c1905. Courtesy Lorna Cummings Howard.

Trulli at Alberobello, Bari Province, Puglia Region, Italy. Courtesy Janet Nemetz.

$2,000 loan from the Consolidated Feldspar Company of Topsham by using his 1927 lease on the Blanche Bennett Quarry in Buckfield as collateral (OCRD v388, p576-577). (One of the companies which had merged to form the Consolidated Feldspar Company was none other than the Maine Feldspar Company, which had had an earlier lease on the Bennett Quarry.) Harold continued to operate the Bennett Quarry. On October 14, 1930, Harold Perham and Blanche Bennett revised their lease on the Bennett Quarry, Buckfield. The boundary line of the limit of mining was firmly established. There was revealed that there was a "Big Pit", implying the existence of a small pit, perhaps what is now known as the Orchard Quarry. Harold also agreed to take on several financial obligations in lieu of settling up the accounts of feldspar removed.

"Harold C. Perham takes over a certain indebtedness of Two Hundred Forty-six Dollars (246) due Alton C. Maxim on a Chrysler car as a part of said consideration.

No other indebtedness of any kind whatever is due Blanche Buck Bennett for any materials mined on the new designated Parcel Number One until after the stated ten year period lapses. …

> Said Harold C. Perham assumes the direct obligation of paying the unpaid accounts due Malcolm Bearce and Earl Hammond & Leon Harlow for work performed in the year 1929 and yet unpaid.
>
> Also this is a release in full to said Blanche Buck Bennett for any and all charges for money, materials and supplies furnished her by said Harold C. Perham since December 19, 1927."

The deal is complicated and implies money owed on account, by both parties and a credit on future feldspar production obtained on the Bennett Farm by Harold. Given knowledge of future events, one wonders which sides of the tally sheet were red or black? Harold paid off his loan from the Consolidated Feldspar Company and the Discharge of his lease assignment was recorded at the Oxford County Courthouse on October 23, 1931 by Norman G. Smith, vice-president of Consolidated feldspar (OCRD v403, p9).

The US Geological Survey, in its annual feldspar report, revealed what appears to have been another proposed business venture in competition with the United Feldspar Company. The lack of fidelity to the West Paris Mill could suggest a split, but one that may not have included rancor. The better established Sagadahoc County feldspar mills may have simply offered "better money" for the crude feldspar or were willing to buy Harold's excess production.

> "The quarries controlled by Harold C. Perham, A. C. Perham, and S. I. Perham are connected with the Trenton Flint and Spar Company, Trenton, N.J. Up to the present time this company has obtained part of its spar from North Carolina but from now on intends to work entirely in Maine. …
>
> The Trenton people are to open a new mill in West Minot to crush rock from the Maine quarries." (*Lewiston Sun,* August 2, 1930)

This newspaper report may have confused Harold Perham's quarries with Alfred Perham's quarry, but the projected new business association with the Trenton Flint and Spar Company suggested continued disaffection with the new United Feldspar Company, even it only involved money. Nonetheless, the projected mill at West Minot was not built, probably because of the Depression, and the Trenton Flint and Spar Company did not work in Harold's quarries. at least for very long, and it attempted to eke out an existence on the lower grade crude feldspar sources in Topsham, but their Maine mill survived barely a half dozen more years, ending its activity with Trenton Flint and Spar's going out of business. The distant observer may wonder if the Trenton Flint and Spar Company went out of business because Harold went bankrupt and no longer supplied them with their needed raw material or if Harold went bankrupt because he was so closely affiliated with a dying company? The above newspaper report further indicated that a "Stinchfield Quarry" was to be operated in West Minot by the "Perham interests",

but the venture does not seem to have materialized, either, although a Stinchfield Quarry is listed without further detail by Morrill et al. (1958). It was also revealed that the new owner of the Maine Feldspar Company, the Consolidated Feldspar Company, would close the grinding mill at Little-field Corner, Danville Junction in Auburn on October 1, 1929. The local Littlefield Corner work force was 23 people, probably mostly quarrymen, "with six at the mill".

The Finnish Influence in West Paris

It should be noted that the Maine feldspar quarries and associated mills had a workforce dominated by new Americans. In Oxford County, many workers and their families were from Finland, and, to a much lesser extent, from other Scandinavian countries. In the Mount Apatite District of Auburn, many workers were from French Canada or Italy, while Topsham workers were frequently from Italy. BILS (1908) described how fortunate West Paris was to have a vigorous young population of Finnish immigrants. The settlement of Finnish immigrants began about 1898 when Jacob McKeen:

> "… purchased an old worn-out farm, grown to brambles and bushes. … Through the influence of Mr. McKeen several of his countrymen were induced to come to West Paris, and one after another they purchased abandoned farms wherever they might be located in that section of country, but for the most part these purchases have been made within from one to three years. … With one or two exceptions these farms are on the mountain tops and from nearly all there are extensive views of the surrounding country. There are hard hills to climb to reach these places but the soil is naturally good, but in nearly all cases they were completely run out, cutting very little hay, and had been abandoned as places of residence before being purchased by the Finns. … About the first move is to contract with the canners to raise several acres of sweet corn, the fertilizer and seed being furnished, and pay for same deducted when settlement is made in the fall. Then on the recommendations of some older member of the colony they may be able to secure a few cows on credit, paying for them in monthly installments. The cream is gathered by someone in the neighborhood and hauled to West Paris station and shipped to Portland. …
>
> The farms usually contain from 75 to 125 acres with buildings considerably out of repair [at time of purchase]. The prices paid vary from $400 to $2,200, not more than two costing so high as $1,000, and when the higher priced farms are bought it is done by two or three men clubbing together. Of the twenty-

three farms already purchased there are at least forty families interested in their ownership. Four or five of those making the earlier purchases have their payments all made, but the larger number have bought quite recently and are still in debt, but so far none have failed to meet their payments when due. In a few cases the [Burnham and Morrill] canning company has advanced the money to pay for the farm and are taking their pay in annual crops of sweet corn."

The blessing of having a supportive community meant that living conditions for many of the local miners was relatively comfortable.

In Georgetown, miners may have recently immigrated and had little access to a support community. Miners lived at the quarries in one room structures. The building, called a trulli, has a rock half-wall and a conical rock roof. There was waste rock at the quarries, but there was no wood to support the roof so the shelter was cemented with mud. The shelters were certainly not intended for long term habitation and the miners probably were able to soon find more conventional rooms to live in after they had a few paychecks to their credit. (Because the trulli is a unique architectural design, it is very likely that the miner who started making the trulli shelters was from near the virtually only town in the world where these structures exist: Alberobello, Bari, Italy.

Other West Paris Mills in the Twentieth Century

For a village, West Paris was remarkably well-supplied with factories. It had a rich history of wood products companies; ranging from saw mills and lumber, custom planing, components to finished merchandise. While "solid" wood is still prized as a raw material, unfortunately, it is also expensive to produce and work and one of the innovations of the century involved finding cheaper substitutes for wood, especially as many uses did not require wood's particular properties. Plastics and composite materials edged wood out of certain uses. Some industries such as wool spinning, required enormous quantities of wooded spools, but they, too, went to sturdier, longer-lasting materials. With the globalization of jobs, a euphemism for getting manufactured goods made more cheaply elsewhere, many industries which utilize solid wood are no longer located in the USA. Some industries just didn't use high quality wood any more and much furniture is now made of particle board. The shift from wood products has changed the County's perspectives on paper making.

No attempt is made here to indicate the detailed history of the West Paris mills, other than feldspar, nor when later companies bought out or merely changed the names of existing mills, nor are their locations noted. What is important here is to show an isolated yet actively industrial community within the "double township" of Paris. In 1957, Paris was split in two parts with West Paris becoming its own

town also governing the settlements of Hungry Hollow, Trap Corner, North Paris (once known as Sucker Harbor), Dean Neighborhood, Forbes District, Porter District, Tuell Town, and Snow Falls areas. Paris retained Brimstone Corner, Elm Hill, Pleasant Valley, Paris Cape, Paris Hill and South Paris.

While this is not a general history of the village's mills, a look at the twentieth century's mills as listed in the *Maine Register* is interesting. The village had a Burnham and Morrill canning factory in 1895 which remained there until 1941. Elingwood's Mill, early known as Ellingwood & Willis, later known as O. D. & S. B. Ellingwood, advertised its cant dogs and pick poles, and, under various designations, was a long time employer from about 1895 to at least 1959. F. L. Willis made spools for machine wool looms, as well as boards, starting about 1890 and may have been combined with Ellingwood's about 1902. Lewis M. Mann and Son's manufactured clothespins, pail handles, etc., starting about 1870, was there until after 1959. N. Johnson Cushman (known for planing boards as well as making ladders) was there from 1895. The L. E. Locke and D. H. Fifield grist mills were in the village from the last year of the nineteenth century (1900) with Fifield continuing 15 years later. The village's grist mills gave way to grain dealers.

West Paris Creamery existed from 1888 until 1907, apparently changing its name. Portland Creamery, Howard Gurney, and Alden Redding variously owned dairies in West Paris at least until 1959. Some wood products companies included: Irish Brothers (lumber and die blocks) 1905-1931, George M. Tubbs (lumber and spools) starting about 1902, with H. G. Brown preceeding them, but they were gone by the 1920s, supplanted by newer companies. Elroy R. Davis started about 1910 manufacturing pick poles, bolt hooks, and welded ferrules. Penley Brothers Company was founded in the mid-1920 and is still in business. Penley's was once known for manufacturing clothes pins.

The branch facility of the Paris Manufacturing Company (wood furniture, toys, and novelties was sometimes known as the "Chair" factory and was listed in West Paris by 1902. Between the two Paris Manufacturing mills, the company's reputation was built of children's toys, sleds, toboggans, wagons, pull toys, game boards, children's roll top desks, furniture, small tables, bookcases, and cleverly designed folding chairs, porch swings, etc. They claimed to be the largest American manufacturers of these novelties. Later, it became Oxford Wood and Plastic Company making boxes and toys

There were also individually owned manufacturers making shoes, carriages, harnesses, etc. for which there was no general workforce. Note that the *Maine Register* is an unreliable source of data concerning names of businesses and when they were in business. Many businesses would suddenly appear in print after years of actual business operations, while some would remain ing the *Register* for years after their closing.

Chapter Four: The Bumpus-Cummings Lease and the Giant Beryls

The original inspiration for writing the story of the Bumpus Quarry was the discovery of giant beryl crystal in 1928-1929. However, the Bumpus Quarry had a very influential role in Oxford County's mining industry beyond the notoriety of an unusual discovery. Although the pre-history of the Bumpus Quarry is summarized, here, with roots going back to the nineteenth century, the Bumpus Quarry, as a business, began on June 1, 1927, when Harry and Laura Bumpus leased the ledge on the Cummings farm for a period of 50 years, from their relatives, Allan E. and Sybil E. Cummings. [305]

"Said land is hereby leased, demised and let unto said Laura J. Bumpus and Harry E. Bumpus and to their heirs, executors, administrators and assigns as aforesaid for the purpose of boring, mining and operating for feldspar and other minerals on said land for the term hereinafter described.

TO HOLD for the term of Fifty (50) years from the first day of June 1927, yielding and paying therefore, the rent of Fifty (50¢) cents for each ton of feldspar mined, payment thereof to be paid by the said Lessees to the said Lessors as payments are made to the said Lessees by the purchasers of said feldspar from said Lessees; and for the further payment of *one-half of the amount received by the said Lessees on any by-products* [emphasis added] and other minerals from the result of operations under this lease to the said Lessors... And the said Lessors doth further covenant and agree to pay all taxes duly assessed on said premises during the term and for such further time as the Lessees, their heirs, executors and assigns may hold the same.

As far as mineral leases are concerned, the terms of this one were much more generous than were usual as the agreement

Bumpus Quarry Lease.

specified a 50% royalty on by-product minerals, not anticipating the value of commodities such as beryl, flint-grade quartz, and rose quartz yet to be discovered. Harry and Laura also relieved their relatives of the burden of having to pay the annual real estate taxes, by agreeing to pay them, although those taxes were about the royalty value of only several shipments of feldspar from the quarry. Although feldspar was the primary target mineral, the Bumpus Quarry was soon to become famous for huge beryl crystals, a mineral that hadn't had a large commercial value previous to that time. Unfortunately, Sybil died before the month was out, on June 28. There certainly was a brief mourning time, but business did not languish for many months. Harry and Laura had to prepare to get into the mining business. During the following "bonanza times" at the Bumpus Quarry, the problems of Sybil's having left no will and testament, influenced events from the background and there were several decisions that seem unusual from this vantage in time. (The complication of an unsettled and tangled estate will be discussed in detail in a later section.) With a mineral lease in hand, but not yet registered, Harry and Laura entered into an agreement with Harold C. Perham on September 9, 1927 to sell him the feldspar, and only the feldspar. [306] No provision was made for other minerals, even as a by-product. The

Cummings house. c1930. Courtesy Ava Bumpus

Bumpus lease was not registered until October 4, 1927, while the Bumpus-Perham lease was registered before it on September 21. (The reader will notice that leases and agreements were frequently registered many months after the fact, with some legal agreements being registered years after they were signed. Sybil's death was certainly a factor in the delayed beginning of mining, although there would have been a number of logistical concerns. Harry and Harlan had to acquire drilling equipment, hire a mining crew, and all concerned had to learn how to become miners. There was no previous local feldspar mining tradition to learn from nor a pool of veteran miners to hire. Dick Nevel was mining for the rare mineral pollucite at Newry and Buckfield, and to a lesser extent in Paris, in the time period and there could have been some input or help from his activity, but that would been unlikely as he was barely experienced, himself. There were also a few miners at West Paris, but they were mostly gainfully employed as miners already. There were experienced miners in adjoining counties, Androscoggin and Sagadahoc, but there doesn't seem to have been much reaching out beyond Albany and Bethel. Probably most of the advice came from the Perhams, who were new to feldspar mining themselves.

The first Bumpus – Perham lease specified:

"1. The said Harry E. Bumpus and Laura J. Bumpus hold a lease of certain feldspar property in Albany, Maine on the so-called Allan E. Cummings [sic] farm, said lease being signed by Allen E. Cummings and Sybil E. Cummings, and has been operated by said Bumpus for several weeks. ...

3. Said Perham agrees to pay said Bumpus for all cars shipped and paid for immediately on the receipt of such returns from the company to whom the material is sold. Each payment shall be accompanied with a statement of correct railroad weights, price material is sold for, car number and initial.

4. Said Perham agrees to furnish market for any amount of feldspar quarried by said Bumpus provided that the total tonnage does not exceed an average of one car of 60,000 pounds per week over any summer working season, such season figured from time of commencing quarry work to finishing same.

5. Said Bumpus agrees that the first 1000 long tons of feldspar quarried on said property shall be sold to said Perham at the above stated terms and that this agreement terminates on the delivery of such stated amount. Said Perham

Earliest known image of the Bumpus Quarry. c. Spring 1928. Workers showing keen interest in the ledge. Note drill steels in the foreground. Possibly Harlan Bumpus in dark hat, Stan Perham in white hat. Courtesy Ava Bumpus.

Bumpus Quarry in November 21, 1929, Portland Press Herald.

allows until January 1, 1930 to deliver such an amount.

6. Said Perham agrees to pay at the rate of six dollars ($6.00) per long ton for feldspar at the loading platform of said quarry and the loading platform shall be in proper condition for loading of spar. Such feldspar shall be of a quality suitable for high grade pottery and acceptable to the company to whom it is shipped. Said Perham assumes expenses and responsibilities of loading from platform to car with one exception: - If said Bumpus quarries until the late fall season and a bad, rutted condition of roads forces said Perham to pay more than usual price for hauling, then said Bumpus shall pay such extra price or assume the hauling re-

sponsibilities himself at a price of eight dollars and twenty-five cents ($8.25) per long ton, f.o.b. cars, Bethel; this price to include demurrage and any other loading expenses. Also such cars shall be loaded to a minimum capacity of 60,000 pounds.

If the present sales price of spar ($8.50 per long ton) now received by Harold C. Perham on spar shipments of this kind of grade stone, should decrease, then, the price paid to said Bumpus shall decrease accordingly. If the present price shall increase, said Bumpus shall receive one half the raise in price per long ton."

Several things are learned from this lease. First of all, the Bumpus Quarry had been in operation "for several weeks". We might imagine that the feldspar was being placed on a stockpile dump to verify that the deposit was more than surficial. It is revealed that Harold Perham would be responsible for transporting the feldspar from the quarry to the railroad station in Bethel for continuing shipment to West Paris. The weight of the feldspar would be fixed by the bills of lading issued by the Maine Central Railroad for the cost of shipping from Bethel to West Paris. The price paid just for the feldspar by Harold Perham was generous, but it may have had to be otherwise Harry and Laura would have had their son, Harlan, do the mining and shipping himself and eliminate the middle man, Harold Perham. It may be surmised that the need to fulfill A. C. Perham's contract to sup-

Bumpus Quarry showing early uncovering of the Giant Beryl ledge. Unidentified 1930? newspaper clipping.

ply the West Paris Feldspar Mill with up to 10,000 tons of feldspar annually was also an incentive for a generous lease. Because the Oxford Mining and Milling Company mill was organized around the feldspar output of the Perham Quarry, it seems that the Bumpus Quarry was regarded as a supplemental source of feldspar, at least in the beginning. Through time, however, the high purity feldspar of the Bumpus Quarry became more and more significant to the performance of the West Paris mill. At least until 1932, the Bumpus-Perham lease was the controlling agreement allowing the Bumpus Quarry a ready market for its feldspar output.

Harlan Bumpus, 1930s
Courtesy Ava Bumpus

Harlan Bumpus, 1930s
Courtesy Ava Bumpus

Small town dirt road (Sodom Road, Norway)
1920s postcard

The Bumpus-Perham lease was in force for two months, however, when it was amended on December 28, 1927. Although the price of crude feldspar had begun to falter, the lease price remained the same at $6.00 per long ton, neither side anticipating more than simple market fluctuations. Of course, Harry and Laura couldn't get away with cutting Harold out and selling feldspar directly to Harold's dad, Alfred Perham and the Oxford Mining and Milling mill. However, flint-grade quartz and rose quartz as well as beryl had been discovered in noticeable quantities and these minerals had to be accounted for as the mineral lease only specified feldspar. The recovery of these latter minerals would have had a royalty of 50% as "byproducts" to Harry and Laura. [307] Harold could have ignored the presence of beryl, quartz, mica, etc. and Harry and Laura would not have been allowed to sell these minerals themselves as there was an agreement in place, even if the by-products were not being purchased by Harold. There might have been the possibility that Harold could have been sued to have the leased annulled based on non-performance, that is, non-purchasing easily recoverable by-product minerals. It isn't certain which side wanted the lease revised, but given the future developments driven by family pressures, it is not likely that Harold anticipated which direction the financial arrangements would go. The large quantity of flint quartz was probably an attractive by-product in the first few months of operations. Beryl was probably becoming apparently abundant, but the high royalty on it seems to suggest that the future bonanza was far from anticipated.

Rose quartz was an uncommon variety of quartz and Harold's brother, Stan, would have been very interested in obtaining it for mineral specimens to sell in his store. There were relatively few commercial sources of rose quartz, the notable locations in the time period being the Bayliss and Kinkle Quarries, Bedford, New York and Scott's Rose Quartz Quarry, southeast of Custer, South Dakota. Of course there were many minor rose quartz locations even in Oxford County. The provision was not just for rose quartz, however, but the rose color certainly attracted attention. Granite pegmatites in Maine and elsewhere frequently have a quartz-rich central zone, which has very glassy and pure quartz, often called "flint" quartz, and which had become a commodity by the time the amended lease was made. Even the Portland Glass Company, a renowned maker of ornamental glass would have been a customer for flint-grade quartz. The *Kennebec Journal* (December 8, 1927) wrote: "The mining of quartz is becoming more and more extensive as new uses are found for the mineral. General Electric Company is working a quartz vein at Paris Hill, and other electrical companies are interested in quartz deposits at Waterford, Norway and other western Maine towns. Fused quartz and quartz glass are still being subjected to much experiment, and quartz cores heated to 2000 degrees (F) are used in experimental radio beacons in aid of aviation." The newspaper article went on to discuss beryl: "Beryllium [sic], another Maine mineral which occurs with feldspar and is now thrown away at feldspar plants is being tested experimentally for airplane use." (The General Electric Quarry in Albany, on Lovejoy Mountain was an excellent source of flint-grade quartz, not far from the Bumpus Quarry on the eastern side of Hunts Corner Road, was first listed in the *Maine Register*, by 1933.)

The new Bumpus Quarry lease specified that it would be in effect for seven years, but this stipulation was one of

Harlan's horses. Veery (black), Tiger (white)
Courtesy Ava Bumpus

the few to benefit Harold. There were additional provisions:

"2. This sales agreement shall remain in force so long as said Bumpus and his wife shall operate or directly control the quarrying operations on the Allen E. Cummings farm in the town of Albany in said County of Oxford; but nothing herein shall hinder or prevent the said Bumpus and his wife from subletting or

Rose Quartz, Bumpus Quarry. 15 x 12 cm, Addison Saunders collection.

sum rental of two hundred and fifty (250) dollars, for each and every year while so operated by said Perham ...

#3. FELDSPAR PROVISIONS. During said Bumpus's operations they agree to quarry a minimum tonnage of five hundred (500) long tons per year, and said Perham agrees to furnish a market for the same up to a maximum tonnage of one thousand (1000) long tons of feldspar. It is also agreed hereby by said Perham that said Bumpus shall not be held in default if he fails to deliver the minimum of five hundred (500) long tons, if ledge conditions have been such that the usual crew of six (6) men could not reasonably quarry that amount of stone during the summer and fall season of any contract year. ...

4. QUARTZ PROVISIONS. Said Perham has full sales rights on all quartz quarried and saved for shipment if the price received by said Perham F.O.B. cars at Bethel, Maine, shall be below five (5) dollars per long ton, then said Perham shall account for the full sales receipts therefore, minus all costs directly due to hauling quartz and also minus a

subleasing the so-called Cummings quarry to any person, firm, or corporation, as they may desire, although the said Bumpus and wife hereby agree not to sublet to the exclusion of said Perham hereunder for a term of Seven years. Furthermore said Bumpus and wife are not holden to continue quarrying operations when the quarrying operations shall prove unprofitable during any calendar year; but if they do not wish to continue such operations, they shall so notify said Perham in writing, and then said Perham shall have the right to carry on quarrying operations up to an average yearly tonnage not exceeding the maximum limit of One Thousand (1000) long tons, for the balance of the term of this contract, provided however that said Perham shall pay said Bumpus and wife such minimum royalties as shall become due for that period of time to the said Allan Cummings and Sybil E. Cummings ... and shall also pay said Bumpus and wife a lump

Beryl, Bumpus Quarry. Perham's of West Paris Collection

Main Street, Perham store eventually located in second building on left, (third story added and tree cut down), West Paris c1920 postcard.

sales commission in favor of said Perham in the amount of five per cent of the said selling price.

If the price received by said Perham F.O.B. cars at Bethel shall be five (5) dollars per long ton, or over, then said Perham shall account for the full sales price minus all costs directly due to hauling quartz and also minus a selling commission in favor of said Perham in the sum of ten (10) per cent of said selling price.

5. BERYL PROVISIONS. Said Perham has full sales rights on all beryl quarried and saved for shipment. If the price received by said Perham F.O.B. cars at said Bethel, be under fifty (50) dollars per ton of two thousand (2000) pounds, then said Perham shall account for the full amount of sales receipts therefore, minus all costs directly due to hauling beryl and minus a sales commission in favor of said Perham in the sum of ten (10) per cent of said price so received.

If the price received by said Perham F.O.B. cars at said Bethel shall be fifty (50) dollars per ton of two thousand (2000) pounds, or over, then said Perham shall account for the full sales receipts therefore, minus all costs directly due to hauling beryl and minus a sales

commission in favor of said Perham in the sum of Fifteen (15) per cent of said price so received.

6. Lapedolite [= Lepidolite] and Pollucite. Said Perham has full sales rights on all lapedolite and Pollucite discovered in quarry operations hereunder. On all sales of such materials said Perham agrees to pay said Bumpus and wife the full amount so received minus all costs directly due to hauling and also minus a sales commission in favor of said Perham in the amount of 5 Percent of said selling price."

The revised beryl and quartz provisions were not very advantageous to Harold. The 10-15% "sales commission in favor of said Perham" was almost equivalent to an 85-90% royalty in favor of Harry and Laura, but with expenses reducing the actual percentage. Previously, the mineral lease provided for 50% royalty on by-products. Income was realized from quartz and beryl, but none for lepidolite or pollucite. Because there was a schedule of payments for the latter two minerals, it has been assumed by some that there must have been discoveries of these minerals. The provisions were "just in case", because pollucite and lepidolite were ores of rare elements and one rare element, cesium, was actively being sought at several Oxford County locations, particularly in Buckfield and Newry, and there had been uneconomic discoveries of the same in Paris and Hebron. No lepidolite or pollucite were ever found at the Bumpus quarry. Most leases did not include provisions for rare minerals and gems unless there already was an awareness of their occurrence. [38] (For example, on September 6, 1951, a lease on the

Nestor and Katie Tamminen Quarry, Greenwood with the United Feldspar and Minerals Corporation mentioned an unusual mineral, spodumene, having a 15% royalty on the gross selling price and "gem stock" having a 50% royalty.) [308]

The Bumpus Quarry was soon a curiosity as it was new and newspaper articles indicated that visitors were welcome. The curator of minerals at the Field Museum of Natural History and probable distant relative, Oliver Cummings Farrington visited the Bumpus Quarry and acquired an interesting crystal in the Autumn of 1927 – a "pocket beryl". The crystal was 3.3 x 2.7 x 2.2 cm, ocean surf green, and was peculiarly cloudy, but with a thin transparent cap. The crystal was in the Field Museum of Natural History's mineral collection until it was de-accessioned about 1975. (See beautiful green beryl crystal in the introduction.) Pockets are very special features in a granite pegmatite. They generally are formed from corrosive late stage solutions by their dissolving away, replacing, or otherwise transforming pre-existing rock. When these "hollow" areas are still subject to an inflow of crystallizing fluids, the open volume may have well-formed crystals in them, sometimes gem minerals. When the openings are small, up to several centimeters, the openings are called vugs. Larger openings, many centimeters to even meters in size, may have such complex fluids trapped in then, or at least "attacking" them, that large crystals often form in a cavity. True pockets usually have a sandy filling between larger crystals and there may be a readily traced history in pockets of one mineral succeeding another in time of formation. Most of the Bumpus "pockets" were simple in their mineral composition and rarely were there other minerals present than quartz and feldspar crystals.

Harlan, at age 24, became a miner at the Bumpus Quarry and was the crew supervisor. Although Harlan suggested he had worked at the family quarry "seven years", he worked, at least intermittently, until 1940 (Earlon "Bud" Paine, personal communication, 2007). When mining was slow, Harlan worked occasionally for Fred Littlefield hauling logs out of the woods. Littlefield sold 1,000 acres of land to the National Forest in 1936 or 1937. In 1929, Harlan Maurice Bumpus, [14] married Annie Elizabeth Cummings Hazelton, second daughter of Harlan's uncle, Wallace Edwin Cummings, on June 5. Annie graduated from Gould Academy in Bethel and had attended, but did not graduate from, Bates College in Lewiston, across the river from Auburn. She was a teacher in the Albany elementary schools for a few years.

Harlan and Annie boarded with Allen and Wallace Cummings. It was on the Cummings farm where Harlan's family grew up. Harlan eventually raised birds such as exotic pheasants for extra income in addition to keeping sheep and a few cows. [402] Harlan and Annie were members of the Round Mountain Grange and of the Congregational Church, both of Albany, and Annie was active on organizational committees and was a 4-H leader. Annie was particularly known as a professional home baker and would bake cakes for weddings, etc. For a number of years, Annie was the Albany correspondent for the *Norway Advertiser Democrat* and the *Bethel Citizen*.

An unidentified 1950s newspaper clipping indicated that: "[Harry] Bumpus opened the quarry that Summer

Bumpus Quarry, Beryls were found to the Right, c. 1928. Possibly Harlan Bumpus in center with dark hat. Courtesy Ben Shaub

[1927] and for seven years worked it by hand. That is, all the pegmatite was shoveled onto horse-drawn carts, after charges of dynamite had loosened the solid rock. The stone was hauled by horses, whence it was loaded into trucks and hauled 23 miles to the West Paris grinding mill." The description, of course, does not mean that Harry Bumpus was ever a miner, he was the business owner. It is not known if there were a loading platform immediately next to the quarry in the earliest years or if the feldspar was brought to the edge of Valley Road. (The above cited 1950s newspaper article indicated the feldspar was trucked to West Paris, but the initial lease indicated that the feldspar was to be trucked to Bethel and then shipped by railroad to West Paris and there is no indication that the original feldspar shipments were shipped by truck from Bethel to West Paris. The newspaper article may have assumed that the current method of shipping had been the only method. Pre-Depression railroad shipping, in full cars, was cheaper than trucking (John Davis, personal communication, 2007).)

Albert B. Kimball, later owner of the Bumpus Quarry, was one of the early Bumpus Quarry miners. He recalled:

"Back in 1927, I was working then for

Bumpus. We done it all by hand at the time. Oh, we drilled it by hand and take a hand drill, ya know, and hold it and a man held a hammer and dynamite and fuse. Now doncha see, they have machinery – air compressors to drill it with and a battery to set off probably 25 sticks of dynamite, where we only set out one at a time.

In the beginnin' all we really took out [to] the mill was feldspar…

Then we commenced to save the mica and then we saved the quartz. The mica – started in by hand, just as soon as I got able enough, I

Albert Kimball, Bumpus Quarry miner, July, 1930
Courtesy Carol Kimball Cox

bought a truck and I had four trucks hauling out of here and hauled to the mill in Keene, New Hampshire, see. And the quartz I loaded onto cars at Bethel…" [249]

The mica may have been sold initially to Harold Perham and was probably added to the total mica sold by him in lots without regard to locations or labeling as that information had little value to the mica consuming industry. Albert could have been referring to a later work episode when mica production was being handled in a different way. The Bumpus Quarry mica seems to have been added to New Hampshire's records rather than being credited to Maine. At any rate, the Bumpus Quarry has received little credit as a mica producing locality. Stan also recovered scrap mica from Black Mountain Quarry, Rumford and his gleanings of the dumps were probably added to the output of the Hibbs Quarry.

The Bumpus Quarry was a constant influence on the family. The quarry was also a playground for Harlan's children and, whenever the quarry filled up with water in the

summer, it was a wading pool. In the late 1940s, when there had been enough scientific talk about uranium and beryllium, by visiting mining engineers, etc., a new Bumpus farm kitten was named "Helium", after a trace gas found in some beryl. In the winter, with the quarry filled with water and was frozen over, the Bumpus Quarry furnished the icebox. The farm had a working quarry, but it was also a home.

Harlan's mining crew first used hand drills and black powder for blasting, but his parents soon up-graded their mining equipment, using pneumatic drills and dynamite. The initial work of removing feldspar from the quarry proper relied on horse-drawn wagons, but it isn't recorded how the feldspar was loaded onto gasoline powered trucks for shipment to the railroad station at Bethel. The rock could have been dumped into a truck, either directly or by chute. It could also have been shoveled.

Distributing the Bumpus Quarry Beryl and Other Newly Mined Specimens: Enter Young Stan Perham

For over half a century, Stan Perham [June 2, 1907 West Paris – December 1, 1973 West Paris] was an influence on mining and mineralogy in Oxford County and beyond. The legacy of those he nurtured in science and mineralogy continues to this day. Stan's love of nature first blossomed at age 7 when Millard Emmons, "an old trapper" and friend of his Father's, interested him in a "small collection of oddities", including some minerals. Emmons told Stan that he could have the specimens, but Stan would have to "earn" the collection. He visited Emmons many times, first working to make a picture of the Washington Monument using tinners' flux on marble. Then Stan had to build a cabinet to house his newly earned oddities. (Linwood B. Emmons of Greenwood was a miner at his quarry of the east slope of Uncle Tom Mountain from 1934 to 1943.)

Natural history ran deep in the Perham family and

Helium, c1950. Courtesy Ava Bumpus.

several interesting incidents sparked excitement (Perham, 1959):

> "About 1917 a meteorite fall was seen by Roland Benson, Harold Perham, Alfred Perham, and Stanley Perham southerly of the old Locke farm in West Paris. An effort was made to find pieces without success."

And

> "In the winter of 1918 the writer saw a meteorite come in over Berry Ledge at West Paris. Its roar was like that of a heavy freight train. It was a blue-white ball of fire and it exploded with a terrific sound 'like dynamiting' just over the 'big birch tree' so called."

By age 12 in 1919, Stan had begun to sell minerals and natural history objects from his bedroom. In that year, "he had secured his first piece of rough [gem] stock, had it cut and received his first order for a finished product (*Lewiston Journal Magazine,* April 3, 1948)." He continued a retail mineral business throughout his life. The Alfred C. Perham farm was located on High Street, about 1 km south of the railroad tracks, on the back road between West Paris village and South Paris.

An interesting job during high school was operating the projector for the silent movies, one night a week. He also delivered newspapers (*Grit*) as well as magazines: "Sold Butterflies and moths cased for gifts[;] went into the selling of gems and minerals when I was 12 with a hired store. Had had a museum, before, first in my home and then in a building out side the home which I purchased by running a sandwich and soda route one summer." (Stan Perham, written communication to Ben Shaub, 1957). During these early days, Stan would go mineral collecting to the best local quarry: "Saturday afternoons, Stan and his friend George Flavin [1907-February 4, 1983 Portland] would hitch up the buggy and drive ten miles to the world-famous Mt. Mica quarry. ... Digging in the rubble heaps, they often found valuable stones, which Stan took to Lorin [sic] B. Merrill, a local gem authority, to cut. Now he had real commercial stock to sell." (Woodbury, 1957).

Stan joined the Boy Scouts of America and in 1924 he achieved the rank of Eagle Scout: "Worked in the mine part time 1923; 1924 when [my mineral] store didn't keep me too busy, it was mostly a summer business then."

Stan graduated from West Paris High School in 1924. His high school yearbook, the *Nautilus,* recorded many activities:

> "Class Pres. (1); Athletic Council (2); Minstrels (1); Class Play (2,4); Operetta (2,4); Pres. Debating society (4); Pres. Athletic Association (4); Vice Pres. Athletic Association (3); Nautilus Board (4); Mock Trial (3); Salutatorian.
>
> How could we ever have gotten through four years without you, 'Stan?' When there is anything hard to do you are always there. Did I say always? Well, perhaps, but this last year a certain member of the Junior Class has taken

Stan Perham. From 1924 Nautilus, West Paris High School Yearbook.

George Flavin. From Nautilus (1924)

Stan Perham store, Maple and Main Street, c1928 Specimen shelves dedicated to (top, L-R) Sturtevant and Bumpus Quarries, and (lower, L-R) Hibbs, Bennett, and Perham Quarries. Note rear door. Photo courtesy Ben Shaub.

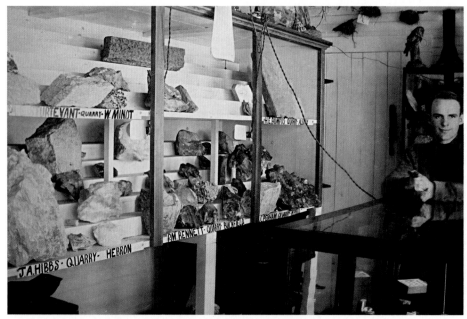

Stan Perham store, Maple and Main Street, c1928 Seashells and stuffed birds on top shelf. Bird eggs in glass case. Note additional minerals on shelves and second rear door leading to Maple Street. Courtesy Sid Gordon.

up considerable of your time."
He was a guard on the basketball team. Stan wrote an essay for the yearbook entitled, "An Amazing Transformation", about the emergence of a moth from a cocoon.

At age 20, Stan: "… moved his business into a proper store located on West Paris' Main Street in 1926." [188, 36] The second outside store was located on the corner of Maple Street with Main Street and the store sign was made of a collage of quartz crystals from Mount Mica Quarry, Paris. The previous occupant to the room was Walter Inman, who was a barber in West Paris, also beginning in 1926 (Milt Inman, personal communication, 2007).

Walter Inman, a mineral collector whose grandfather was noted mineral collector, mineral dealer, and bear hunter, Edgar D. Andrews, may have alerted young Stan that the barbershop room was going to become available. The accompanying interior pictures show Stan's Main and Maple Street store.

Stan needed a genuine store because of the new production of minerals from Harold's quarries which provided enough stock to pay for the location. The specimens were probably on consignment. Perham (1987) noted: "When Harold Perham began his mining venture, Loren Merrill handled the distribution and sale of the gem material and many minerals which were found in conjunction with the mica, beryl and feldspar. This task was then turned over to Stanley Perham who continued to cull the massive amount of material produced by the mining operation." Woodbury (1957) noted Stan's early popularity: "The Harvard scientists who came to Mt. Mica to dig formed the habit of dropping over to the Mineral Store… Alton Bicknell, a man who knew something about mines, told Stan's father, 'I'd put more stock in what that boy [Stan] says than in many of the city experts.'"

While he was attending Bates College, Stan was Bates Outing Club president in 1929, but he had to discontinue his studies because of the Depression and it was then his chosen career became full-time. Stan had entered college with the Class of 1931, but he completed two and a half years of study. The first appearance of Stan's Maine Mineral Store, in the *Maine Register,* was in the 1930 edition under jewelers. It was while he was a student that he had the initial influence on the preservation of the beryls in the ledge.

Although the big boom in cesium prices had passed, by 1930, Stan worked with Benjamin B. Burbank [June 5, 1901-June 6, 1979 Brunswick] of Brunswick, chemist for the Bath Iron Works, on how to manufacture cesium salts from pollucite as there was still a market for cesium, although no longer for radio tubes. Stan was interested in mica mining as well and worked, undoubtedly intermittently, "from 1931 to 1938" culling mica from the dumps of the Black Mountain Quarry, Rumford with a lease from the Maine Products Company. [64, 140] In 1939, the Black Mountain Quarry was being operated for mica by Frank D Pitts who had a sublease from Cesare Trusiani [March 30, 1897 – December 1972] of Brunswick. A newspaper account erroneously stated that Black Mountain was being worked by a New Jersey company, although they may have bought any mica produced (*Lewiston Journal,* October 20, 1939).

In 1931, Stan joined International Order of Odd Fellows of West Paris and was president in 1935-1936 of the West Paris High School Alumni Association, as well as member of the Bates College Alumni Association. Stan was a founding member of the Oxford County Mineral and Gem Society (1948) and was elected an honorary member of the Boston Mineral Club, Maine Mineralogical Club, and the Almeco Gem and Mineral Society. He was a member of the Geological Society of America, the American Association for the Advancement of Science, and Maine Association of Engineers. He was active in the West Paris Universalist Church and was State treasurer of the Church Young Peoples Group in 1924. He was president of the West Paris Chamber of Commerce in 1959 and 1960 and was active in the Norway-Paris Kiwanis Club (December 7, 1948; *Portland Press Herald,* "S. I. Perham named to Norway-S Paris Kiwanis Club membership committee"). Throughout his career, Stan was sought as a lecturer on gems and minerals. Nearly a full page of the *Lewiston Journal* (April 2, 1936) was devoted to a synopsis of a lecture he gave in Augusta. In the "Gorham News" column, November 3, 1950 *Portland Press*

Herald, it was announced that S. I Perham "spoke on Maine Gems and showed a collection in the home of Mrs. Francis L. Bailey." This was presumably a meeting of the Cosmopolitan Club. A short list of guests was mentioned. Many other such lecture notices have been found. He was supervisor of the local Soil and Water Conservation District, while his hobby was square dancing and he and Hazel were members of the *Swinging Bears.*

Mineral Specimen Dealers and Independent Lapidaries as New Maine Industries

Mineral specimens are of interest to naturalists, museums, and educators. Due to these interests, Maine minerals have been sought since at least the discovery of Mount Mica

Maple Street from Main Street, Perham store in first building on left, room at rear lower window, Second building was the livery stable, West Paris. One-legged boy (left with crutch) was Raymond Dunham, 1915. Courtesy West Paris Historical Society.

Main and Maple Street intersection, Perham store at rear lower window, West Paris, c1920. Courtesy West Paris Historical Society.

Rose Quartz with beryl crystal, Bumpus Quarry. 1 x 0.6 meters, Merriam-Logan specimen.

Quartz parallel-growth crystal, 4.0 x 1.8 cm, Perham Quarry. Specimen and photo from Joseph A, Freilich Collection.

in Paris which likely occurred in 1821 (not 1820 as is commonly reported). The earliest known Oxford County mineral sales seem to have been made by Cyrus Hamlin in 1822 or slightly afterwards. He sold minerals that he and his younger brother Hannibal had found at Mount Mica. Brother Elijah seems to have held on to his early discoveries, at first. Professor John Webster collected minerals in Sanford in 1848 to raise money, but he sold his specimens in Cambridge, Massachusetts. By the early 1860s, there was a part-time mineral dealer, Nathan Perry [1830-1890], originally in North Woodstock, suggesting that there were enough customers to support more than occasional dealing in minerals. There were several prominent Maine mineral collectors in the mid to late nineteenth century: Samuel R. Carter of Paris, Ezekiel Holmes of Paris and Winthrop, Elijah and Augustus Hamlin of Paris and Bangor, Nathaniel T. True of Bethel, Elron Chadbourne of Lewiston, Loren B. Merrill of Paris, Edgar D. Andrews of Stow, Edmund Bailey of Andover, et al., but there was undoubtedly a large number of occasional collectors whose names have never received publicity. By the 1890s, many more collector/dealers became active, including George Howe, Shavey Noyes, and Charles G. Andrews, all of Norway.

The ebb and flow of collectors has been, of course, related to the availability of new mineral specimens. Before the Civil War, the number of Maine mineral localities was small. Mount Mica was the shining star, if not somewhat dulled by its inactivity. The rise of mining at Mount Rubellite in Hebron about 1860-1861, on the future farm of Daniel Bumpus, inspired a rise in scientific specimen mining, and certainly influenced Augustus Hamlin when he returned to Maine from the Civil War. With an explosion of discoveries of gem and specimen producing localities in Oxford County and adjacent Androscoggin County, in the 1880s, there were many mineral collectors who were also part time mineral dealers. By 1880, Nathan Perry became Maine's first full-time mineral dealer. (See King (2000) for additional details.) There was even a multi-dealer mineral show in Paris in 1882, undoubtedly the first such mineral show in the USA. By 1889, Loren Merrill and Kim Stone had begun to devise their own gem cutting and polishing equipment which led to an eventual cottage industry of gem cutting in Oxford

County. Loren Merrill was teaching people to cut gems at least by 1902.

Nathan Perry did not have a store front, but did business from his barn first in Woodstock and, in 1880-1890, on Drift Rock Farm, Gothic Street in South Paris, later home of Ben Conant. Perry's widow, Estella Robbins Perry married George H. Davis on February 7, 1892 and continued to sell the residue of Nate's mineral stock until at least 1904 under her new name. She also advertised in the national magazine *The Mineral Collector*. Other mineral and gem dealers, such as Loren Merrill from about 1889 until his death in 1930, sold gems and minerals from their homes and barns. Some may have had a dedicated storage shed.

Apart from minerals that may have been sold on a few shelves in general stores, music stores, post offices, bank windows, etc., the first mineral specimen business in Oxford County, and probably in Maine, with a dedicated store front seems to have been that of Robert "Bob" Bickford's (b. 1873) in Norway. Robert Bickford was listed in the *Maine Register* as a mineral dealer in 1906 and as a lapidary as early as 1912. He also had a classified advertisement offering his services as a lapidary in 1912. (Brothers, Ross Bickford (b. 1875) and William Bickford (b. 1866), had lapidary businesses in 1906 and William was the first gem cutter in the family in 1903. William was listed as a mineral dealer in 1910. However, the transition from just mineral dealing to include lapidary and gem work was indicated for only William and Ross in 1906.)

Smoky Quartz, Bennett Quarry, Buckfield, 10 x 10 cm, found c1920s. Jim Mann collection.

Knox Bickford (b. 1889), was learning to be a gem cutter in 1910. He cut gems for members of the jewelry trade and probably cut the most gems of any lapidary in the family, although he was the last of the brothers who learned the art (Ben Shaub, Bickford notes and interviews, 1958).

In 1914, Robert Bickford's storefront address was at 25 Main Street in Norway and by 1917, there were four mineral and gem dealers in Norway: Robert F. Bickford, John H. Fletcher, Charles E. Bradford, and Freeland Howe, Jr. The other Bickford brothers probably put all of their specimens on consignment with Robert. (Freeland, who was listed as a mineral dealer as well as a chemist in 1908, had a music store on Main Street by 1917 and probably sold minerals collected by his older brother, George, as well as George "Shavey" Noyes, and others.) George Howe, an insurance agent, was listed as a mineral dealer, at least in 1898-1908. Shavey Noyes was a listed mineral dealer in the *Maine Register* 1898-1907, although he sold minerals through the mail and may have sold from his family's business, Noyes Drug Store, on Main Street.

Within the context of a widespread mineral selling environment, Stan Perham opened his business. His sources of supply were, initially his own discoveries, but later his inventory came primarily from the quarries his brother Harold worked. In Stan's store, there were dedicated shelves to the Bumpus Quarry, Albany; Hibbs and Sturtevant Quarries, Hebron; and Bennett Quarry, Buckfield. The Sturtevant Quarry was probably a poor specimen producer as it yielded reasonably quartz-free commercial feldspar. The above pictured reddingite was mined at the Bennett Quarry by Harold's crew and probably bought from Stan by Andover mineral dealer, Dick Nevel [1887-1938], and then resold (King and Mann, 2008). There were undoubtedly specimens obtained from other miners and collectors and it is common knowledge that miners would sell minerals to visiting mineral collectors and dealers, even if they were expected to

Earliest known specimen collected at the Tamminen Quarry, Greenwood, Quartz crystal 20 x 16 x 13 cm, found 1930. Addison Saunders collection.

Giant Beryls at the Bumpus Quarry. 1928.
Wallace E. Cummings (bottom) and possibly
Harlan Bumpus (top). One of Maine's
Most Famous Mineral Photographs.
First Published by Merrill and Perkins (1930).

Beryl, Wardwell Quarry, Albany. 1-3 cm lengths

turn them over to the quarry foreman. The role of the metal lunch in carrying mineral specimens home is also well-known. Although the early photos of Stan's showroom show shells and taxidermied birds on display, the time required to obtain replacement stock may have meant that less and less emphasis was being placed on general natural history objects and that minerals and gems became the primary materials offered by the Maine Mineral Store. Stan had a single cut gemstone in stock when he opened his doors for business in

1919. By the mid-1930s, gems, particularly ornamental rose quartz from the Bumpus Quarry, had become an important part of his stock. Ironically, the Perham Quarry was never a good mineral specimen producer. The most interesting specimens were quartz crystals with extra faces in parallel growth. Shaub (1961) wrote an article about a provocative-looking quartz crystal from the Perham Quarry, but he fought off the urge of saying that it looked like anything, but a quartz crystal. (See facing page.)

The Huge Beryls Discovered

During the earliest mining, beryl was abundant in the Bumpus quarry. By the winter of 1928-1929, some large beryl crystals had already begun to be exposed in the south face of the quarry. An unidentified newspaper clipping writing about the history of the Bumpus Quarry revealed:

"In December of 1928 workmen were puzzled by a formation of soft, green-colored stone imbedded in the face of the rose quartz and feldspar. As the pegmatite was cleared away the formation grew to a whorl of five huge crystals. Once again young Perham was called in. Beryl, he said, the largest find of its kind ever discovered. When processed, he told the owners, it made beryllium, a lighter-than aluminum metal. ...

In a few days, Professor [William Berryman] Scott, head of the geology department at Princeton University, stood beside the green mineral and said 'nothing like this had ever been found before in the entire world'."

Of course, the miners were not puzzled and, as will be seen, there were already several written agreements in place concerning the beryl. While there may have been uncertainty of the fate of the crystals, beryl was a mineral that Harlan and the mining crew were familiar with. It may also be mentioned that beryl is one of the harder minerals known from Maine. It is not "soft".

The various photographs of the beryl crystals in the ledge show that much of the enclosing quartz had been carefully chipped away from the crystal faces. Some photographs show a progression of "dressing up" the ledge through this careful trimming and exposing of the giant crystals.

"Big Beryl Crystal is Found in Maine. Find in New Albany Mine May Be Biggest Yet. Albany, Me, Nov. 20 [1929?] – A beryl crystal, believed to be the largest in the world, has just been found in the Bumpus feldspar mine in Albany.

The crystal is about four feet in diameter, but its length is unknown as it has yet been but partially uncovered. From what has been exposed it bids fair to exceed in length any crystal taken from this mine, and some big ones have been found here. Geologists who have examined the Bumpus mine are almost unanimous in the opinion that the beryl crystal deposit is the largest known in the country, it being estimated that there is not less than 200 tons of this crystal and there may possibly be 500 tons. ...

Harry E. Bumpus of Auburn, owner and operator of the mine several years, came here Wednesday, remaining until today, overseeing the excavation of the huge crystal."

Giant beryls were, historically, a New England phenomenon [1] and the news media were widely contacted when the enormity of the Bumpus beryl crystals was realized. Mineralogists, both amateur and professional, local residents, and tourists swarmed to Al-

bany to see the sight.

Edwin Gedney, [2] religious geologist, divinity student, and later Reverend of the Adventist Church, along with Harvard University mineralogist, Harry Berman, reported in 1929 on the giant beryl discovery, in the western end of the south wall of Bumpus Quarry, in the then leading USA amateur mineralogist's magazine, *Rocks and Minerals*.

"The beryl occurs in great, six-sided, log-like crystals embedded in the quarry wall. A few plates of white mica are found coating

Sturtevant Quarry dump. Lewiston Journal Magazine, September 30, 1963.

World's largest reddingite crystals, Bennett Quarry, 5 x 3 cm, sold by Dick Nevel in December, 1931. Jim Mann collection.

Quartz crystal, Perham Quarry. Hazel Perham Collection. From Shaub (1961).

other minerals, and one or two masses of a beautiful rose quartz six or eight feet across are found associated with the feldspar and beryl. …Three crystals are well exposed by the mining operations and are each about seventeen feet long and a yard or more in thickness. The largest has a length of about eighteen feet, a diameter of four feet, and an estimated weight of about eighteen tons. [9] Beside these observed crystals there are exposed portions of seven others which give indications of being fully as large.

One of the more prominent of these has a diameter of four feet and is exposed for a length of fourteen feet, although neither extremity of the crystal has as yet been uncovered. Another of equal thickness has been exposed for a distance of eight or nine feet without reaching either end. Others have been truncated by the mining operations and are only exposed in roughly hexagonal cross-sections which are all from three to four feet in diameter. In addition to the ten large crystals, many smaller ones ranging from six inches to a foot or two in thickness are noted in the feldspar and quartz. …

The larger crystals appear to be in groups radiating outward from a common center … This indicates that the crystallization of the beryl began at one or more centers and proceeded outward from these centers in all directions forming star-like groups. The attitude of the crystals in the quarry wall seems to indicate that there are at least two, and perhaps three, of these radial groups partially disclosed. …

They seen to be more or less localized in their occurrence and all of those so far noted could be enclosed within a circle of a radius of thirty feet or less. The beryl manifests little or no preference in its mineral associates. It is found with either quartz or feldspar, and in

Bumpus beryl cross-section, rod may be a four-foot drill steel.
Lewiston Daily Sun, November 23, 1929

Dr. Edwin Gedney, c 1930.
From Gordon College Alumnus (1974).

Bumpus beryl crystal side view. Possible four-foot drill steel placed diagonally through section.
Lewiston Daily Sun, November 23, 1929

places a single crystal may cut directly across both of these minerals. Some of the larger crystals are somewhat cracked and fractured, and plates of mica up to an inch across are formed along these fractures.

In general the beryl is a light apple-green in color, and more or less milky to opaque. It occasionally varies to an aquamarine variety which, however, is quite commonly badly fractured. …

The exposed portions of the crystals were measured with a view of estimating the amount of beryl observed on the surface in the quarry and on the dumps. About a hundred tons were found exposed to view, and the evident similarity of size between the partly exposed crystals and those large ones more fully revealed, together with their general radial grouping indicating many still uncovered crystals, suggests that at least an equal amount may yet be revealed by future mining operations."

Stan Perham examining the Giant Beryls before they were fully exposed. c1928. Courtesy Jane Perham.

Edwin Gedney maintained a lifelong interest in the Giant Beryls (See Chapter Seven). He was a trained geologist and began his career in that field. Gordon College Alumnus (1974) summarized:

"Following graduate studies at Harvard University, Edwin Gedney joined the New York firm of Fleetwings, Inc. as its Director of Research and Development. To aid in his trek through the States in search of mineral deposits the company built him a trailer house, long before they were produced commercially. Dr. Gedney believes that 'It could have been one of the first in America.' An airplane company that had an interest in the firm supplied a hollow aluminum tube from which his living quarters were fashioned. This mobile home was pulled by Edwin Gedney's 1929 Ford and he lived in it with his wife Dorris, their baby, and a dog for a year and a half. …

Gold was discovered in East Africa in the early 30s. Geologist Gedney was invited to the African continent and was given as job of Director in charge of gold exploration for Ventures, Ltd. His responsibility was to locate the gold so that the company could set up and man the mines. The firm had exclusive rights to an area of land two-thirds the size of Massachu-setts in parts of Kenya, Uganda and Tanzania."

Gedney also wished to pursue his spiritual interest and in 1934, he joined the faculty of Gordon College of Theology and Missions in Wenham, Massachusetts: "Sitting on a warm, sunlit mine shaft in the bush of Kenya, waiting for his partner to emerge from the dusky depths, a young geological engineer made a decision that dramatically changed the direction and purpose of his life." Gedney had met with the College President, Dr. Nathan Wood, and accepted the challenge of creating a new science department for the College. Gedney taught Earth Science, Biology, Chemistry and Physics. In the ensuing years, Gedney started so many new programs for the school that he became known as "Mr. Gordon", in loving recognition of his dedication to educational development.

Regarding the Bumpus Quarry, Hugh S. Spence (1929), of the Department of Mines of Canada, wrote a "letter to the editor" of Engineering and Mining Journal, also reporting the same crystal dimensions as Gedney, including having his picture taken alongside the giant beryls.

A dizzying array of numbers has been attributed to the sizes of the Bumpus Quarry beryls, frequently without reference to when they were measured. A contemporary account by the Maine State Geologists [145] also provided measurements of these beryl crystals, presumably measured by themselves and presumably from the giant cluster: "Eight or ten very large crystals are exposed. The four largest measure: 18 ft. x 3 ½ ft., 13 ft. x 3 ft., 12 ft. x 3 ft., and 10 ft. x 3 ½ ft. The crystals are ordinary opaque bluish beryl. In addition a little aquamarine and golden beryl has been found. There is no evidence or other indicators of rare minerals." (Small amounts of cleavelandite were found later.) Based

Bumpus Quarry Beryl Ledge, note pick ax at left
Boston Herald, November 20, 1929

on newspaper photographs with a rod leaning against the beryls from the giant cluster, the Merrill and Perkins (1930) measurements are probably the most precise of all that have been reported and were made when the crystals were exposed at their maximum lengths. While diameters of some of the crystals were sometimes reported in the 42-48 inch range, it appears that the larger diameters occurred on shorter crystals, but the inquisitive reader, of later articles,, may find all of the largest dimensions combined into one huge "dream crystal".

The Attempt at Preserving the Albany Giant Beryls

The fame of the huge beryls attracted not only naturalists, tourists, and reporters. The discovery of a beautiful ledge of rose quartz and the enormous cluster of well-defined, variously light blue, light blue green, to light sea-foam green beryl crystals must have inspired many conversations of how beautiful they were and "Wouldn't it be nice if they could be placed in a museum for all to see?" The earliest mention of preservation, in print, may have been by Alice Frost Lord writing for the *Lewiston Journal* (June 1, 1929):

> "Asked if these specimens are likely to find any permanent home in Maine, [Stan] Perham shook his head doubtfully. They would make marvelous accessions to any of the college collections or to some institution in a city like Portland where visitors would thrill at the sight, if they have any love for mineral beauty.
> 40

Richard M. Field, c. 1935
Geologist, Princeton 1923-1950

The difficulty in these places, is to find quarters where such a shaft of beryl could be exhibited properly. Eighteen feet of crystal is a bulky and heavy bit of room decoration.

William Berryman Scott. c1930
Paleontologist at Princeton 1884-1930

Mr. Perham has had negotiations with Harvard and other educational institutions where these might find a home. Several universities are anxious to obtain one. But the arrangements are as yet not completed for any."

The *Boston Herald* (November 20, 1929) had an article "Giant Beryls in Quarry at Albany, ME. Harvard Experts Confirm Reports of Maine Beryl Find" The article notes that two Princeton University professors inquired of Charles Palache at Harvard University if there was anywhere "interesting in New England" where he could take their summer geology class on a fieldtrip. Field was directed to go to the Bumpus Quarry to see the fabled beryl crystals that had been discovered. After the trip, the two Princeton professors had an idea:

"Prof. Field, on viewing the quarry, became very enthusiastic about the beryl crystals – larger than he had ever seen. His associate, Prof. W. B. Scott, was even more enthusiastic and thought that something should be done to preserve the deposits. [Professor Scott was a descendant of Benjamin Franklin.] Hence his suggestion at the meeting of the scientific body [National Academy of Science, autumn meet-ing] at Princeton that Maine authorities be apprised of the value of the crystals and make some attempt to save them from destruction."

It may have been that the enthusiasm by the new-to-the-scene academics was just what was needed to do something towards preservation. The idea of saving the beryls was not a new one, even to "Maine authorities". As will be revealed, Palache had already had a similar idea, but his "feathers had been sufficiently ruffled" and he had already abandoned hope: "… geologists and mineralogists at Harvard University are not very excited about the story [of preservation] as reported from the autumn meeting of the [National] academy of sciences meeting at Princeton University."

In the December 6, 1929 issue of *Science,* it was simply reported of the proposal:

"Remarkable Beryls in the Maine Pegmatite: W. B. Scott. In the Albany Quarry, near Albany in Maine, has lately been uncovered a group of crystals of beryl which are quite unprecedented in the matter of size. … Two lantern slides, illustrating them from different points of view, are shown. The object in presenting this brief statement to the academy is to ascertain if some way of saving this wonderful group and keeping it *in situ* can not be devised."

Who Would Buy the Beryl?

The December 31, 1927 amended lease gave an official status to beryl production at the Bumpus Quarry. The idea of having a special exhibit of beryls at the American Museum of Natural History was also taking root, probably, because of the appeal issued by Scott at the scientific meeting. George Kunz, a member of the National Academy of Sciences and a frequent attendee to its annual meetings, may have been present as the meeting was conveniently located near New York City. (Coincidentally, the American Museum of Natural History owes the impetus for its existence to

West Minot, Maine from an old postcard.

Abner Kimball and boy on Beryl, Bumpus Quarry. One of the last photos known of the beryl in the ledge. Unidentified newspaper, Winter 1930-1931

Maine native, Albert Smith Bickmore, who was born in St. George, Knox County, just south of Rockland [March 1, 1839 Tennants Harbor-August 12, 1914 Nonquitt, MA]. Bickmore was co-founder and first director of the American Museum of Natural History. [29]) Despite the appeal for academics to preserve the natural wonder of the Bumpus beryls, there was also a small growing market for beryl as an ore of beryllium and therein laid a conflict.

By the end of 1929, there had to have been plans already in motion to enlist Stan's help disposing of the beryl. Stan had just left Bates College and was trying to make a living. Stan had been working in Harold's quarries, but Harold also made a deal with Stan. The December 1929 agreement was barely a month old, but there had to be pressure to resolve the giant beryl crystal problem, quickly. The heirs of Sybil Cummings had seen little, if any, of their inheritance. While few anticipated what the Black Tuesday Stock Market Crash would mean to them, it was still necessary to scrape out a living.

The agreement between Harold and Stan was an option for Stan to buy the beryl from Harold and was signed on February 8, 1930 (OCRD v. 389, p152-153), but was not registered until October 7, 1930 and the agreement had been approved by Harry and Laura Bumpus. The option agreement was for 120 tons of beryl shipped in string double bags, probably burlap, "...suitable for usual handling by ocean and/or rail transportation companies without loss ... on or before March 1st. The remaining 60 tons to be delivered at buyer's option before August 1st." [309] Provision was made to have the beryl sampled and assayed to verify that the beryl contained at least 10% BeO and an analytical laboratory, Ledoux and Company of New York City, was listed as an example. Stan's purchase price was $35.00/ton F.O.B. Bethel. A special note is important. The agreement required: "Payment within five days from date Stanley I. Perham receives payment for beryl and said Stanley I. Perham shall speed *as fast as possible* [emphasis added] all such deliveries

mentioned and financial returns." Stan was also "obligated" to "accept delivery of an additional 130 tons ... between now and February 1, 1932" Although there were a few other details, this sale may have been the first big deal of Stan's long and illustrious career as a miner and mineral dealer. Nonetheless, it is probable that the actual amount of beryl produced from the Bumpus Quarry could not meet the projected weight of the contracts' requirements.

The Taylor Beryl Contract
A Pivotal Development

Unbeknownst to the beryl preservationists, a beryl ore sales contract also had been negotiated by Stan with the David Taylor Company, of New York City, on February 14, 1930. The Taylor Company specialized in "Ore and Products of Tungsten, Beryllium, Vanadium, Columbium, Tantalum". Taylor was to pay Stan $50/ton F.O.B. Bethel for beryl that contained 11% BeO or more with a provision for an extra $3.50 for each percentage point higher of BeO per ton. There was also a penalty for lower grade beryl and there was also a release available if the beryl was discovered to contain important amounts of cesium, but Taylor had the option to buy the beryl at a premium according to the added cesium value.

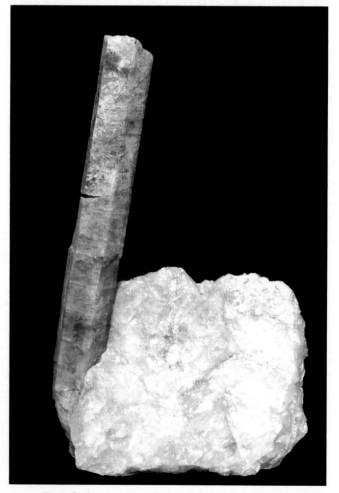

Beryl, Songo Pond Quarry, 29 x 3.8 cm. Addison Saunders specimen.

Tamminen Quarry, Greenwood. c1938. Courtesy Ben Shaub.

There are no available quantitative analyses to suggest that Bumpus beryl actually contained the cesium provided for and the market for cesium was still thought to be strong, but with the abandonment of the cesium radio tube within a few months from the Perham-Taylor contract date, the issue wouldn't have affected the actual shipments. [38] The Perhams were encouraged to recover beryl from all of its quarries and they probably bought beryl from various miners and mineral collectors. The price in the contract meant that the usual 13% BeO of Bumpus beryl would sell to the Taylor Company for at least $57.00 a ton. Expenses would come out of Stan's payments to Harold. Section #9 of Contract #2 had a relevant provision: "It is specifically agreed that seller may retain for his use or sell [to] jewelers, small manufacturers, etc., up to 5 tons of Beryllium ore between the dates February 3, 1930 and February 2, 1932 at a price not less than $90.00 per short ton of ore containing 11% oxide, himself retaining the proceeds and not rendering any account thereof to the buyer hereunder." This contract was registered at the Oxford County Court House on November 13, 1930. This "clause" permitted Stan to sell a few tons of beryl to the preservationists without a penalty.

The Bumpus Quarry could easily have been the world's largest beryl supplier in the time, mostly because of the huge crystals exposed and "waiting" in the ledge. The contract specified: "It being recognized by both buyer and seller that this contract constitutes an unusually large Beryllium ore transaction and that buyer hereunder may have to hold the material for as much as two years, before he can distribute it in the market, it is agreed that there will be no additional Beryllium ore sold, offered for sale, or shipped from the properties controlled by you and your brother be-

fore February 2, 1932, except with the consent of the buyer hereunder, but that after delivery of the 120 tons herein called for, the buyer herein is obligated, between now and February 1, 1932, to accept delivery of an additional 130 tons under the terms of Contract #2, attached hereto. ... As requested by you, we have already sent you, as evidence of our good faith, check for $500.00, receipt for which is acknowledged by you, as advance payment on the first delivery of Beryllium ..." We might imagine that company president, David Taylor, had already personally inspected the huge beryls in place at the Bumpus Quarry. Although the Taylor contract sounds lucrative, there have been hints in various interviews that the contract was never started due to outside concerns. Stan Perham's address was given as Lewiston, Maine on the contract and that may have been the address of his lawyer or a local bank with offices also in New York City.

Harold was managing feldspar shipments from various quarries spread from the Bumpus Quarry in Albany, which more or less "ran itself" as far as his direct supervision was concerned, to his own quarries in Buckfield, Hebron, and Minot. Because the concentration of his attention was in southern Oxford County and western Androscoggin County, Harold moved his family to the tiny village of West Minot. Stan lived in West Paris and could better manage the beryl issue, which undoubtedly had become a severe headache for Harold and a headache for almost all who were involved. The giant whorl of beryl crystals was exposed in the Bumpus Quarry for two years. No beryl was being sold and no heirs were getting their share. The giant beryl crystal deal offered to Palache in late 1928 or early 1929 was at a premium price: $100 per ton. The 1930 contract with Taylor

Abner Kimball, c1940. Courtesy John Kimball.

Of this amount, $7,000.00 would be paid by Stan Perham to Harold Perham (@ $35/ton). Stan's margin would have amounted to $3,000.00. Harold Perham would have kept 10% or $700.00, as sales agent; leaving $6,300.00 residue for payment of 200 tons of beryl to Harry and Laura.

Harry and Laura would get $3,150.00 as they had a contract requiring them to pay 50% royalty. The remaining $3,150.00 would be split equally with Allen Cummings and the heirs of Sybil Cummings. Each of the full-share heirs would get one fifth of $1,575.00, equaling $315.00, while the one-fifteenth interest heirs would get $105.00. (Laura, as a full-share heir, would also get one fifth of the royalty, in addition to her share according to the contract.) (The one fifth share ($315) was equivalent to 6 months' wages if you had a job in a Maine town and may have equaled a year's wages if you lived in Albany, Maine.) Their sister, Cora Emma, died in 1909, but her three daughters, Adelia, Laura, and Edith, were entitled to split her one fifth share and would have each received one fifteenth shares. Of course, there would have been shares in the sales of feldspar, quartz, and mica.

Everything Changed

Up to this point, there are not many public indications of trouble brewing. The giant beryls were still in the ledge and were still garnering admiration. Unfortunately, the excessive publicity the beryls had received would prove a disaster for all involved.

Also in 1930, Stan, age 24, moved his formal retail and wholesale mineral business, the Maine Mineral Store, to Trap Corner in West Paris on the better traveled State Route 26 and was listed in that year's *Maine Register* under "jewelry". He lived in the majority of the building, while the Maine Mineral Store occupied the south end of the building. It is important to note that Stan Perham would eventually become the most important mineral dealer in Maine and it was principally due to his diligence and connections that so many Maine minerals have been preserved and not discarded or damaged. Miners recognized through Stan's educational efforts that minerals had scientific and monetary value and that he was willing to buy them. The importance of his influence for half a century cannot be over-emphasized and hardly an official Maine or Federal geological and mineralogical report dealing with pegmatite mineralogy failed to mention a consultation with Stanley I. Perham. An early, probably 1930, advertisement for Stan's store read:

> "The Maine Mineral Store has the first choice of everything from the Harold C. Perham Feldspar Quarries and those of the Trenton Flint and Spar Co., the largest group under one control in the Pegmatites of Maine. Included in the group is the famous Bennett Ledge in Buckfield, Mt. Marie, So. Paris, The Hibbs Quarry in Hebron, The Bumpus Quarry in Albany. Each a noted locality in which every collector should be interested... Your cutting

may have been a long time in the negotiating and it may have been that convincing a customer to buy "an unusual amount of beryl" had already had a protection clause stipulating that if anyone bought beryl from Harold Perham that it would be at a premium rate so that Taylor would not be disadvantaged by any competitor getting beryl at a rate close to theirs. The implication of the various negotiations implies that the revised Bumpus-Perham contract of late December 1927 meant that Harold was able to entertain offers for the beryl crystals. The Palache deal was a small one. The Taylor offer was a large one and Taylor may have been courted as a buyer as early as sometime in 1928. The Stan Perham-Harold Perham option was signed February 8, 1930 and the Stan Perham contract with David Taylor Company signed only six days later, has to imply a previous long term negotiation with Taylor.

As pressure increased on Harlan, Harry, and Laura due to their relatives, they in turn certainly pressured Harold to do something about selling the beryl and getting it out of the quarry so feldspar could be more easily quarried. If the contract for 200 tons were fulfilled, the total sale would have amounted to $10,000.00 paid by Taylor to Stan (@ $50/ton).

trade in gem and rock work is solicited. ... When in Maine be sure to call at the New Store with its two floors – downstairs, public display; Upstairs, private display."

Stan's store may have been a part-time or seasonal venture, at least before 1933. The weekends and summer months were the time to make the sales and he had to make sure his shelves were stocked. It was worthwhile for Harold or his mining crew to collect display-quality mineral specimens, but Stan kept an eye out for the more casual specimens that he could sell on a regular basis, as well. Nonetheless, when good specimens were discovered, they sometimes got publicity. One press release by Stan on August 7, 1929 to the *Lewiston Journal* read:

> "A pocket of Royal Purple Apatite was opened at the Hibbs Quarry in Hebron, at 3:00 P.M. today. It is the first to be opened in Maine in many years. There were 20 to 30 crystals in the pocket, the largest of which is a magnificent specimen, true royal purple in color, fully 7/8 of an inch long by ½ an inch in diameter, has a perfect termination and is in a beautiful white matrix. Every crystal found except one or two sets in a fine white base.
>
> The Quarry is owned and operated by Harold C. Perham for feldspar. The very choice crystals have been placed on display in the Maine Mineral Store, West Paris, Maine, where they may be seen by any one."

The press release had a penciled note that the magnificent piece was purchased by the Boston Museum of Natural History. Formerly, Stan was a summer time miner, going to Bates College, as well as store owner. After leaving school, he continued working for his brother. One evening he made a still famous discovery of rose quartz, which had a deep strawberry red color (*Lewiston Journal*, July 11, 1930):

> "Looking around after work, Stanley Perham ..., employed at the Hibbs Quarry, discovered a rose quartz weighing about 50 pounds. It is a deep rose color and is the best museum piece ever collected at the quarry... The piece is valued at $100. The stone will be on display at the Maine Mineral Store in West Paris for the next week."

Harold was proud of his Hibbs Quarry minerals. On August 2, 1930, the *Lewiston Sun* reported:

> "A layer of feldspar 40 feet wide, 100 feet long, and between 15 and 20 feet thick, has been located in the Hibbs quarry. The value of the layer is estimated at $35,000. That is not the only find as alongside of this feldspar was located a good-sized quantity of mica. ... It is claimed that the best mica in the United States is taken here. Last year 120 tons were sold from this same quarry, some of it worth as high as $20 a pound.
>
> Aquamarine stone worth $100 was dis-

covered, the first of any amount found this year. It was located near crystal rock and is of clear bluish hue. In a few days more this stone will be removed since there are many crystals showing. This was found near the spot where the rose quartz was taken a few weeks ago.

> The first large crystal of columbite was also found Friday. This is a very rare mineral and found occasionally in the ledges and when in crystals is of value. Work has just been started in the bottom of the quarry and it is believed that throughout the summer a considerable value in rare minerals will be removed. The quarry is operated by Harold C. Perham of South Paris who sells feldspar to the Trenton Flint & Spar Co., of Trenton, N.J."

War of the Scouts

Stan was motivated to fulfill the beryl contract. He was sure of the huge beryl exposure at the Bumpus Quarry, but every ton of beryl found elsewhere was more money in his pocket as he didn't have to share the $50/ton with anyone. In order to maximize beryl production, Stan hired a scout about late July or early August of 1930 to obtain beryl and other mineral leases. Given the timing, one might suspect that the "company", probably the Cummings heirs, were threatening to stop honoring the beryl provisions and the hired scout could as easily have been a desperate measure. The scout's agreement was not registered with Oxford County Clerk until September 11, 1936, however: "I, the said Loris M. Rollins [b. ~1902, resident of Kennebec Co., ME 1910 and 1920; probably a resident of Berlin, NH in

George F. Kunz, c1930

Charles Palache, c1930

Rollins and Donahue were at the County Clerk's office on the morning of September 18, 1930 and the first Rollins lease and the first William Donahue lease are on the same page in County records! If we are to interpret the "received" time as the actual time the paperwork was handed to the clerk, Rollins recorded a lease, signed the day before, with Errold Donahue of Albany at 10:35 AM, with both Errold Donahue and Rollins appeared before the clerk. William Donahue's lease, signed on September 12, with Rodney Willard of Woodstock was recorded at exactly high noon, on the 18th, also with both signers present. Donahue returned two hours and eighteen minutes later recording a lease seemingly with the ink still wet from Nestor and Kattie Tamminen of Greenwood, who were also present for the filing. The UFC leases were from places close to the West Paris mill. A beryl quarry in Albany didn't need to be close to West Paris, if there were no intention to market by-product feldspar.

Not long afterwards, one of the Rollins' leases was for the Kimball Ledge, just south of Songo Pond, Albany, made with Abner Kimball, Albert's father. [340] The lease was seemingly written to specialize in:

"1. Gem material: 50% of worth in rough at quarry or mine as determined by a disinter-

1930], have been in the employ of the said Stanley I. Perham for a period of about three months prior to the date of this instrument [October 21, 1930], and under his instructions and under his pay I have been procuring certain leases in connection with minerals, gems and similar operations within the limits of the aforesaid County of Oxford." [310] The beryl excitement had spread with consequent prospecting and "cruising" of ledges in order to discover additional beryl sources.

United Feldspar Corporation also had to have been desperate. We may infer that the Cummings Family feud was threatening the continuation of all mining at the Bumpus Quarry. In the early Depression, national demand for ground feldspar was plummeting, but UFC had to open new sources for its supply of crude feldspar. It still had some share of the business that existed and the loss of the Bumpus Quarry would have been potentially ruinous. With this epiphany, it is plausible that the Bumpus feud was manipulated, or at least complicated, by the United Feldspar Corporation. We can only conclude that the family feud and pending court action had shut down the Bumpus Quarry, temporarily. Neither beryl nor feldspar could come out. UFC hired William B. Donahue as their scout. Although Stan's scout seems to have been in the field longer than UFC's, we only have a vague inkling of the timing. In an interesting coincidence, both

Harry Berman, c1939

Large Bumpus Beryl fragment, c1930 from postcard.

Rollins was active at other locations in Stan's behalf. A very interesting lease was obtained from Albert B. Kimball on September 25, 1930 (OCRD v389 p165). The lease was on the D. A. Cummings lot, also known as the Pine Hill Pasture. There is, of course, a question if Albert were still in the employ of Harlan Bumpus? It may have been that Stan's brother Harold was fully aware of Rollins' activities and may have been contributing to his ability to supply feldspar to the West Paris Mill. Nonetheless, Albany was known, at least to mineral collectors, as a beryl-rich area and the prospecting may have been completely directed toward finding beryl, rather than for feldspar. Also on September 25, Rollins leased the adjoining lots to the Allen Cummings Farm where the Bumpus Quarry resided, also from Abner Kimball (OCRD v389 p166). On October 10, 1930, Rollins leased the Upton Farm

ested party chosen jointly or by arbitration process by the lessor and lessee.

2. All other marketable metals or minerals other than feldspar, mica, beryl, and quartz, 10% of value at quarry or mine as determined by joint agreement or disinterested party...

The following royalties shall be paid in the event of removal of any forms of feldspar, mica, beryl, or quartz:

Feldspar .50 per ton of 2240 lbs.
Mica 5.00 per ton of 2240 lbs.
Beryl 40% of value
Quartz 50% of value per ton of 2240 lbs."

The "lease" sounded more like a sale of mineral rights: "It is further understood and agreed that all rights and terms of this lease shall [endure] to the benefit of, and pass on to, the heirs, successors, and assigns of the parties hereto." The permanent nature of the agreement might still appear generous given that percentages were used for beryl and quartz royalties. The prices for feldspar and mica are not particularly low even with three-quarters of century's inflation. The Kimball Ledge Quarry was prospected and a few specimens were obtained, but even by 1955, Morrill only noted in his mineral collector's guidebook: "Kimball mine – Beryl". In 1958, this was upgraded to read: "... small pit and old vertical face on SE toe of knoll. Operated for aqua (Deep Blue Xls, terminated);

Verry and Tiger hauling huge beryl crystal fragment away from the Bumpus Quarry. Courtesy Ava Bumpus.

along Crooked River in Albany from Helene Bruce of Bethel (OCRD v389 p168).

On September 26, 1930, Rollins obtained a renewable lease on the "Millard Lord place" in Albany from Fred R. Littlefield using a lease with the standard royalty provisions (OCRD v389 p150). On September 30, 1930, Rollins had a ten year lease of the "Alton Fernald lot" in Albany from Rodney F. Willard (OCRD v389 p151) and this lease was transferred to Stan Perham on October 16, 1930 (OCRD v416 p373). Perhaps the only contender among the leases, as a true beryl producer, was the Donahue Quarry, the land being leased from Errold O. Donahue on September 17, 1930. Because the lease documents were all signed within a short period of time, it is likely that Rollins and Stan conferred over time concerning the merits of each property. The battle of the mineral scouts may have been the subject of considerable interest on the Oxford County mineral and mining grapevine. While Stan was the employer, Rollins' leases on potential feldspar properties would have complemented Harold's business, even if Stan were the real manager or superintendent of any quarries activated.

Beryl Crystal Fragment on sled, c1930 from postcard

American Museum of Natural History, c1930, 77th Street archway entrance, New York City.

Negotiations for the Giant Beryls

With many influences at work, but before any of the beryl problems openly emerged, George Kunz [September 29, 1856 NY, NY – June 29, 1932 NY, NY], then vice-president of Tiffany and Company jewelers, realized that there was an opportunity to obtain a prize for the American Museum of Natural History in New York City. Kunz had long been associated with the American Museum and Maine minerals, since about 1880. Kunz also had begun purchasing Mount Mica tourmalines a half century before from Augustus Hamlin, as well and rare minerals from Nathan Perry and others. [126] Kunz frequented the Poland Spring Inn, located just one train station before South Paris, whenever he came to Maine to buy minerals and gems. Kunz also had been instrumental in obtaining many worldwide treasures and donations to the museum. (It is an interesting coincidence that Hermon Carey Bumpus [May 5, 1862 Buckfield, ME – June 21, 1943 Pasadena, CA], who was a distant cousin of Harry Bumpus, was a zoological curator and later director of the American Museum of Natural History from 1900-1911. By the time the Bumpus Quarry beryls were in the news, Hermon Bumpus was organizing museum exhibits for the National Parks Service, having had a prestigious career, including his work as a zoologist and as president of Tufts College.

Kunz and George Howe assuredly knew him.) As Kunz died only two years after visiting the Bumpus Quarry, his trip to Albany may have been one of the last he ever made into the field to buy important mineral specimens.

A series of letters passed between Kunz and Charles Palache and Harry Berman at Harvard University concerning the giant beryls. (Dates stamped on the reply letters indicate being received the day after the letters were dated by their sender; normal mail service in the time period.) On April 23, 1930 (HU Mineral Museum), Harry Berman wrote to George Kunz:

"I agree with you that if action is to be taken, it must be done in a hurry. I am not sure, however, but what it is too late already since they have been exposed all through the winter

Stan Perham, Summer, 1930 on crate used to ship one of the Bumpus Beryls.
From Perham, 1987

and may be already not suitable for Museum purposes.

  However, I will probably be in the vicinity of Albany within the next week and I can easily run up to Albany to take a look at the crystals again. If there is anything that you wish to have me say in your behalf to Mr. Perham, who has charge of the sale of the crystals, kindly let me know within the next few days."

Of course, the beryls had been through two Maine winters, before the days of global warming, and certainly had received rough treatment from rain, ice, snow, freezing, and thawing. The crystals do not seem to have been exposed to their maximum length until after the winter of 1928-1929, however. Probably after a reply to Berman, Kunz wrote to him (May 19, 1930, HUMM):

  "Mr. Perham, of West Paris, Maine writes me a long letter about the beryl, but says absolutely nothing about the price. He says that there is danger of it being broken up, and also states that you recently called on him.

  Can you let me know as soon as possible what you said to him? I would surely like to see these preserved, but his letter does not sound very encouraging in that line. P.S. Does he own it? Who owns it?"

There may have been a confusion of Perhams in the following letter. Stan had no interest in selling the beryls until two and a half months before. Harold would have been the one to negotiate with before that time. Palache wrote back (May 20, 1930, HUMM):

  "I have seen your letter to Mr. Berman and take the liberty of answering it myself since I have had enough dealings with Mr. Perham to give you some idea of the situation about the beryls.

  I first saw these crystals two years ago in June when they were just being uncovered and I was at once interested in the possibility of securing a group. We watched them through the summer and Mr. Gedney, one of my students, went up in the fall of '28 and came to the conclusion that it would be possible to quarry a block of feldspar containing the beryls which was then sound enough to hold together; that the block would weigh upwards of 20 tons and that transportation would not be impossible in the winter time to the railroad. Having in mind some of the block might be set up in the courtyard of the museum, I negotiated with Perham, who does not own the quarry but owns the mineral rights on its product. Perham was holding the beryl at $100. a tone [sic] as an ore of beryllium. He is absolutely devoid of any interest in the scientific aspect of the mineral but is out for all he can squeeze out of it in the way of money. I tried to make a contract with

"Visitors" to see the Giant Beryl. Note that the published image is reversed. Unidentified newspaper.

Giant Quartz Crystal, Auburn.
26x19x13 inches 66x48x33 cm
253 pounds = 115 kg

him on the basis of $1000 for the right to take the specimen out, I to pay costs of extraction and take all risks. However, as I had no desire for a broken specimen, I stipulated that if the material went to pieces in extraction, I would not be responsible for more than a small payment as a guarantee and that the beryl would

Giant beryl under archway of the American Museum of Natural History. Courtesy Ava Bumpus.

remain as his property. He insisted on full payment in advance, offering enough to attempt to dispose of the beryl should I fail in my efforts. I considered the matter so risky and was so utterly unwilling to enter into a contract on this basis that I gave up the idea definitely. Since that time he has made repeated approaches to me to reconsider my decision. Extraction of the feldspar has extended in such a way that the group is now very much in the way of the development of the quarry. Leaving it where it is without protection from the elements means certainly that it will be destroyed by the frost. It is already injured by exposure of two winters. On the other hand, to leave it practically stops the future development of the quarry. As far as I can see, the matter of the preservation of the group in place would mean the purchase of the whole quarry and undoubtedly this would involve a considerable sum of money. I of course have little interest in its preservation where it is as far as spending money on it goes. If the state would take it over it would be a fine thing. That suggestion has, as you know, been made before by the National Academy of Science. Knowing Perham as I do as a grasping young man with no sense of scientific value, I do not think there is the remotest chance of any such plan going through.

You have seen the pictures which Gedney took. As a whole the group is magnificent but the individual crystals are not so sharp and clean cut that they would make good Museum specimens."

The agreed price of only a $1,000 meant that only 10-20 tons of beryl were being paid for, a bargain as the crystals were estimated at higher weights. The deal did mean that Harold wouldn't have to pay someone to do the slow work of extracting the huge crystals. Harold may have realized his offer was not a very lucrative one and could not afford to negotiate a low price with academics who frequently felt that anything "scientific" should be theirs for the asking.

Undaunted, but in a desperate sounding letter, Kunz wrote to Palache again (June 4, 1930, HUMM): "Cannot something be done about the beryl up in Maine? I could arrange to have somebody meet us there and I believe it is possible that one of these specimens could be presented to Harvard, as well as one to the American Museum of Natural History."

Remnants of the Beryl Whorl, Merrill and Perkins (1930).

Palache replied (June 9, 1930, HUMM) that he didn't believe that a deal would be successful and that he didn't want to be involved further, especially as Palache had to go to California for most of the summer. This letter was soon followed up by a letter from Kunz to Harry Berman at Harvard (June 12, 1930, HUMM):

> "Could you or Mr. Gedney arrange to go up to Albany, Maine with me sometime in the very near future to see if anything can be done about the beryl? …
>
> Please let me know about this as soon as possible as I should like to go up there within the next week or two."

Berman replied the next day (June 13, 1930, HUMM):

> "Mr. Gedney, I understand, is not available since he is shortly to go into the field for commercial work.
>
> I could go with you to Albany, preferable after June 26th, since we are in the midst of important work. It may be that you will not want to go through with the beryl proposition but I am sure that it would be worth your while to take a look at them."

Patiently, Kunz waited and wrote back to Berman (July 18, 1930 HUMM):

> "Could you go to Maine next week? Con-

Janice Wood Jung during the summer of 1934 with a Bumpus beryl crystal section. Courtesy Seabury Lyon.

> fidentially: I think that I may be able to arrange something with regard to the beryl that might be satisfactory all around."

The "confidentially" remark suggests that Kunz had reached an agreement with American Museum mineralogy curator, Herbert P. Whitlock, and the Museum's administration to use money from John Pierpont Morgan legacy funds to finance the beryl's acquisition. (Although Morgan died in 1913, he had established a number of endowment funds for the museum to continue to purchase important specimens, including minerals and gems, in his name. Note also that there have been many attributions of various museums receiving sections of the giant beryls. The only museums to acquire portions of the giant beryls were the American Museum of Natural History in New York City and the Field Museum of Natural History in Chicago, Illinois. Oliver C. Farrington may have contacted a benefactor, William J. Chalmers, to buy a section. Chalmers was chairman of the Allis-Chalmers heavy equipment company as well as trustee of the Field Museum. Regarding reports of other institutions (*Lewiston Journal,* February 18, 1956), Stan Perham was quoted as saying that the Carnegie Museum of Natural History in Pittsburgh, Pennsylvania was a recipient of a giant beryl section, however, the curator of that museum's mineral collection found no records or reports indicating that such a beryl had ever been in their possession (Marc Wilson, personal communication, 2007).)

Destruction of the Giant beryls

An unidentified newspaper clipping had the following continuation to the story, which doesn't seem to have been published elsewhere.

> "Bumpus waited months [actually over a year] for someone to decide what to do with the precious crystal. Finally he announced a deadline. If a decision wasn't reached by a certain day he would blast it into smaller pieces and remove it. Kunz waited one day too long and Bumpus carried out his promise. Kunz ar-

Stanley I. Perham, 1930. Lewiston Evening Journal

rived to find the original whorl of 100 tons reduced to four and five ton chunks."

The destruction of the giant beryls, the day before Kunz was scheduled to arrive and inspect them for purchase seems an astonishing act. What would prompt such action? Was the destruction of the beryl crystals a spiteful decision? Had the feuding relatives decided that they were not getting their fair share of the sales prices of the beryl? It is possible that Kunz's intentions of visiting were unknown to Harlan.

We know the family feud was blazing in 1930. A parade of famous, and not so famous, people had viewed the spectacle of the ledge, most vocalizing not only awe, but mouthing the "value" of the crystals, either monetary or scientific. The two-year old revised lease with Harold Perham was signed when the amount of exposed beryl was small. After the giant crystals became known, family members must have come to the ledge time after time, staring at it and wondering what more lay hidden. The ledge became the metaphorical "Goose" and the beryl was the "egg". Just as in the fairy tale, when the ledge was broken open, beryl ceased to be found. The Bumpus-Perham lease was in force at least until August, 1930, after the large beryl chunks had been shipped off to New York City.

The beryl "whorl" probably weighed less than the widely quoted 100 ton estimate and the U. S. Bureau of Mines noted that the whole Bumpus Quarry beryl produc-

Edson Bastin, c. 1906
USGS Geologist in Oxford County

tion, up to that time in the early 1950s, was only 255 tons. Perham Stevens (1972) wrote:

"By the week of July 25, 1930 the last of the remarkable beryl specimens had been removed from the quarry wall. Among the largest of these crystals was a specimen a little more than four feet in diameter and fourteen in length. Two beautifully crystallized specimens of golden beryl, about three inches in diameter and five inches in length, were set at angles to this green beryl. This pair of golden beryls were the only specimens of this sort found at the Bumpus Quarry during this era. They also discovered a massive beryl which was a maximum of four and one-half feet in diameter and about twenty feet in length. The largest of the beryls mined by the Perhams was a colossal crystal which was a little more than six feet in diameter at the base and twenty-two feet long. "

Thus, the recovery process at the Bumpus Quarry resulted in yet another record size for Maine beryl, albeit unpublicized in the time period. It should be remembered that Harlan was in charge of the actual mining at the Bumpus Quarry until 1932, after which time, OMMC was the company of record, at least in 1933. However, Harlan seems to have been

Field Museum's giant beryl crystal, Bumpus
Quarry. Weight = 1,000 pounds.
From Science News Letter, January 17, 1931.

the Bumpus Quarry foreman for OMMC resuming in about 1935. This record size beryl, 22' x 6', was cited by Rickwood (1981). [37] Merrill and Perkins (1930) illustrated a pile of "detached crystals of beryl" from the beryl whorl.

Berman does not seem to have accompanied George Kunz to the Bumpus Quarry in June as Kunz wrote to him (August 20, 1930, HUMM)"

"I have been very busy since my return, but I am still waiting to hear from Albany further. I have secured one large imperfect beryl, for which I made a good offer which was accepted, and the beryl is to be shipped at such time as they notify me. Have you heard anything further from them?

Confidentially – If a good beryl can be obtained there I am sure that I can have it given to the Harvard Collection. Naturally, I refer to a beryl weighing from two to five tons, if they can get [one] of that size out with some idea of perfection. Do not go near them. P.S. Am surprised at the non contact on the five things given."

The comment, "Do not go near them." certainly referred obliquely to the family feud. Apart from the feud, there would have been many impediments to preserving the huge beryls in place, the most important of which was that the mine was very active. A shelter would have been required to protect the beryls from the weather and the width of the quarry, about 15 meters or 45 feet, was not wide enough to allow the blockage made by a building. Additionally, the quarry was also going deeper. The method of working an open pit is to remove rock for a convenient distance and when the face can not be widened, then the work must start a deeper level, or "bench". Quarrying requires that the forces used to remove rock have a path of least resistance. Even when granite is being quarried, the wedges which are used to separate the block from the ledge must have a free direction away from the confining rock. The successive deepening of the quarry would have left the huge beryls "high and dry".

We might imagine that all of the beryl that was to be shipped from the huge beryl whorl were shipped about the same time. The three big fragments known to have gone to museums were prepared as soon as possible and were shipped in crates in August. The small pieces of beryl were bagged as required by the contract and large pieces were broken up so the would fit in the shipping bags. Woodbury (1957) indicated somber details rarely revealed:

"It was shortly after this that a national magazine printed an article which called Per-

Beryl crystal, about 6 x 20 cm, Bumpus Quarry. Perham's of West Paris Specimen.

ham 'the youthful beryllium millionaire,' which Stan stoutly denied. ... Suddenly the mining company laid claim to the beryl itself. All that Stan finally got was 300 tons of rose quartz."

The Bumpus Quarry may have been closed sometime in the autumn, perhaps even as payments were being received from David Taylor Company. Payments had to be disbursed within five days after Stan had received them and it would have been then that inflammatory discussions began of closing the Bumpus Quarry down. The expectations of the amount of money resulting from the sale must have impossible to meet. Why couldn't the Cummings Family be the "Beryllium millionaires"?

No documents have been found to verify that a court order was issued nullifying the beryl contracts, but that detail may have been in the court transcripts that were discarded by the Oxford County Court House in the 1960s. The Bumpus-Perham lease was certainly cancelled: both the original 1927 lease and the amended 1927 lease. The Perham-Perham beryl lease and the Perham-Taylor contract were probably also suspended. However, more problems came into play resulting in even more complications.

The American Museum's Beryl Sections

The American Museum purchased its beryl sections in the summer of 1930, before the legal proceedings between the Cummings Family members and the Perhams began. When the beryls were installed at the museum, there was a small presentation ceremony. The published American Museum budget for 1930 merely noted that $3,000 was

The Two 'trunk-less" beryls in Dog Run, 1989

transferred from the J. P. Morgan Fund and without explanation to deny the supposition, that at least most, if not all of the transferred money was expended on purchasing and shipping the beryl crystal segments to New York City. [255, 256] There was no mention of the beryl crystals in the Mineralogy Department's section report of new acquisitions and the later AMNH annual reports were abbreviated due to budget shrinkages caused by the Depression. The giant beryls started their exhibition life prominently placed, as mineral sentries one on each side, in front of the main entrance of the American Museum of Natural History, under the archway of the museum facing 77th Street and for decades were proudly set there. Lucas (1936) in his guide to the Museum revealed that the beryl crystals were positioned under the archway and they were joined by two large specimens demonstrating the effects of glacial action: one was a rock with a pothole eroded into it while the other was a boulder of fossiliferous limestone that had been scratched and smoothed by glacial movements. The Guidebook contained only one and a half pages of text dedicated to the very popular Morgan Memorial Hall of Minerals and Gems, although it did have two full-page photographs: one was a general view of the exhibit hall and the other was a quartz crystal from Auburn, Maine weighing 115 kg.

The view of the large quartz crystal demonstrated that at the time, there was an appreciation for giant crystals. A short notice in 1937, the first in the Annual Reports of the Museum mentioning the Bumpus beryl crystals, merely said: "The exceptionally large beryl crystals from Albany, Maine,

which, due to their great weight, were not suitable as floor exhibits, have been installed under the archway at Seventy-seventh Street." [256] By the 1980s, the huge crystal sections had acquired the dinge and grime of the city air and countless hands touching them and, while mostly protected from the elements, the two giant beryls had progressed from being the elephants every circus had to have, to becoming white elephants no one knew what to do with. The crystals ceased to hold their attraction and when crowds of school children were being funneled through the museum's gate, the crystals were somewhat in the way. There was a large poster, encased by glass, near the giant beryls explaining: "LARGE BERYL CRYSTAL From the Bumpus Quarry, Albany, Maine One of the largest crystals of beryl ever taken from a quarry. Beryl is a silicate of aluminum and the rare metal beryllium. It is principally found in granite, of the variety pegmatite characterized by coarse crystals of the minerals comprising it. Beryl is at present the source of beryllium. The gem varieties are known as emerald, aquamarine, golden beryl, and morganite. Presented by J. P. Morgan, esq." The dry scientific explanation may have excited a few intellectual people, but the original intention was to have an exhibit which called out; "Hey, look at me world! Gaze at my mighty size and despair!" (with apologies to Ozymandias and P. B. Shelly). [23] By the time the sign was erected, the museum was featuring both crystals, one on either side of the archway. In 1989, the two beryl crystals were in a fenced outdoor enclosure used as a "dog run" on the north side of the Museum along 80th Street. The difference in the sharp peak on the beryl crystal in the picture under the archway and the picture of the same crystal in the dog run suggests that some of the crystal was broken off by souvenir hunters.

Ideas which have not been widely considered were the thoughts of the miners as they gradually uncovered the huge beryl crystals during 1928 to 1929 as well the subsequent dismantling of the beryls in 1930. Obviously, the enclosing matrix was lovingly chiseled away from the crystal faces as were the quartz masses between the crystals. The crystals were certainly prepared for "exhibition" to visitors to the quarry. Two big sections had been crated and shipped away, but what happened to the rest of the beryl? Were there only two big sections with no third section destined for Harvard University? Did Palache refuse a huge beryl section because of his disappointing history with the Giant beryls? How long did it take to disassemble all of the beryl from the ledge? To be sure, most of the beryls were removed from the hanging wall because there were valuable feldspar and rose quartz near them and the presence of the outcrop was an obstruction in the quarry. Even today, the pit is relatively narrow. The deepening and lengthening quarry had to be cleared of a rock mass that was becoming an obstruction. How much of the beryl was sold as specimens in the Perham mineral store? (Alice Frost Lord said in the *Lewiston Journal,* June 1, 1929: "… Stanley I. Perham … handles all the beryl thru his unique Maine Mineral Store." In the summer of 1930, Stan issued two, now very rare, postcards showing fragments of the giant Bumpus Beryl.

Perhaps because of his early interest in the acquisition of the beryl from Bumpus, Palache (1932) wrote a short article entitled, *The Largest Crystal,* in which he tabulated well-established records for sizes of crystals. Interestingly, the world's record crystal with the smallest dimension (12 ½ x 10 x 14 cm), mentioned in Palache's article, was a pyrite from Colorado that had been purchased about 1927 from Andover, Maine mineral dealer and miner, Wallace Dickerson Nevel. [123] The Bumpus beryl was the second largest well-measured crystal on Palache's list, being surpassed only by spodumene from the Etta Quarry, near Keystone, South Dakota at 42 feet by 5 feet 4 inches (12.802 x 1.626 x 2.469* m; *third dimension calculated, with a stated weight of 90 avoirdupois tons (81.72 metric tons [given the size of the calculated third dimension, it is probable that the reported weight was grossly in error or was the weight of more than one spodumene crystal]). (Palache did not cite Schaller's (1916) report of a spodumene crystal 47 feet long.) Jahns (1953) summarized USA giant beryl discoveries. [21] Palache's list is very curious as he cited Bastin's (1911) description of Maine pegmatite texture: "The pegmatites show remarkable differences in coarseness, some, especially the narrower dikes and sills, being little coarser than medium-grained granites, though differing strikingly from the latter in texture, and others containing single crystals of nearly pure feldspar up to 20 feet across and single beryl crystals the diameter of a hogshead." Although the Maine microcline locality was not specified, Palache mentioned it as he did not know of another microcline record

even approaching the poorly "measured" Maine crystal. Bastin (1911) also stated of the Maine Feldspar Quarry, Mount Apatite District, Auburn, Maine:

> "Crystals of light bluish-green beryl also occur rather abundantly, embedded in the solid pegmatite. One hexagonal beryl found about 1898 is reported by J. S. Towne to have been 4 feet in diameter and 20 feet in length, but the majority does not exceed 1 foot in length and a few inches in diameter. Near the gigantic beryl mentioned occurred several pockets bearing the finest crystals of herderite ever found on Mount Apatite; the form and composition of these have been described by Penfield [in 1894]."

Had Penfield's article come out four or five years later, the report of the world's record beryl from Auburn certainly would have been mentioned by him and would have received full credence by Palache. The only reasons that can be suggested for Palache's not recognizing or mentioning the Auburn giant beryl are that he was 1) unaware of it, although he cited from the still standard work on Maine pegmatites for his approximate world record for microcline, 2) he didn't believe it, despite the definitely stated dimensions and the fact that it was "hexagonal" and not just a mass, or 3) as no claim of a <u>record</u> for the huge beryl was made, the actual size of the Auburn beryl crystal was dismissed, as an exaggeration. The latter option is quite plausible as there is a report which may have referred to the same crystal as it fits the time period. BILS (1904) related: "A few beryls have been found of enormous size. One was found at Mount Apatite a few years ago, that was 12 feet long and 20 inches in diameter." Bastin may have been told a "fish story" with the beryl's size increasing to 20 *feet* in the retelling of the discovery. Bastin, a Unitarian, received his Ph. D. in 1909 and was still a relatively young employee of the U. S. Geological Survey (Anderson, 1954). Bastin wrote as accurately as he could and without bravado, typical of scientific reports. (Most of Bastin's pegmatite observations in Maine were made in 1906.) Rickwood (1981) wrote an article also titled, *The Largest Crystal,* mentioning a beryl found sometime before 1976 that was 18 x 3.5 meters and which had an estimated weight of 379.5 metric tons from Malakialina, Madagascar (equivalent to 59 x 11.5 feet and weighing 307.5 avoirdupois tons) and this crystal currently ranks as the largest reported crystal of any mineral species, although it is perplexingly unphotographed and undescribed and, therefore, a very poorly established oral record. [41]

Mica – Another Bumpus Quarry Mineral without a Specific Royalty Rate

The Bumpus deposit was unusual in that it had a third mineral occasionally providing income, but there was no royalty rate mentioned in the various leases. This third mineral was mica. Harlan must have been responsible for this mineral's mining and sales. Mica could have been removed from the Bumpus Quarry at the rate of 50% royalty under the general by-product mineral provision, although that is an exorbitantly high rate to pay for an inexpensive material.

Mica comes in a variety of grades, but the Bumpus Quarry yielded only the lowest grade, scrap mica, at least in any quantity. Scrap mica has had a wide variety of uses. When finely ground, it has been an important ingredient in all-weather house paint and used as a filler in roofing tar. The use as "glitter" in wallpaper, greeting cards, etc. did not consume much mica. Mica mining is more labor intensive per kilogram because of the difference in mining style between feldspar mining and sheet mica mining. Most Maine pegmatites were known to be worked for one item at a time, particularly for efficiency's sake. Nonetheless, some mineral locations had royalty provisions in their leases for three or four items including "other minerals". Mica was abundant enough in big pieces to permit stockpiling at the Bumpus Quarry, but occasionally big concentrations of mica were found: "The quartz body contains green beryl, small quantities of plagioclase, and large whorls of wedge muscovite." [64] On November 6, 1935, a particularly large "whorl" of muscovite mica was discovered. [190] The whorls consisted of nearly 100% mica in a roughly spherical aggregate of radially splayed plates. Although size measurements were not reported, the probably 7-9 foot whorl yielded "about forty tons of mica". The going rate for crude scrap mica in 1935 was about $17.74/ton. [107] It is perhaps during the 1930s that Albert Kimball trucked mica to Keene, New Hampshire? A second giant mica whorl was uncovered in 1965.

Top photo: Stowell Wood products Mill, Woodstock. From a 1920s postcard.
Center photo: Ore cart at Perham Quarry.
Photo by Vi Akers. Courtesy Sid Gordon.

Lake Front Store, c1941, Norway where
Lester E. Wiley and Lucy A. Cobb first met about 1936.
Courtesy Laura Wiley Ashton

Chapter Five: Economic Changes

Oxford Mining and Milling Company Mortgaged to United Feldspar Company

There were many forces at work affecting the fate of the Bumpus Quarry's giant beryls, as well as the health of the mining industry in the County. The presence of the Mill

Front and back of Oxford Mining and Milling Plant c1928, from postcards

was the economic driving force, and without the local mil, there might have been a small mining industry, perhaps being serviced via railroad by the Littlefield Corners grinding mill in Auburn, at least until 1929. The absence of an Oxford County mill may have extended the life of the Littlefield Corner location, although that mill was certainly technologically obsolete when it closed. It is improbable that the Oxford County feldspar industry would have developed before the Depression, but for the timely efforts of Alfred Perham, C. Alton Bacon, J. F. Cheney, William L. Adams and others. Feldspar prices were falling when the mill was being built and no other investment would have occurred before WWII. There might also have been a small boom in mica exploration during the war time price subsidies (q.v.), but in light of the great expansion in mica and feldspar mining in North Carolina, even a post-Depression development of an Oxford County feldspar industry was unlikely. The

mid- to late-1920s were the only years that could have spawned a feldspar industry.

Whether West Paris would have become a town, separate from Paris in 1957, is another point. West Paris eventually had a number of wood working mills and the addition of a feldspar grinding mill provided diversification: Lewis M. Mann & Son's (made pail handles and clothespins, in *Maine Register* by 1931), Penley Brothers {in *Maine Register*, at least by 1934} (made clothespins, etc.); and Ellingwood & Son's (in *Maine Register* by 1950; made wooden handles for tools such as peavey handles, etc., ladder rounds, pick poles, etc.). Diamond Match Company was listed only in 1951. (A detailed list of mills and factories follow.) For a small town, the labor market was unique: "You could be fired from one factory at 8 A.M. and have a new job by 9 A.M." (Milton Inman, personal communication, 2007).

Originally, the Feldspar Mill was a local effort. Alfred C. Perham had a feldspar-bearing ledge and he convinced investors to invest in a feldspar grinding mill. The expansion to include the Bumpus Quarry and Harold Perham's quarries was a natural consequence. The important factor to remember was that feldspar prices were plunging immediately after the Mill was built. Plans of output budgeted against price per ton for the finished product quickly meant that OMMC was being financially stretched. The Mill had been financed and then re-financed by Maine money. In order to meet its bills and payroll, the Mill had to increase its output, but a third round of investment was hard to come by. One must imagine that OMMC courted potential out-of-state investors who were already in the feldspar grinding business. The sale of the company would probably mean only a "break even" return, but the alternative was less attractive.

In 1929, Oxford Mining and Milling Company was bought out, including stockholders such as Alfred Perham, by Charles H. Pedrick, Jr., of New York City. It is true that several of the feldspar-producing areas in the eastern USA were declining in production. The Bedford, New York District was nearly finished as was southeastern Pennsylvania. Maryland, Virginia, and even Connecticut were not as productive as was necessary. United Feldspar had an interest in most of these districts and would have wanted to maintain its base of feldspar supply. North Carolina was a relatively new producer and, unlike most feldspar-producing states, was increasing rapidly in production. Maine had been a historical leader in feldspar production and Oxford County was a new area opening up, promising to raise the State's output dramatically. The promise of the district was the amount of feldspar being produced from relatively few quarries associated with OMMC, in order of their probable production size: Perham, Bumpus, Bennett, Hibbs, Sturtevant, Mount Marie, Bessey, and Corbett Ledge.

All of Alfred's mineral rights in West Paris, including land not before considered as feldspar sources, were purchased on May 1, 1929 (OCRD v388 p317-318). Relatively soon, Pedrick finished organizing the United Feldspar Company, ranging from New England to North Carolina operations, and on January 20, 1930 the new conglomerate purchased the Oxford Mining and Milling Company and all of its holdings and agreements for $100,000 by five-year mortgage.

After the Huge Beryl Flurry

In answer to the question, "What happened to the Bumpus Quarry?", the short answer is that Bumpus Quarry beryl production rapidly ceased to be significant after the early 1930s. The huge beryl clusters were depleted and there were no new big discoveries for a while. Certainly beryl continued to be found, but the discoveries were significantly smaller and less worthy of note. Even a new ten-foot beryl would not have received press attention. The Bumpus Quarry rose again in notoriety and importance in the 1940s.

As will be seen, 1932 was also a pivotal year in the Bumpus District's development as Harold Perham stopped his personal mining business. Harlan Bumpus continued to

Railroad looking southeasterly towards Bryant Pond village with "Colby Grade" beyond, c 1920 postcard.

quarry feldspar and he hauled feldspar by truck to the West Paris grinding mill.

The Economics of Bumpus Quarry Mining

The Bumpus Quarry dropped out of the mineral news for many years. The famous source of beryl seemed exhausted. Neumann (1952) reported that the total Bumpus Quarry beryl production from 1930 until about 1950 was 255 tons. In 1929-1930, the estimates publicized suggested that there was about 100 tons in the famous giant beryl cluster and that amount may have been produced before cancellation of the Perham contracts. The remaining years of production, 1931-1950, would have only required a few tons produced per year, agreeing with generalized production figures reported by the U.S. Bureau of Mines. The Bumpus Quarry originally had localized rich pods of beryl, but they

did not continue throughout the pegmatite.

Neumann (1952) calculated, based on the size of the Bumpus Quarry, that only 18,000 tons of pegmatite and 3,000 tons of overburden had been removed from the Bumpus Quarry to that time, an average of 19 tons of pegmatite per week assuming a 38 week mining season over 24 years. The map made in 1945 published by Cameron et al. (1954) suggests that the volume removed from the Bumpus Quarry had a weight nearly twice as much as this.

As will be seen, there may have been 11 idle years during the lifetime of the Bumpus Quarry and the average feldspar, quartz, mica, and beryl totals could have been about 36 tons per week. In 1934, Harlan may have ceased mining at the Bumpus Quarry. Harold Perham may have been briefly in charge of the mining at the Bumpus quarry and removed about 1,600 tons of mineral-bearing pegmatite, an average of about 42 tons per week, also assuming a 38 week season. (The difference in the weekly production averages may have been partly due to mining technique, size of the mining crew and whether or not the pegmatite was all marketable minerals or contained a higher percentage of waste pegmatite to be tossed onto the quarry dump. If the 18,000 tons of "pegmatite" were all marketable feldspar, Harry and Laura Bumpus may have paid Allen and the Sybil Cummings heirs up to $700 per year plus the cost of real estate taxes, a good amount of money in the time period. The Neumann (1952) figures do not reveal what was meant by pegmatite and whether or not he included known feldspar shipments received by the mill. (Overburden usually means the unmined waste rock and soil that has to be removed from the surface of a mineral deposit which will expose the deposit.) We know that the "pegmatite" contained a significant amount of waste rock and that payments were probably lower than the $700.00 the calculations permit, possibly only $350. Given various figures to choose from, it is evident that there was a very high percentage of usable feldspar in the Bumpus Pegmatite, a conclusion supported by the very small waste dump around the quarry. It is possible that there were sales of dump material over the years, as a replacement for gravel, for road building projects along Valley Road, but there is no known record of any such sales.

The amount of beryl produced from the Bumpus could have been high in 1930-1932. Because beryllium metal had few applications until the mid to late 1920s, the USBM did not keep records of USA beryl production until 1932, but even then no USA occurrences or detailed production figures were specified. It was revealed that two manufacturers were making beryllium copper alloys: American Brass Company of Waterbury, CT and Riverside Metal Company, Riverside, NJ. [235] (David Taylor Company only bought and sold the ores.) The only USA beryl production in 1932 was limited almost exclusively to the Black Hills pegmatites of South Dakota and some localities in Colorado. [236] Much of the USA's beryl needs were supplied from India. By 1935, it was announced, [237] that the consumption of beryl in the USA amounted to 35 tons a month, with predicted increases in demand. In 1935, beryl brought $25-$35/ton, according to the beryl's richness in BeO. As beryllium became significant as a new technological metal, Maine State Geologist, Freeman Burr, announced that there would be a beryl survey of 50 Maine pegmatites, but no USA production was specified by the USBM for that year. [238]

In 1936, beryl was only $30/short ton for 10% BeO beryl and $35/short ton for 12% BeO grade. Tyler (1937) noted: "The revival of lithium and feldspar mining in South Dakota has stimulated domestic production [of beryl]. Colorado, Nevada, and perhaps other States promise to yield substantial supplies, and small amounts already have been obtained in New England, New York, Virginia, and North Carolina." Tyler (1938) estimated that the entire beryl production of the USA was 500 tons a year, and, although there were no specific production figures by state, Maine was not listed as a producer. Tyler (1938) also made what today would be a humorous observation, at least to mineralogists: "Although the beryllium content may reappear as a constituent of other minerals, such as bertrandite, herderite, or beryllonite, none of these minerals is heavy enough to be separated by the sorting action of streams…" The humor lies in the extreme rarity of these minerals in the time period. Bertrandite was found in a remarkable deposit is Utah and bertrandite had replaced beryl as the major ore of beryllium.

In 1939, Tyler reported that in January of that year, laboratories in Copenhagen, Denmark and "almost simultaneously in the USA, 'atom smashing'" had been achieved and as a result of these fission experiments, "atomic energy" might yield enormous amounts of power as well as the creation of rare elements: "The practical importance of these discoveries is yet to be learned." Also for 1939, the same source reported: "Beryl occurrences have been reported in various parts of the United States, but few mines outside of the Black Hills, S. Dak., have produced as much as a carload, and most domestic production has been obtained as a byproduct of feldspar, lithium, or rare-metal ore mining. … A substantial part of domestic production has been supplied by imports…" The apparent malaise in Maine beryl production was probably real. Domestic beryl cost $30-55/ton, while imported beryl cost only $3.25-$3.50/ton "f.o.b. United States Atlantic ports," [234] but imported beryl prices soon rose to approach domestic prices (Matthews, 1943). In 1940, domestic beryl production was estimated at 100-150/tons, while imported beryl was about 459 tons. The very first public records of Maine beryl production were not available until Matthews (1943). He tabulated production, although ranking of states was not necessarily in the order of their mention: 75 tons in 1937 (CO, ME, SD); 25 tons for 1938 (only ME); 95 tons for 1939 (CO, ME, SD); 121 tons for 1940 (CO, ME, SD); 158 tons for 1941 (ME, NH, SD): "… South Dakota contributed 96 percent in 1941 … compared with 61 percent in 1940." From these data, we may draw several conclusions. The first of which, as there were no obvious sources of Maine beryl production other than the Bumpus Quarry, the principal Maine location accounted for less than 47 tons in 1939, quite probably only a few tons over that reported for 1938. However, Maine production could not have been over 6 tons for 1941, based on reported proportions. To be sure, during the late 1930s and early 1940s, there may have been a minor contribution to Maine beryl production from Emmons Quarry, Greenwood, Nevel Quarry, Newry, and the Black Mountain Quarry, Rumford. There even may have been some beryl produced elsewhere in Albany. Harold Perham was out of the mineral business and he essentially said so in biographical notes in his various Perham family genealogy books. Stan Perham was the essentially the sole agent for Maine beryl shipments and he has been frequently mentioned as a "government agent" for beryl in WWII.

Bethel railroad station, c 1940s -1950s postcards

The interest in beryl by the Federal Office of Production Management in pre-war times was noted by Matthews (1943) and the price of beryl rose to "$60-$72" per ton. Because of the relatively high cost of beryllium alloys, their uses were specialized. The aircraft industry principally focused on moving parts of engines subject to the greatest rates of wear and, occasionally, fuselages required beryllium, etc. Properties relating to high strength versus weight were required in making parachute harnesses. Another property of beryllium alloys included freedom from metal fatigue, such as for springs in measuring devices. Other very important alloy properties were non-magnetic and non-sparking characteristics. However, in Germany, beryllium was beginning to be used in iron and steel armor plating on military vehicles.

Thompson Pottery Company c1930
East Liverpool on Ohio River, from a postcard

What Happened to Maine Ground Feldspar?

The ground Maine feldspar had different destinations depending on which mill did the grinding. The feldspar was sold partly by contract and probably partly on spot demand and shipments would have been sometimes commingled with batches from other states, although there was said to have been a preference for Maine feldspar by some manu-

Potteries in East Liverpool, 1906 from a postcard

facturers. The potteries shown may or may not have used Maine feldspar. The Maine Feldspar Company, with quarries and mills in both Auburn and Topsham, sold much of its feldspar to Charles M. Franzheim Company of Wheeling, West Virginia. Franzheim was influential in the politics of the American Ceramic Society and worked on a number of their technical committees establishing standards for raw mineral products for the industry. Golding and Sons later merged with Trenton Flint and Spar Company, with quarries and mill in Topsham, and they sold their ground feldspar principally to potters in Trenton, New Jersey. The Consolidated Feldspar Company, absorbing the Maine Feldspar Company, continued to sell to Franzheim.

Much of the Oxford County feldspar, ground in West Paris, went to potters in East Liverpool, Ohio, and nearby manufacturing towns, the largest pottery and porcelain pro-

ducing center in the nation. Ben Shaub photographed china supposedly from the Mount Apatite District, Auburn, but the attribution has to have been made on the china owner's knowledge of what potters purchased the district's output. [123]

There is an unidentified Summer, 1938 quote in a clippings book at the Maine State Library that claims: "The china, made by a Trenton, New Jersey, factory for use in the White House during the administration of Woodrow Wilson, was made from 'hand-picked' spar from Keith's Quarry, Mount Apatite, Auburn, Maine."

The following graph shows how the Depression affected Maine feldspar prices and production. The price of crude feldspar dropped from a high of $9.19 per ton in 1924 (a price that convinced the Oxford County miners that feldspar was a good business to enter), to $7.11 in 1930, reached $5.02 in 1933, and sunk to a low of $4.10/ton in 1939, even as demand was rising. As has been mentioned, the price of ground feldspar was artificially high in 1924 because of a long labor strike in the porcelain industry. Reserve stocks of ground feldspar were very low as distributors of the raw materials did not buy ground feldspar and the resumption of production in the ceramics industry quickly depleted available stocks. The available ground feldspar temporarily became more valuable and it was about a year before prices began to decline. Unfortunately, ground feldspar production capacity increased all across the USA and by decade's end, compounded by the advent of the stock market crash and its consequences, prices for ground feldspar declined.

The price of ground feldspar did not fare as badly as the price of crude feldspar. The 1925 price of ground feldspar, at all Maine processing mills was $18.76, down from $19.05 from the inspiration year of 1924. By 1930, the average price of ground feldspar was $15.90. The Depression production low was in 1932 but the average Maine price was a robust $16.79, a price that was significantly higher than the national average of $13.18. Orders may have been dropping off and the mill owners offered the miners less for crude feldspar the following year, not knowing what might come when the Depression was lingering with no end in sight. Before the Depression, Maine miners typically got almost half the value of the ground feldspar. Except for 1933 when miners only got a quarter of the value of the ground product, the Depression era crude feldspar brought only from a third to 40 percent of the ground feldspar value. In 1938, despite the increased use of ground feldspar, excess capacity of the nation's grinding mills dropped the price received in Maine to $12.55/ton.

The production of crude feldspar mirrors the output of ground feldspar, of course. The loss in weight was roughly equal to the difference in how crude feldspar was purchased (by the long ton) and how ground feldspar was sold (by the short ton). The profit the mill owners received was based on giving the miners less money. In effect, the mill instituted a form of economic slavery: hard work yielding low pay. The mill hedged its bets by successively offering lower prices in

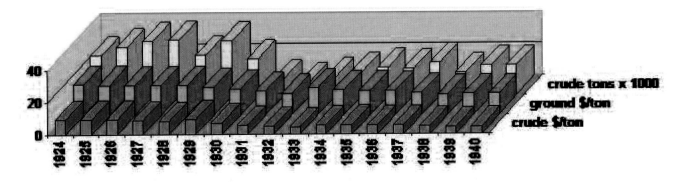

All Maine Feldspar: Crude and Ground

crude tons x 1000
ground $/ton
crude $/ton

40
20
0

1924 1925 1926 1927 1928 1929 1930 1931 1932 1933 1934 1935 1936 1937 1938 1939 1940

advance of the lower prices they received, rather than in parallel with prices they received for their product. The predatory payment schedule kept the Oxford County mining industry from expanding and, as the modern era approached, the work of mining became less and less attractive in Oxford County. The price of crude feldspar remained fairly flat even after the Depression era low. After 15 years, Maine feldspar miners were getting less than half as much money for their mineral despite the increases in costs of mining and living.

The high point of Maine feldspar production, during the era considered, was 33,897 tons in 1929. Maine ranked as the second crude feldspar producer in the nation, behind megaproducer North Carolina. In the early part of the century, Maine was frequently the number one producer, but North Carolina feldspar mining had undergone a rapid expansion from the early part of the twentieth century. In 1929, the Mount Apatite District dropped out of production and the State's annual output was reduced accordingly.

The West Paris Mill's Mineral Impurity Concerns

Good ground feldspar had to be free of impurities, but some impurities were more important than others. Missed beryl was not too significant for the West Paris mill's output of ground feldspar: "It does not discolor the fused feldspar but makes it opaque and in small amounts it lowers the deformation temperature of the feldspar. Its chief disadvantage is its tendency to reduce the translucency of the ware." [253] Impurities such as garnet, tourmaline, and the black mica, biotite, were another matter. Beryl fused quite readily and if it were in small enough percentage and well-mixed, might pass unnoticed by inspectors of the freshly kilned products, whether porcelain fixtures, tableware, bottles, art objects, etc. Individual specks of garnet, tourmaline, or biotite would show up as blotches down to pepper-like spots in the finished products, which either sold as factory seconds or had to be discarded: "An addition of 1 percent of garnet to feldspar that [otherwise] fuses to milky white causes it to fuse to an intense yellow brown, and the presence of one-tenth of 1 per cent of garnet would render a

feldspar unmarketable as a white porcelain feldspar. Garnet grinds at about the same rate as feldspar, and its presence in a feldspar imparts a faint flesh tint to the powdered rock. As many feldspars are naturally flesh colored, the presence of garnet can not be detected by this color and generally is not suspected until it is shown by the mass fusing to a yellow color in finely pulverized feldspars or by brown specks in coarsely ground feldspar." Low amounts of biotite could be tolerated in products "… where a white-burning feldspar is not required.", but such feldspar was not #1 grade. Black tourmaline was not very easily fusible and "Being black, it is easily detected [in white porcelain or china] and its complete removal is imperative if the feldspar is to be marketed for pottery uses." While rarely encountered even in small bits in Oxford County, the Sagadahoc County feldspar could have black magnetite mixed with the feldspar: "Its presence absolutely condemns the feldspar for ceramic uses … In the fused feldspar the magnetite shows as black specks surrounded by yellowish or brownish zones of ferric silicate."

Up to this point, the reader will have read about feldspar, but many readers need to know that there are two common kinds of feldspar in Maine: potash feldspar and soda feldspar, named after common chemicals in them. Potash feldspar, the feldspar usually referred to in these pages, fuses to a glass, while soda feldspar forms a crystalline texture when it cools. Tableware, porcelain, glass, art objects, etc. would become crazed if the fused feldspar did not keep its glassy form. The development of crystals would ruin the finished materials. Although potash feldspar always has some soda content, if the soda content is low, e.g. less

China made with Greenlaw Quarry, Auburn feldspar. Photo by Ben Shaub

than 3 weight percent, the fused material will still make a glass and will be of pottery grade. Quarry foremen and millers had to know how to tell the difference between the two kinds of feldspar and when there was a large percentage of soda feldspar present in a part of a quarry, the direction of work had to change or a lot of waste rock had to be removed which could not be sold. If the potash feldspar content did not eventually return to a high enough percentage, the quarry had to be abandoned. High soda feldspar could be sold as an abrasive or in other applications, however. Maine feldspar grinding mills would test various lots of ground feldspar for fusing characteristics in a kiln or would

Typical piece of Bumpus Quarry waste pegmatite rock showing tight intergrowth of feldspar (F), quartz (Q), and tourmaline (T).

have specific lots of feldspar chemically analyzed. Many mills would have stockpiles of pure potash feldspar to mix with soda feldspar-bearing lots to keep their output up to the required specifications.

The more difficult to judge impurity was not so different appearing than the feldspar: quartz. An impurity up to 10% could be tolerated, but Maine's ground feldspar was usually very pure as the miners and millers were very careful. Quartz impurity greater than 10% resulted in the feldspar being classed as Number 2 and was unsuitable for tableware and good porcelain, although it was saleable at a lower price. Number 3 grade, with innumerable impurities, was only suitable for abrasives, fillers, chicken grit, and other low value products. The Bon Ami Company favored Maine #1 feldspar in the manufacture of its cleansers because of its reputation of being very low in free quartz. Bon Ami's scouring powders were sold with the slogan: "It hasn't scratched yet!" Quartz is harder than typical bathroom porcelains and would ruin their high lusters, while feldspar is softer and would be less destructive.

Undoubtedly because of the success of the local feldspar quarries, the new owners of the West Paris feldspar grinding mill, the United Feldspar Company, up-graded and expanded the mill. Charles Pedrick, Jr. was active in the American Ceramic Society, particularly committees estab-

lishing specifications on feldspar purity and he was a leader in installing magnetic separators in his feldspar grinding mills to remove iron-bearing minerals from the ground ore. The improvements mentioned shortly after the United Feldspar acquisition of the West Paris mill must have included these impurity separators.

The Depression and the Cummings Family Feud

The New York Stock Market Crash of Black Tuesday, October 29, 1929 is a marker date leading to shrinkage in wealth and business throughout the world. The initial loss of real and "paper" stock wealth rippled through the business world. The immediate loss of some companies through bankruptcies was a prelude to continued increase in numbers of personal, corporate, and bank failures. Great numbers of people became unemployed as business diminished and their consequent absence of buying power further fed a downward economic spiral. The year 1930 was filled with uncertainty for many families and as the economy worsened, with more people loosing their jobs, each subsequent year saw more financial ruins and disasters, notably banks containing people's savings and checking accounts. The loss of wage-paying jobs was drastic, even in Maine. Fortunately, only a few bank losses were experienced in Oxford County. The famous "Bank Holiday" was not enacted until the day after Franklin D. Roosevelt took office, March 5, 1933. The former president, Herbert Hoover, had been trained as a mining engineer, but he had little business or economics training and instituted few programs designed to relieve economic stress.

On June 24, 1931, the Casco Mercantile Bank merged with the Paris Trust Company, becoming Casco Bank and Trust, while the South Paris Savings Bank was still in business. Comparing listings in the *Maine Register,* between 1929 and 1935, Oxford County's banks had only minor losses. Bethel's two banks survived: both the National and the Savings banks. Buckfield's branch of the Paris Trust, of course, became a branch of Casco Bank and Trust. Dixfield retained its branch of the Rumford Falls Trust Company. Fryeburg's Fidelity Trust Co. was replaced by the Casco Bank. Both the Norway National and Norway Savings Banks survived, as did both of Rumford's: The Rumford Trust and The Rumford National Banks. No other towns in the County had a bank or branch listed.

There were social crises across the nation due to the new poverty. The descent into the Great Depression affected the Cummings, Bumpus, and Perham families, in its own way. The author remembers an elementary school teacher telling the class that Maine may have suffered less from the effects of the Depression as the State's rural inhabitants had never experienced the urban prosperity attributed to the 1920s. Nonetheless, although there were few bread lines, soup kitchens, or obvious symbols of poverty in Maine's historical legacy, the affects were very real. Rural poverty from any cause is diffuse and is always more prevalent than apparent. Many Depression period residents held out as long as possible before they would be known to have "gone on the town" – meaning having had to accept the shame associated with municipal welfare for their survival. When the Depression was at its height, the so-called Poor Farms, had ceased to be useful and it was not practical to incarcerate all of the poor. In a small town, your financial condition, real or imagined, was always part of local gossip and the associated shame not only affected you, personally, but also all of your family living in the area and all of your close friends.

Origins of Bumpus Quarry Strife?

The circumstances leading to the Cummings family feud began with an executor's problems trying to deal with a stressful set of expectations from the heirs of Sybil Cummings. The feldspar mill would have cared little for family matters so long as good quality feldspar continued to come in from the Bumpus Quarry. Unfortunately, the change of ownership of the mill changed the delicate balance of motivations.

The original purchase price of feldspar in West Paris may be thought to have been high, given Harold Perham's lease with the Oxford Mining and Milling Company. The Perham – Bumpus contract did allow for readjustment of price of feldspar as the national market price fluctuated. Although hardly a mining boom, there had to be bonanza feelings among some members of the Cummings family. By 1930, the re-organized and expanded West Paris Mill was actively seeking new sources of feldspar. New sources would provide cheaper costs for feldspar.

Harold Perham had been selling some of his feldspar to other Maine feldspar grinding companies; first to the troubled Trenton Flint and Spar Company and then to the Consolidated Feldspar Company, successor to the Maine Feldspar Company. On June 24, 1930, Harold borrowed money against the value of his Bennett Quarry lease with Blanche Bennett. The agreement was a sublease of the Bennett Quarry to the Consolidated Feldspar Company in return for their paying him $2,000. On October 14, 1930 Harold and Blanche Bennett signed an agreement as to the location of the boundary pins for the lot on which the quarry was situated and it was agreed that:

"Harold C. Perham takes over a certain indebtedness of Two Hundred Forty-six Dollars ($246) due Alton C. Maxim on a Chrysler car as a part of said consideration.

No other indebtedness of any kind whatever is due Blanche Buck Bennett for any materials mined on the now designated Parcel Number One until after the stated ten year period lapses.

Said Blanche Buck Bennett hereby acknowledges full satisfaction for all money or demands due her for any and all quarrying operations and material since December 19, 1927, to date hereof.

Said Harold C. Perham assumes the direct obligation of paying the unpaid accounts due Malcolm Bearce and Earl Hammond & Leon Harlow for work performed in the year 1929 and yet unpaid.

Also, this is a release in full to said

Worker at a porcelain factory assembling a sanitary fixture mold.
Journal American Ceramic Society, c1926.

Bin tag used at the Littlefield Corners Feldspar Grinding Mill to identify ground feldspar lots tested by the Maine Feldspar Company. Pre-1928.

Blanche Buck Bennett for any and all charges for money, materials and supplies furnished her by said Harold C. Perham since December 19, 1927."

The total value of the assumption of debt was probably over $300.

The Tangled Cummings Estate

There may have been an earlier feud in the Cummings Family, than any concerns relating to the Bumpus Quarry.

Upper: Knowles, Taylor, and Knowles pottery, East Liverpool, Ohio. 1930s postcard. Lower: Sterling China plate made in Liverpool, Ohio

As the surviving spouse, on the death of her husband, Joseph, Abigail Cummings was the sole owner of the Cummings Farm, until her death on February 11, 1921. Abigail's Last Will and Testament of November 2, 1918 was an unusual one. She left half of her estate, including 120 acres of land, equally to her unmarried children and executors: Allen, and Sybil:

"KNOW ALL MEN BY THESE PRESENTS, that I, Abbie W. Cummings, ... being of sound mind and memory, do make, publish and declare this my last Will and Testament.

First. I direct my executor, hereafter named, to pay my just debts and funeral charges after my decease.

SECOND. I give, devise and bequeath to my daughter, Sybil E. Cummings, and my son Allen E. Cummings, all the property both real and personal of every name and kind of which I shall die seized.

THIRD. In making this will I have not forgotten my daughters Laura J. Bumpus and Viola E. Cummings or my son Wallace E. Cummings, or the three children of my deceased daughter Cora [Emma] Cummings, but I purposefully omit devising or bequeathing anything to either of them, not, however, from any lack of affection for them but because I deem it wise and just to devise and bequeath all of my property as above indicated."

The will was witnessed by Albany's famed Civil War veteran, Amos G. Bean, as well as Angie Bean, and Benjamin Inman. From this disposition we might surmise that there was belief that at least one of her children, Laura, was "well enough off" that she did not need an inheritance. However, her other descendants were not well-off and probably would have felt themselves disadvantaged even for a rural family. Allen and Sybil remained unmarried and were loyal children remaining at home and taking care of their mother. Wallace was also living in Abigail's household and his disenfranchisement was probably a matter of controversy. Without regard to motivations leading up to the disposition, the after effects must have created hurt feelings among those disinherited. Allen and Sybil, as sole inheritors of the Cummings Farm, had full legal right to lease the farm's mineral production to their sister and brother-in-law, Laura and Harry Bumpus.

Interior of a Trenton, NJ pottery. c1915. Courtesy Nathaniel Edwards.

Oxford County Court House, 1930s postcard

United Feldspar Mill with addition. Note expanded office/laboratory. 1930s postcard

Unfortunately, Sybil Cummings died on June 28, 1927. Her brother Allen became executor of her estate, by applying as next of kin to the Oxford County Probate Court during the November term, 1927 of Oxford County Superior Court, indicating that the Cummings Farm had an estimated value of $1,500. As Sybil left no will of her own, the inclusion of the various formerly disinherited heirs was dictated by State law, but was seemingly contrary to Abigail's previously written wishes. Allen Everett Cummings, Wallace E. Cummings, Viola Cummings Ballard, and Laura Josephine Cummings Bumpus, each received one fifth of the estate. The remaining child of Abigail and Joseph Cummings was Cora Emma Cummings Cummings, was deceased. She had three children, Laura Etta Cummings Pinkham, Edith E. Cummings Stearns, and Adelia Abbie Cummings Waterhouse, who split her one fifth share amongst themselves and each received a one fifteenth share. (Laura E. Cummings married Clifton S. Pinkham of Kennebunkport May 8, 1921. Adelia Cummings married Clarence Waterhouse on September 30, 1927, while Edith married Hugh W. Stearns on October 1, 1927.)

The context of probating the estate involved a new found source of income for the heirs, although no mining had begun by the time of Sybil's death. Nonetheless, the formerly disinherited heirs were facing the possibility of dramatically increasing their annual incomes through mining royalties by the time Allen became executor. The hullabaloo of the giant beryl discovery, beginning in 1928, claims of world-record size, etc. had to raise expectations. The estate was not settled quickly, perhaps because there was an interest in seeing what bonanza was going to arise. The first year of Allen's executorship came and went without action.

Allen was re-appointed executor for another year on December 3, 1928 as he had yet to find a "suitable" buyer "who would pay what the administrator considered an adequate price" for the land. It is interesting that the role of executor seems to have implied the liquidation of both Sybil's and Allen's shares in the farm. It further seems that Sybil's half share in the Cummings Farm was being treated as a full share, at least by the wording of the 'final decree" of the probate court. Allen actually "owned" 60% of the farm: his original half inherited from his mother plus his inherited one fifth of Sybil's one half. It may never have occurred to him that his share amounted to this much. Nonetheless, Allen's license as executor expired a second time, but he continued to act in that capacity without re-application to the court, until October 3, 1930 when it was routinely granted. On October 6, 1930 (see OCRD v389 p364), Allen entered into a sixty-day option agreement with an "independent" exploration geologist, William B. Donahue, to sell just Allen's share of the Cummings Farm and Bumpus Quarry, including interest in any unpaid royalties from minerals being quarried on the Cummings Farm. Two days later, Stan registered the beryl agreement with his brother, Harold, at Oxford County Records and Deeds. This sequence of three legal actions seems to be the consequence of informed people.

Donahue's role in the Cummings feud may have been far darker than has been previously suspected. (It appears that Harry and Laura put the feldspar royalties they owed in escrow, if not in their bank account, awaiting the formal distribution of money to the various heirs. The unpaid royalties would have been an incentive for some heirs to want a quick resolution.) Allen received $50 earnest money for the option to sell his share of the land for $3500 (OCRD v389 p179). Allen's later actions got him into trouble. We might imagine that Donohue did not keep in contact with Allen or that Donahue was known to have been in New York City, "out of town", etc., but Allen acted as though he believed Donahue was not going to exercise the option to buy his share of the land and he didn't wait for the option to expire. Forty-six days after signing the option with Donahue, on November 21, 1930, Allen sold "his" share of the Cummings farm, containing the Bumpus Quarry, to his sister, Laura J. Bumpus, for $2500 and forgiveness of a mortgage, still a large sum for a poor farm in Albany, but not for a money-making business. Judge Shaw indicated that Allen sold the Cummings Farm acting "against advice of disinterested counsel". Again, there is an implication: Allen had asked a lawyer or appraiser about the selling of the farm, while an option had not expired.

Allen and Sybil had previously mortgaged the Cummings farm to their sister, Laura, on May 25, 1925, two

Androscoggin County Court House, Auburn.
Stereoview c1895.

Oxford County Court House and Railroad
Station, South Paris. Postcard about 1920.

years before Laura and Harry obtained a mineral lease from Allen and Sybil. Of course, Laura Bumpus would want to buy the Cummings Farm as she and Harry would not have to pay royalties on feldspar and other minerals quarried. The purchase might have paid for itself in just a few years. On the morning, of the sale, Laura discharged half of the mortgage (the total amounting to $206.50 according to the account of the final decree) at the Oxford County Court House in South Paris:

> "I, Laura J. Bumpus of Auburn in the County of Androscoggin and State of Maine mortgage owner of a certain mortgage given by Allen E. Cummings and Sibyl E. Cummings (in her life time) to Bethel Savings Bank dated May 25th A.D. 1925 and recorded in Oxford County Registry of Deeds, Book 352 Page 354, do hereby acknowledge that I have received full payment and satisfaction of the

same mortgage upon the one-half interest in common and undivided of the property of Sibyl E. Cummings and of the debt thereby secured, and in consideration thereof I do hereby cancel and discharge and release unto said Sibyl E. Cummings, her heirs and assigns forever the premises therein described. Meaning and intending to discharge the mortgage upon the interest of the said Sibyl E. Cummings in and to said real estate; the share of the debt due upon said mortgage from her estate having been paid by her administrator." [312]

Four hours later, in Auburn, Androscoggin County, Laura purchased Allen's half of the Cummings Farm.

It is, of course, uncertain what the true motivations were, by either side. Money pending was different than cash in hand and Allen may have decided that cash in hand was the way out of the pressures being exerted on him by expectant and, probably, arguing family members and a probate court that also expected action, although it seemingly had not previously questioned Allen about his lack of progress. There is the appearance of a self-serving motivation on the part of Allen as he sold his share, but not the other shares belonging to his siblings and nieces. There are many possible speculations concerning who was willing to sell what and for what minimum amount of money. The transcript of this and other trials were discarded by the Oxford County Superior Court in the 1960s, so there is even room for speculation as to the details of the litigation. Perhaps Allen may have intended selling the others heirs' portion to Donahue, if and when Donahue came forward to exercise the option? Alternatively, Allen may have presented the idea of selling all of the Cummings Farm to Donahue and there may have been the idea among some of the heirs that Donahue's offer was still too low, given all the publicity that was being generated. Allen may have been willing to sell, but the other beneficiaries were not. A very probable alternative was that Laura and Harry did not want to loose control of the farm and their portion of the feldspar and beryl business to Donahue, whom they may have known or suspected was United Feldspar Company's agent? Donahue's actions may be attributed to

Hakala mining at Mount Marie, Paris, ~1949.
Courtesy Wayne Ross.

Stan Perham, examining Bumpus rock, circa 1930. Perham (1987)

the same motivation. Donahue and the United Feldspar Corporation did not want Harry and Laura end up with a 70% interest in the Bumpus Quarry minerals: Harry and Laura's one fifth of Sybil's half interest plus Allen's 50% and Allen's inherited one-fifth of Sybil's half. The pressure on the United Feldspar Corporation was a need to preserve its feldspar production, although it was attempting to become more independent of the Bumpus Quarry's feldspar production. By opening new quarries.

Donahue did come forward on the fifty-seventh day of his sixty-day option and started his objections. He also purchased Edith Eleanor Stearns' and Adelia Abbie Waterhouse's portions of their inheritance in the Cummings Farm land and royalty payments, but Donahue did not record his purchase of Adelia's share until July 13, 1931, almost a month after the Oxford County Court's "final decree" (OCRD v401 p118; OCRD v401 p220). (One would think that Adelia would no longer have had any interest in a lawsuit, if she had already sold her interest?) At any rate, Sybil's and Allen's niece, Adelia A. Cummings Waterhouse, [22] along with mining engineer, William B. Donahue, successfully sued Allen, Laura, and Harry to nullify the sale of the Cummings Farm with the final decree on June 15, 1931. [313]

Ostensibly, Adelia's interest in the lawsuit was the increased money she would have gotten from the estate settlement if the farm were sold for $1000 more than it had

been. As one of three heirs of Cora, who was entitled to a one-fifth interest, Adelia got one-fifteenth of all of the estate's value. The one-fifteenth interest was, of course, in Sybil's one-half interest.

The contention of the heirs was that the sale was agreed upon when Allen's license to act as executor had expired and that the new extension of the expired executorship was improper. There is the implication that the family contended that an interest, undivided, of a piece of property could not be split into "yours" and "mine". Allen could not get all of the money from that sale, because each heir had an interest in every sale. Allen's accepting a new offer that was lower than a pending offer was unacceptable. Donahue's position was that he had a valid option that he had chosen to exercise.

Allen was ordered to make an accounting of his financial matters relating to the estate, as the suit alleged that he had yet to do so. Allen's account included: Sale of Sybil's personal items = $140.00; sale of a cow = $50; sale of one half interest in the Cummings Farm = $2,500; and sale of minerals (unspecified) = $1,233.00. The total was $3,923.00 with various payments to settle Sybil's debts (including funeral expenses, cemetery monument, prior doctor bills, lawyer's fees, Allen's executorial fee, etc.) at $1,656.13. Allen's account had a net value of $2266.87. Interestingly, Allen included all of the sale of feldspar and all of the Cummings Farm taxes due for 1927-1930 ($215.00) in the account of Sybil's estate. The lease of the mineral rights clearly indicated that Harry and Laura were to pay these taxes, but no tax-specific payments from Harry and Laura were indicated by Allen. Laura and Harry may have paid the taxes to Allen, but then, Allen could not claim those payments against the estate without also accounting for the payments to the town. At any rate, the financial mess is more or less an experiment in creative accounting.

A 1931 an Oxford County newspaper clipping revealed:

"BERYL MINE FIGURES IN EQUITY CASE

The beryl quarry in Albany, widely known as the Bumpus mine was brought into a legal controversy at the court house this week. This was a specialty equity hearing before Chief Justice [William R.] Pattengall in which William B. Donahue, a mining engineer takes action against Allen E. Cummings administrator of the Abbie W. Cummings estate. The mine is on the late Joseph Cummings homestead farm, occupied many years by the widow, now deceased. A distribution of the estate was made and Donahue alleges he was given an option by Cummings on the sale of a certain share and placed in trust $3500 as the sale price. Cummings has not fulfilled his agreement according to Donahue and the question of Cummings' equity is one of the matters involved.

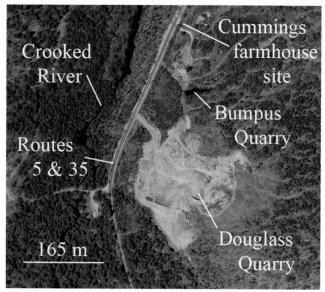

Crooked River

Routes 5 & 35

165 m

Cummings farmhouse site

Bumpus Quarry

Douglass Quarry

A mass of documentary evidence has been introduced and many witnesses heard during the hearing from Monday afternoon until Wednesday noon. The history, chemical composition and market value of beryl figured strongly during this time.

Alton C. Wheeler and E. W. Abbott were attorneys for Donahue; Berman & Berman for Allen Cummings."

The fact that the Bumpus Quarry beryl figured in the case may suggest that there was an expectation that the huge beryls were extremely valuable and the beryl that had been exposed was merely the tip of an iceberg. By this time, the Cummings heirs would have been aware of Stan Perham's allegedly becoming a "beryllium millionaire". In fact, Woodard (1957) cited Stan Perham that it was that assertion in a national magazine which led to the family feud. The only similar quote found so far was in a 1930 issue of *Business Week* which said: "Of the few hundred tons of beryl in the world, Stanley I. Perham, 23-year old student at Bates College, Maine, owns 150 tons." The news item claimed that owning the beryl was "better than [owning] an oil well".

Additionally, the court decree made no specific mention of any payment of beryl royalties. The missing details are whether or not Stan Perham was actually able to complete shipment of beryl to David Taylor Company. If about 100 tons of beryl were sold, the calculations previously made on a theoretical sale of 200 tons are to be reduced by a factor of two. Allen's accounting of minerals sold, $1,233.00, is tantalizingly close to the calculated royalty of 78 tons of beryl sold to David Taylor Company. There are deductions which could be imagined at this far vantage point in time, where approximately 100 tons of beryl could have been sold to the Taylor Company and the labor and other costs could have reduced the payment to the Cummings' account to the amount of $1,233.00. This accounting would imply that there were yet to be paid royalties from Harry and Laura Bumpus on the mining of feldspar.

There were certainly enough indications that the

Bumpus Quarry had yet to produce its full potential of minerals. In order to prepare the way leading to the final decree of the litigation, Allen had to specifically assign to William B. Donahue's purchase option, all of his interest in the <u>un-paid</u> royalties from the sales of minerals from the Bumpus Quarry in effect giving Donahue a discount for purchasing the land. This deeded assignment (OCRD v389 p364, June 6, 1931) had the effect of allowing the release of Donahue's $3500 escrow held by the Clerk of Courts for sale of half of the Cummings Farm and the Bumpus Quarry. The $50 option check was never accounted for. The sale of the half interest in the farm was made by quit claim deed on July 6, 1931 and registered on July 14. The litigation did not affect the mineral lease of Laura and Harry Bumpus. The court decree undoubtedly involved the contracts made with Harold Perham and the beryl sub-contract with Stan Perham: "Suddenly the mining company laid claim to the beryl itself." Technically, Harry and Laura were the "Company" before the litigation. UFC could be called "the Company", afterwards. As a result of the court decree, the United Feldspar Company acquired substantial control of the Bumpus Quarry and the Cummings Farm, but there were still heirs to contend with.

After estate bills were paid, according to Allen's figures, the one fifteenth share was worth $82.20, while a one fifth share was worth $164.40, plus, of course, the residue of the estate and future payments. It isn't known who paid the lawyers' bills, but if Adelia were responsible for part of them (totaling $150.00), it is difficult to imagine that she had much money left over. One would have thought that love of family was worth more than that? Of course, if there really were a projected fortune at stake – "Better than an oil well", the motivation might have been greater. If Donahue paid for all of the lawyer bills, then it may have been the United Feldspar Company which *de facto* paid the lawyer and court costs, because Adelia's participation was necessary to strengthen the lawsuit. William B. Donahue transferred his interests in the Cummings land and mineral royalties to OMMC by registering documents on June 26 (OCRD v401 p251), July 24, and July 31, 1931. [314, 356, 357] (Coincidentally, two days before the first Cummings land documents were registered, OMMC leased the portion of Mount Marie, not then also leased by Harold Perham, from Victor Piirainen and Charles Edwards, in effect preventing Harold from further expanding his mining efforts on the hill (OCRD v389 p392-393). If Harold was not going to sell feldspar to the West Paris mill, the West Paris mill company was not going to permit Harold to expand into fertile areas and sell feldspar to its competitors.)

After all the litigation and unknown behind-the-scenes conflicts, the remaining control of the Cummings Farm was not sold for more than a decade. United Feldspar did obtain a controlling interest in ownership of the Cummings Farm. They added Allen's previous 50% ownership the purchase of 6.6% from Adelia and of 6.6% from Edith, equaling 63% interest. Allen's inherited 10% seems to have been lost in the shuffle. This was still enough for the United

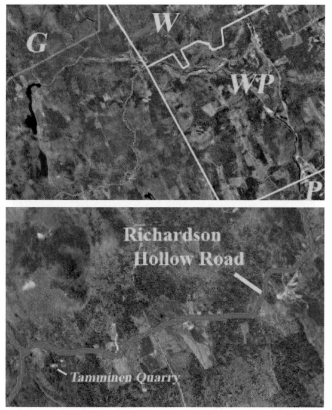

Upper: G=Greenwood, W=Woodstock, WP=West Paris. Lower: Richardson Hollow Road Area detail, white patches are outcrops or quarries, 2008. Courtesy Google Earth.

Feldspar Corporation to control the fate of the Bumpus Quarry. Laura kept her share, Wallace kept his share, and Laura Pinkham retained hers.

Harold Perham entered an agreement with United Feldspar Corporation, through its subsidiary, ten days after the "final decree" relating to the Bumpus Quarry. Specifically, he sold his two Bumpus Quarry contracts to UFC for $1,000.00:

> "Whereas. The said Harold C. Perham has an interest in certain sales agreements made between himself, Harry E. Bumpus and Laura J. Bumpus, both of Auburn, … one being dated September 9, 1927, and the other December 28, 1927, which he has agreed to transfer, sell and assign to said Oxford Mining and Milling Company for the consideration mentioned herein.
>
> And Whereas, the said Harold C. Perham has another contract with said Harry E. and Laura J. Bumpus for the sale of certain Beryl from their mine at Albany, Maine, and said agreement being dated November 29, 1930, which he has agreed to transfer his interest in to said Oxford Mining & Milling Co. …
>
> THIRD: As a further consideration for this agreement, the said Oxford Mining & Milling Company agrees to purchase from

said Harold C. Perham a maximum of sixty tons per month or seven hundred twenty (720) tons per year during the year commencing July 1, 1931, and ending on July 1, 1932, this feldspar to be delivered approximately two-thirds from the Hibbs mine located in the Town of Hebron, and one-third from the Sturtevant mine, located in the Town of Hebron, it being understood that these proportions may be varied with the consent of the Oxford Mining & Milling Company, said feldspar to be of a good commercial grade and not to contain not more than ten per-cent of free quartz and foreign material, said feldspar to be delivered at the mill of said Oxford Mining & Milling Company… and … agrees to pay said Perham seven dollars and fifty cents ($7.50) per long ton of 2240 pounds for said feldspar…

> It is further understood and mutually agreed should said Perham to desire he may deliver during the month of June feldspar which he now has on hand at both the Hibbs and Sturtevant quarry, not exceeding sixty ton, this to be in addition to the amount called for as above…"

The "Final Decree' obviously was a compromise for the Perham's, although it must have been humiliating. Feldspar could still be mined by Harry and Laura Bumpus, and they presumably would still get the $6/ton for mining feldspar. Harry and Laura would still pay their royalties to the land owners including the UFC, for the feldspar, divided according to their shares. No mention was made if they still were responsible for taxes on the land. The Oxford Mining & Milling Company would presumably be responsible for transporting the feldspar to West Paris, but would no longer have to pay Harold his "cut". It may be further surmised that it was at this time that trucks began to be used to transport the Bumpus Quarry feldspar to West Paris instead of using the Grand Trunk Railroad out of Bethel. As Harry and Laura had a contract that specified a 50% royalty on "other" minerals as by-products, any beryl encountered could be sold by them, again with appropriate distribution of payments. The compromise deal for Harold included a good rate of pay for Hebron feldspar, but Harold had to pay to transport it to West Paris, however, and the quantity purchased was not really a huge amount.

No matter what the outcome of the family litigation, it started during the time that the giant beryls were uncovered and must have been a serious concern to those associated with the quarry. The financial expectations never matched reality. The unfettered publicity in newspapers across the nation caused most of the problems. The court actions irreconcilably divided the family and there is little or no detailed tradition in the family concerning the feud as the litigation became a taboo topic. Other footfalls were to be heard within a few years in the attempt by the "Mill" to fi-

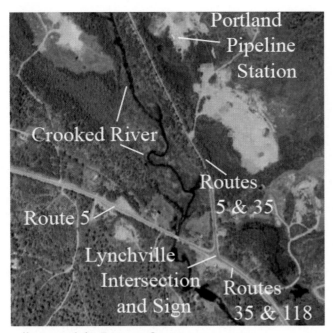

Portland Pipeline Station

Crooked River

Routes 5 & 35

Route 5

Lynchville Intersection and Sign

Routes 35 & 118

nally control the Bumpus Quarry.

William B. Donahue's Other Leases

Two months before Allen's fateful decision to sell his share of the Cummings Farm, the *Norway Advertiser* had an interesting personals item on August 1, 1930: "Wm. B. Donahue of New York City was in town a few days last week. He is interested in mines of different kinds." There was also a personals item about "Mr. and Mrs. E. O. Donahue" of Albany, but no family connection was suggested or implied. As has been indicated, Donahue may have been an entirely independent exploration geologist or may have been in the employ of the United Feldspar Company. The latter seems more likely, especially as Stan Perham simultaneously had Loris Rollins acting on his behalf as an exploration "geologist". As an independent agent, Donahue would have been freer to act as a roving geologist rather than the identified instrument of a predatory out-of-state corporation.

Donahue's expertise as an exploration geologist was excellent and he particularly recognized the value of feldspar deposits along Richardson Hollow Road in Greenwood. Three important quarries were soon located on the south side of that road including, from west to east, the Tamminen, the Waisanen, and the Nubble Quarries. The Nubble, a small knoll on the south slope of Noyes Mountain, was owned by the world-renowned author Charles Asbury Stephens. Stephens signed a mineral lease with Donahue on September 12, 1930. (Stephens was cousin of locally noted mineral collector and internationally known biologist Addison E. Verrill, who in turn was Vivian Aker's father-in-law. Stephens, himself, was a mineral collector and created the famous and undying myth of *young school boys* discovering Mount Mica.) Matti and Aino Waisanen signed a mineral lease on August 13, 1930 (OCRD v). The land immediately east of Waisanen's', owned by Peter and Este Kuvaja, was leased September 4, 1930 (OCRD v388 p614; cancelled August 3,

1931, OCRD v389 p377). Donahue approached Kusti Nestor Tamminen in August, 1930 to explore his farm for its feldspar potential and through perseverance, the Tamminen Farm was leased for five years on September 18, 1930. The land of Anna J. Noyes of Greenwood, bounded to the east by land of Kusti Nestor Tamminen, was also leased, August 16, 1930 (OCRD v388 p603-604). It should be noted that the mineral leases usually provided for a minimum annual royalty or rent and so any leases not resulting in feldspar production would be cancelled, lest the minimum annual payments became excessive. For example, the Donahue lease with Rodney Willard of Sept 12, 1930 was cancelled by mutual agreement on July 31, 1931 (OCRD v389 p417). Donahue also leased mineral rights on Richardson Hollow Road on Jack Mustonen's (October 9, 1930, OCRD v389 p155-156) and George Dewey Verrill's lands (September 23, 1930, OCRD v389 p140). William B. Donahue was later an employee of Feldspathic Research Corporation and he was an author of a Canadian patent (#411473) to produce magnesium-free beryllium on March 30, 1943.

The Oxford County mining excitement did not remain just the OMMC's, Perhams', or Donahue's realms. For example, the Henry K. Stearns farm on the Hunts Corner Road in Albany was leased by William M. Daniels of "Hebron and Minot" (OCRD v388 p618-619). Daniels either read the Oxford County mineral leases being filed or was at least familiar with details of royalty and rent terms, and his leases, while differently worded, were on essentially the same terms as others being let. Other people followed suit.

Effects of the Great Depression on Oxford County's Mining

Partly, because of the great contraction in the feldspar, mica, etc. markets which were directly affected by the nation's housing construction industry, Harold Perham mortgaged his mineral leases and agreements to the Paris Trust Company of South Paris on May 16, 1931. His need for money must have been exacerbated by the closing of the Bumpus Quarry during litigation. The mortgage included a deed for Corbett Ledge land and Quarry (Paris); mineral rights of the Kalli Piirainen property (the Mount Marie Quarry area in Paris), a lease in Peru from Anthony Z. Mockus and others, probably of what is now known as Hedgehog Hill; mineral rights in Canton on the Charles W. Walker property; a mineral agreement with Blanche Bennett on the Bennett Quarry (Buckfield); mining lease on the Doughty Farm (Hebron); mining rights at the Hibbs Quarry (Hebron); mining rights on Effie Bessey's land (Hebron), mining rights at the Sturtevant Quarry (Hebron); mining rights at the Harry W. Bearce property (Hebron); as well as several rights of way. [315] The mortgage permitted Harold to continue to exercise these rights as he had been.

On July 22, 1931, OMMC also negotiated a lease, probably to the Mount Marie Quarry in Paris (formerly in Hebron), on land owned by Victor Piirainen. [316] The economic conditions of the Great Depression were unrelentingly

Albany pipeline section north of the Bumpus Quarry looking northwest, 2007

severe and Harold Perham sublet his mineral lease to the Bennett Quarry on September 1, 1931 to the Whitehall Company. [317] The sale of the mineral lease still granted Harold the right to take quartz and "second grade feldspar" from the dump of the Bennett Quarry for free, even including that placed there by the Whitehall Company. In fact, the Whitehall Company was prohibited from removing second grade feldspar from the mine area and selling it itself. A specific provision included: "The Whitehall Company agrees to use reasonable precaution to save and lay aside as said Perham's property, Beryl, Mineral specimens of recognized commercial value, and contents of Mineral Pockets. Said Perham agrees to pay a reasonable wage cost for the saving of these materials." Harold also received a royalty of 50¢/ton on feldspar removed by the Whitehall Company. The payments were made directly to the land's mortgage owner, the Paris Trust Company which would, in turn, would pay Blanche Bennett the overage. This sale apparently provided the Whitehall Feldspar Company its first exposure to mining in Oxford County.

Bankruptcies

Depending on how you view it, United Feldspar Company defaulted on its mortgage in 1932, seemingly not even having paid interest on its mortgage. The United Feldspar Company previously had acquired the Oxford Mining and Milling Company by five-year mortgage on January 20, 1930 through the Paris Trust Bank. On March 23, 1932, Francis Edgar Haag (OCRD v v389 p566-567), bought the

mortgage of OMMC from UFC. Haag appears to have been engaged in acquiring feldspar grinding companies including UFC. The new mortgage was for $100,000 and the wording of the new mortgage suggested that UFC was in arrears. It is because the Oxford Mining and Milling Company still existed as a company, only bought by mortgage, did its name continue in the *Maine Register.*

"... a certain mortgage, made by Oxford Mining & Milling Company, a corporation organized under the laws of the State of Maine given to secure payment of the sum of ONE HUNDRED THOUSAND ($100,000.) - - dollars and interest, dated the 23rd day of December, 1929_ recorded on the 23rd day of January, 1930 in the office of the Registry of Deeds of the County of Oxford, State of Maine in Book 388, of mortgages, at pages 499 – 501, covering premises in the Town of Paris and in the Town of Bethel, County of Oxford, State of Maine, as therein more fully described,

TO HAVE AND TO HOLD the same unto the assignees and to the successors __, legal representatives and assigns of the assignees forever, as security for the payment by United Feldspar Corporation, its successors and assigns, of its six certain notes on which there remains unpaid to the assignees the sum of Twenty seven thousand five hundred seventy five ($27,575.) dollars with the interest,

Albany pipeline section, 2007

and to secure any further notes which the assignor may deliver to the assignees in extension of or in addition to the said notes.

And the assignor covenants that there is now owing upon said mortgage, without offset or defense of any kind, the principal sum of One hundred thousand ($100,000.) dollars, with interest thereon at six (6%) per centum per annum from the 23rd day of June nineteen Hundred and thirty-two."

Through the acquisition of this note, the transition began whereby Charles Pedrick's United Feldspar Corporation was able to become Francis Haag's United Feldspar and Minerals Corporation. As a formality, on October 14, 1932, United Feldspar Corporation assigned their interest in the Oxford Mining and Milling Company to a Mr. A. E. Haven of New York City, further suggesting UFC's poor financial condition. Although some of the transitions relating to the various companies and corporations involved actions away from the prying eyes of Maine's citizens, on February 21, 1938, Oxford Mining and Milling Corporation was further conveyed to F. E. Haag and Company, United Feldspar Corporation, and Feldspathic Research Corporation by $40,000 paid by Francis E. and Winifred L. Haag doing business with the above names.

The further shrinking of the economy relating to mining eventually took its toll and Harold had to declare bankruptcy August 15, 1932: "In 1929, the Stock Market Crash arrived, and his business was cleaned out, — lock, stock, and barrel." [171] By February 1, 1933, Harris M. Isaacson, lawyer of Lewiston, was acting as trustee, arranging for the liquidation of Harold's assets, specifically mineral leases at the Sturtevant and Hibbs quarries in Hebron, on behalf of the Androscoggin County Courts [318] with issuance of a trustee's deed to all remaining rights and agreements still held by Harold Perham. [319] On April 24, 1933, Isaacson filed

papers officially releasing Laura and Harry Bumpus from any mineral lease agreements negotiated with Harold Perham although that lease already had been reassigned to OMMC. [320] There remained a few economic issues, but Harold's mining buildings on the Hibbs Quarry area were assessed taxes that went unpaid in 1934 and a lien on them was issued on April 5, 1935 as well as on January 25, 1936. [321, 322]

The struggle between Harold Perham and the "mill" was officially over on May 5, 1937. The UFMC registered the purchase of his mineral leases held by Casco Bank and Trust on August 28, 1937. The Casco Bank was then labeled as "insolvent". The mineral lands or leases included: two lots purchased from Victor Piirainen, rights to Lobikis Quarry, rights to Walker Quarry, Canton, rights to Kalli Piiraininen's land, rights of way to the Walker Quarry, Canton and to the Hibbs Quarry, Hebron, rights to the Doughty Farm, and rights to the Bearce Farm.

CCC, WPA, and Portland Pipeline Important Construction Projects

Cold Water CCC Camp, Stow.
From Schlenker et al. (1988).

Portland Pipeline welding

During the Depression, there were still investments and construction projects in Oxford County including those by Federal Works Progress Administration (WPA), Civilian Conservation Corps (CCC), and the Portland Pipeline Company. The first two were Federal, so-called "make work programs", while the latter was commercial. The CCC was one of the first Federal projects that provided jobs in Maine, although many of the workers were from out-of-state. CCC began in 1933 and two of the Camps were located in Oxford County: Cold River Camp (#152, #160) at Stow and Wild River Camp (#156) in Gilead. [251] The CCC worked on various forestry projects, including building fire towers. They also constructed the Evans Notch Road in 1933-1934, along the Maine-New Hampshire boundary between Gilead and North Chatham, New Hampshire: "The company used more than thirty tons of dynamite to cut through granite ledges and to uproot stumps." CCC also worked on portions of the Appalachian Trail. The WPA improved many bridges and roads in Oxford County after its inception in 1933. Wild River Camp was active May, 1933 to 1937, while the Cold River Camp had two periods of activity: May, 1933 to 1936 and October, 1938 to October, 1939. As Maine mining shrunk and miners lost jobs, some turned their skills to construction projects.

"In 1932, the [Harold Perham] family moved back from West Minot to West Paris, where they lived on the Mine Road. Those 'hard time days' were a conglomeration of 'dickering cows', 'swapping labor', and raising garden crops, cutting firewood, picking up odd jobs, making Apple Butter, bucking up cordwood at $1. per cord. {Hens for eggs, and for eating, were kept for many years.} His sixth child was born in the Fall of 1932. Eventually, the family was to number Thirteen.

... he was lucky to hit a bunch of work as Blaster when they were building road base during the winter months. Later on in 1934, he

got a regular job working at the Feldspar Grinding Plant, and worked there until the summer of 1941, when he started work on the Pipe Line as a Jackhammer Operator, and helped build the first Pipe Line from Portland, Maine, to Montreal, Canada." [171, 34]

The Portland Pipeline Company was organized by Standard Oil of New Jersey, now the Exxon-Mobil Company. The pipeline surveyors completed their work in the summer of 1941 and construction began immediately. The route of the Portland Pipeline, for petroleum oil and natural gas, extends northwesterly from Portland through Raymond and along the northeastern margin of Sebago Lake. Waterford has an important pumping station along the pipeline route and the pipeline enters Albany Township near its famous Lynchville village intersection of Routes 118, 5, and 35. At Bethel, the pipeline follows westerly and generally along the southern shore of the Androscoggin River into New Hampshire.

"Originally the right of way held three pipelines: an 18 inch line that carried natural gas from Canada to the US, a 24 inch line that transported oil to Montreal, and a third 12 inch line that was cleaned and retired in 1984. Currently [2007] the 18 inch line has been returned to oil service as is the 24 inch line. It now takes 43 and 36 hours, respectively, to pump a barrel of oil through the 18 inch and 24 inch lines to Montreal. The 12 inch line has been abandoned. The three lines are about three feet beneath the ground surface, pass through approximately 236 miles of countryside in Maine, New Hampshire, Vermont and the Province of Quebec. ... The objective was to accept oil year around in Portland and delivering it to four refineries in Montreal East operated by Imperial Oil Limited, Shell Oil Company of Canada, McCall Fontana Oil Company (later known as Texaco) and the British-American Oil Company (later named Gulf). Previously these Canadian refineries shut down in the winter when the St. Lawrence River was frozen and only operated six months a year. In 1946 Standard Oil of New Jersey sold the Montreal and Portland companies to the four companies operating the Montreal refineries. In December 1950, the system changed to a three-station operation resulting from ever changing and more efficient equipment." (Holmquist, Wayne R. 2006, www.raymondmaine.org/historical_society/portland_pipeline.htm)

The first "use" of the pipeline occurred in the Autumn of 1941. Harold continued to work for the Portland Pipe Line Company at the Waterford station for 24 more years, retiring in 1964. He became an inveterate hiker and genealogist. As part of his duties in the Appalachian Mountain Club, he helped maintain trail sections in Andover.

Reg Ross sitting holding drill; standing man with sledgehammer. Nellie Hibbs, foreground, c1928. Courtesy Irene Card.

Joseph Alton Hibbs' home
[Morris Bumpus homestead]. c1921,
Lower views = Joseph Alton Hibbs [May 1, 1865
-1934]. L c1895, R c192 Courtesy Irene Card and
Wayne Ross.

Harold had a lifelong dedication to community service beginning as an Eagle rank Boy Scout in the 1910s and was later a Boy Scout troop leader and was on the regional executive board. He was on the Paris School Board for ten years, president of the West Paris Alumni Association, secretary of the National Association of Universalist Men, and was active on the Republican Town Committee and served two terms, 1929 and 1931, as a Maine State Representative. Harold worked to have a gymnasium built in the West Paris high school and to establish a public athletic field which was eventually named in his honor. He was also a "Granger" in West Paris. In 1953, Harold wrote a book: *The Maine Book on Universalism* and later wrote a history of the Ledgeview Memorial Nursing Home. When West Paris wanted to secede from Paris, Harold was one of the leaders who helped it happen. In his retirement years, Harold prepared an extensive manuscript genealogy of the Perham and related families.

The Bumpus – Perham – Hibbs Family Connection and the Hibbs Quarry, Hebron

The patterns of immigration and migration have a strong bearing on family relationships. Western Maine experienced a strong influx of settlers after the Revolutionary War. By March 4, 1805, there was a successful movement to establish a new county carved out of the formerly existing York and Cumberland Counties. The new Oxford County was later reduced in size when towns from it (Berlin, Carthage, Jay, Madrid, Weld, Letter D, and Letter E, and Number Six North of Weld) were re-allocated to form part of the new Franklin County on March 20, 1838 and the Oxford County towns of Livermore and Turner were transferred to the new Androscoggin County when it was formed

Miner (L), Nellie Hibbs, Reg Ross (R), c1928.
Courtesy Wayne Ross.

on March 18, 1854. Although new residents were constantly entering Oxford County, this county, as well as Maine, itself, did not begin to receive large amounts of new settlers until much after the Civil War. Consequently, families present early in the county's history have many family ties just because they were there and their children married locally. Nonetheless, in recent times, it has been the job of genealo-

Wheelbarrows and ramps, Hibbs Quarry, c1928.
Courtesy Wayne Ross.

gists to re-discover old ties as most families are unaware of their family history further back than their great grandparents, and frequently only back to their grandparents.

The reason why Harold Perham worked at relatively far-off Hebron may not have been by chance. Harold's experiences, first at the family's quarry in West Paris and later at the Bumpus Quarry brought Harold in contact with the Bumpus family. As has been mentioned, Harry and Laura Bumpus first contacted Harold's younger brother, Stanley, to identify their feldspar. Harry Bumpus was Daniel Bumpus' [b. October 10, 1800 Hebron – June 29, 1889 Hebron] great grand nephew. Daniel purchased the land where Mount Rubellite quarry in Hebron is located on January 24, 1868. [200] (The mineral rights to Mount Rubellite had been sold, previously, in 1862 to Samuel R. Carter and Forrest Shepherd [October 31, 1800 Boscowen, NH – December 8, 1888 Norwich, CT]. Shepherd was a professor of "agricultural chemistry" at Western Reserve College (now part of Case

Evelyn Hibbs and Reginald Ross at Hibbs Quarry.
c1928, Courtesy Irene Card and Wayne Ross.

Western Reserve University, Cleveland, Ohio), 1847-1856, but he was very well-known as a free-lance consulting geologist/mineralogist. He was a resident of New Haven, CT in 1862 and may have learned of unusual Hebron minerals from Hebron native and Yale student, later Yale professor, Oscar Dana Allen [1836-1913], who had discovered the presence of the newly discovered element, cesium, in Hebron and Mount Mica lepidolite. Samuel Carter was Augustus Hamlin's brother-in-law. Carter's interest in minerals may have come from his association with Eliphalet Nott, noted mineral collector and president of Union College, from which Carter graduated in 1852. The scientific excitement associated with Mount Rubellite is a long and interesting but technical tale and will be published elsewhere.)

L-R Nellie Hibbs in front of daughter Evelyn, unknown miner standing, Reg Ross sitting, unknown miner at right. c1928.
Courtesy Irene Card and Wayne Ross.

Harry Bumpus's presence on Auburn City government and presumed familiarity with the feldspar mining and grinding industry in Auburn, as well as family ties, explains much of his interest in the feldspar business. (Harry was alderman 1906-1907 and on the Auburn City Planning Board, at least 1924-1935, while his father, George Washington Bumpus, was Auburn City Clerk and Pensions Agent, at least 1901-1935.)

The Hibbs Quarry was originally opened in 1906 by Winfield Scott Robinson and J. A. Gerry and was re-opened by Harold C. Perham in 1926. Joseph H. Hibbs [1835-1904], Joseph Alton's father, had married Dulcina Rebecca Bumpus on February 9, 1862. Interestingly, Harold Perham Bumpus [March 21, 1894 Turner, ME – May 16, 1967 Coventry, CT], son of Raleigh Martin Bumpus and Mabel Loverna Perham, was born five years before Harold C. Perham.

On August 12, 1925 Harold, signed a five-year, renewable, mineral lease for the J. Alton and Nellie J. Sawyer Hibbs [b. June 25, 1876 Mechanic Falls] property with the provisions: "Operations hereunder are to begin in the early part of summer of **1926**." [352] A one year lapse of mining would allow cancellation of the agreement. The lease specified payment of a $0.50/ton royalty on the feldspar and quartz removed, and a 50% royalty on the selling price of "mica, gems, and other minerals". There was a $100 advance on royalties required with monthly accounting and payments. Because of the generous royalty for the "other minerals" besides feldspar, it is reasonable that the original intent was to have a feldspar quarry rather than a mica quarry, as Harold was also then foreman of the nearby Bennett Quarry in adjacent Buckfield. The Hibbs' lease specified: "After beginning operations said Perham is to keep three men or more employed on said operations at all suitable seasons; winter

Hibbs Quarry visitors, man with hat on center right possibly George Howe, c1928.
Courtesy Irene Card and Wayne Ross.

work not being obligatory." The delayed date for the start of mica mining may have been related to Phyllis' illness, Harold's first wife. She died of pneumonia within three weeks of the signing of the Hibbs' lease.

No Maine mica production was reported in 1926, however, and it might be assumed that the mica was sold to a Boston firm with the mica's weight being assigned to New Hampshire or remained unassigned. It is unlikely that Harold stockpiled his mica for sale in the next year, although it is possible. (Reginald W. Ross [c1901-March 27, 1969] who worked for Harold Perham, met his future wife, Evelyn Hibbs [January 2, 1913-August 8, 1999], while working at the Hibbs Quarry. Evelyn and Reg were married on December 1, 1928. Earlier, Reg had worked for Dick Nevel, at the Dunton Quarry, Newry, at least in 1926 and 1927.)

Perham (1987) wrote about the Hibbs Quarry:

"Mica produced by this locale tends to be very flexible and the Hibbs Quarry has produced what is believed to be the choicest 'cigarette' mica found in the United States. (It is so-called because it may easily be rolled about a core.) The Huse Liberty Mica Company of Boston purchased several carloads of this unique form of mica during 1927 and they were most impressed with the size of the sheets and with their flexibility as well.

During 1929, the mining uncovered a tremendous crystal of mica which weighed seventy-five pounds. This material was of the finest amber color and it yielded single flawless sheets of mica fifteen inches wide and two feet long."

Cigarette-mica was used in the manufacture of sparkplugs used for aircraft engines. Sparkplugs made of porcelain, or the then known ceramic materials, could not perform to the same high standards requiring the shielding of 20,000 volts. The extra shielding, afforded by mica, also effectively reduced the noise received by conventional radios. The cigarette-mica was split to a thickness of about 0.0004 mm around the central electrode of the sparkplug. By about 1950, the use of cigarette-mica was being phased out. The Hibbs Quarry also produced a few tons of beryl, but the grade was only about 0.5% compared to the amount of feldspar produced. It is notable that some of the Hibbs' microcline feldspar crystals were up to 1 x 0.65 meters in length. Additional details of Hibbs Quarry mining activity and production will be found in Appendix C.

The Famous Lynchville Intersection Sign
Historical Review

Although not tied to the Bumpus District's mining history, a sign post at the Lynchville intersection deserves to be mentioned. Lynchville is the closest to a business dis-

trict Albany township has ever had and is one of several such tiny "business districts" along the road between Norway and Fryeburg. Occasionally, some of these business districts did not have a single operating store.

The sign's location is just over 8 km south of the Bumpus Quarry where State routes 5, 35, and 118 intersect. In the early 1940s, the sign was erected to call attention to the many local towns with international names. It may be supposed that there were precursor signs at the location which listed only the towns nearby and that there was always some tourist attention caused by the sign's curious names. Certainly Bethel, Albany, and Waterford were also listed. It is certain that the international names only version of the sign was erected because curiosity increased as tourism rose after the great Depression. The sign has become a popular tourist attraction and may be one of the most photographed spots in Maine. Over the years, the sign has had to be replaced and/or redesigned due to the ravages of time, automobile accidents, and thieves. The design changes have mostly involved using one or two support posts, frequently of wood, but sometimes of metal.

The earliest usage of the Lynchville sign on a post-

Lynchville sign with "Believe It or Not" caption,
1942 postcard

card seen by this author is July 5, 1942 and showed the black and white single masted sign. A second black and white version featured the phrase "Believe It or Not" and tried to capitalize on the famous and very popular newspaper cartoon of the same name authored by Robert Ripley. Interestingly, the black and white cards have numerous examples where a creative sender would paste some tape with their own town name on the postcard or they would simply pencil it in. One scarce black and white card had a shadow of someone pointing to the sign.

By the mid to late 1940s, the sign appeared on so-called linen postcards and were issued with fauvistic colors. These cards were still being used by 1960. There were several versions of these cards and they mostly varied in the details of the foreground where there might be colorful flowers drawn in or not.

At least by 1953, the Lynchville sign appeared on "chrome" cards and the sign had a short support bean suggesting it had suffered from an accident. These postcards were smooth and multicolored. The sign had some human interest with a child holding a pole with crude cloth duffle bag over his shoulder, in hobo style, suggesting he was choosing a town to go to while he was running away from home. There was a pose with a high angle to the pole and one that was horizontal, presumably so that a different image could be sold to several printers without one violating the copyright of the other printer, although several identical images were not owned outright and were published by the different companies and one image was seemingly published by five different companies. One image had a child on a tricycle trying to figure out which way to ride. There is another contrived card where the photographer brought in corn

Postcard from mid 1940s

stalks, pumpkins, and gourds to decorate the sign while autumn colors were blazing in the background. During the 1950s, the postcard was sometimes issued with scalloped edges. An intrepid company even had the audacity to issue the sign image in a horizontal format, while another cropped the sign to its bare essential names without showing supports or background.

About 1960, the sign was placed on two supports and cards from then on until the 1990s had this feature. Of course, out of the way gift shops continued selling old postcards and would not have the new style until they'd run out of the old. The cards were sold all over Maine and there are several dozen images available to collectors, although the linen cards had a weird change in them. Artists would airbrush out vegetation and draw in new backgrounds or foregrounds. The foliage might be enhanced, recolored, or

Postcards showing Harold "Hank" Rolfe of Waterford with alternate poses, from 1960s.

Horizontal postcard from the 1980s

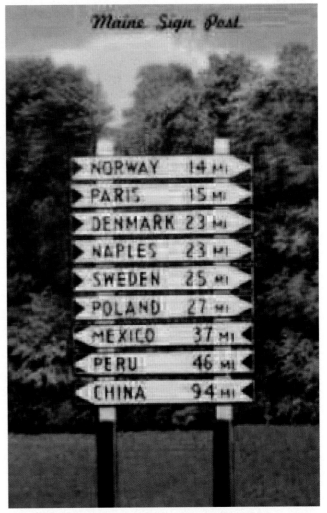

Corner art details of "identical" Lynchville postcards.

Decal from 1950s with extra town to appeal to an all Maine tourist market.

Postcards from late 1940s to early 1950s, note added town, probably Woodstock, NH not Woodstock, Maine, see also pointer's shadow.

Color postcard from the early 1960s

subdued, depending on the press run. Big trees in the background would appear or vanish from one card to the next. Early prints of cards tended to be sharper and brighter than later editions.

In recent years there have been several signage imitators in the towns of China, Norway, etc. The Norway sign appeared about the year 2000. While this new sign featured international capital cities, Maine residents noticed that there were many missing municipal candidates, some of which are also capitals at various levels: Edinburgh, Cambridge, Jerusalem, Bremen, Avon, Detroit, Lexington, Concord, Windsor, Troy, Princeton, Sorrento, Fryeburg, Oxford, Carthage, Bath, Hebron, Gilead, Dallas, Dresden, Washington, Mount Vernon, Hartford, Palermo, Canaan, Frankfort, Limerick, Exeter, and, of course, Albany. A more complete

sign would be more expensive to maintain or replace. However, the international sign in Lynchville could include Wales and Lebanon. The South China sign is an imitation of the Lynchville "Classic" design, complete with three tiered cap. (Note: Egypt, Maine is a settlement, not a town.)

As tourism increased in Maine, decals, whiskey glasses, and bumper stickers were not far behind, each displaying Lynchville's pride. The various manufacturers of tourist memorabilia recognized that they could sell more souvenirs if they had an inclusive character. The Lynchville sign was marketed with an extra town on the sign – the far

Lynchville signs, 1960s-1990s.

Reed Grover on tricycle from the 1960s

*Ellory and Collene Lawrence of Stoneham
from the 1980s*

Upper left: Lynchville sign, note ax cuts on supports, 1991.
upper center: Van King at South China's sign. Photo by Nathan King 1991
Upper right: Norway sign, 2007

eastern Maine town of Calais (pronounced Kal-iss, mispronounced by tourists after the French town of Kal-lay). The mileages were not altered from the Lynchville location, nor was there a deviation from the ascending order of mileage.

One of the interesting points concerning the sign is that the souvenir postcard was rarely labeled as to location, lest the inquisitive mobile tourist might attempt to visit the sign for free. The land around the sign is currently being decorated with figurines, etc. by a local citizen.

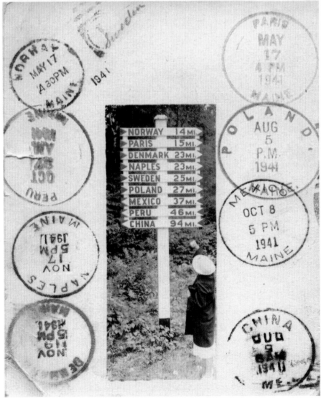

Photograph of the Lynchville Sign with postmarks of the various post offices of the towns listed.

The dates suggest that the Lynchville Sign was erected in its modern form in the Spring of 1941.

Lynchville Sign souvenir made of cedar sticks and a postcard fragment glued on to a backing.

Tag simply says: "A Sign Post in Maine"

There is a wide variety of Sign souvenirs that have been produced from ashtrays to key chain tags and sold to possibly over a million tourists - all unaware of the sign's actual location.

Chapter Six: Transitions from the Depression to World War II

The Depression made many people change their career paths. Harold Perham changed from miner to handyman and eventually had lifelong work with the Portland Montreal Pipeline Company. Some such as his brother Stan changed the emphasis of their livelihood's approach. Stan had been frequently a miner, but he also loved minerals in and of themselves. When mining work diminished, Stan shifted to direct sales.

Rise of Family Mining

The Depression was also a time when individual farmers were scraping for a subsistence living. The farms

feldspar on our land [on the Koskela Road] in South Woodstock. My father, Edward [Etvarti] Koskela, started a quarry in our field in front of our house. Later, we had another one down the hill into the woods near a brook. We had no mechanized equipment. I was the oldest child and it was my job to hold the drill, while my father and a hired hand would alternately strike the drill with sledgehammers. The deposit was not rich, but when we could bring several tons of feldspar into the mill, we had earned some desperately needed money." [404]

In the time period, crude feldspar would be worth about $4-

Stan Perham in front of his store. c1933. Courtesy West Paris Historical Society.

might be able to feed the family, but there were few chances to make hard money. Part-time lumbering was common and any extra work would help pay the mortgage or taxes, but outside purchases were rare for some families. Some Oxford County families knew that the West Paris feldspar mill would buy feldspar and the farmers would sometimes try to eke out a small feldspar crop of a few tons per month, although several local families were comparatively successful from their mineral leases: Tamminen, Waisanen, Emmons, and others. One might imagine that the mill expected a higher grade of crude feldspar from the occasional producers, otherwise they would have not purchased from them. Taisto Koskela related:

> "During the Depression, there was very little cash to be earned and bartering was common. However, during idle periods, we would mine

5 per long ton. Hand drilling was slow work, especially if a family had only one short drill. Black powder fortunately had little shattering power so each blast would have had more large pieces per blast than better equipped regular quarries which used dynamite. It might be imagined that a week's work would have yielded up to three or four tons, a good week's wages in the Depression, but spread over the effort of three or four people.

Rose Quartz Gems from the Bumpus Quarry

Money for college was getting tight and Stan Perham opted to leave Bates College half-way through his Junior year. He worked for a while at the Perham Quarry and soon afterwards, for his brother, Although Stan opened a new store on Route 26 in 1930, he still held several jobs. He was also working as an hourly-wage miner at the Hibbs Quarry for his older brother, Harold. His contract to sell beryl from the Bumpus Quarry promised a good combined income and led to yet another contract: marriage. Stan had met his future wife, Gwendolyn Louise Wood at Bates College and she graduated with the Class of 1927. They were married January 3, 1931. By the time of the marriage ceremony, however, the previously described courtroom drama had emerged and the future mineral business must have suddenly become less certain.

Deanie (center), Stuart Martin (Left, Chairman of Rumford Board of Selectmen), and Thurston Cole (Right, amateur mineralogist) at Newry overlooking Whitecap Mountain. c1958.
Photo by Ben Shaub.

An unpublished manuscript (Stan Perham, ~1958) indicated that the Trap Corner store "opened" in 1933. Although the store was located there in 1930, we must believe that it was yet to become a full-time business and was probably open only on week-ends, perhaps full-time during the Tourist Season. Stan frequently mentioned in interviews that Bumpus rose quartz had made a difference for him in his early days at the Trap Corner store:

"One memorable day, Mrs. Edward Bok, [Louise Curtis Bok, widow] of the famous editor of the Ladies Home Journal, saw some of the rose quartz and ordered $100 worth made into paperweights. That was the event that marked the tide's turn." (*Country Correspondence,* Haydn S. Pearson, nationally syndicated column, September 2, 1960).

When Harold went through bankruptcy in 1932, Stan couldn't have continued mining at the Hibbs Quarry any longer and naturally would have opted to expand his store hours. Stan's son, Frank, was born in 1934. Tragically, Gwendolyn died on her and Stan's wedding anniversary in 1935, while she was at her parents' home in Waterbury, Connecticut. Nearly a year later, Stan married Hazel Louise Scribner [March 23, 1912 Melrose, MA – December 14, 1991 West Paris] on November 28, 1935. His new wife, Hazel, helped as shopkeeper while Stan was out mining.

The Maine Mineral store's expansion in hours was necessary to survive and the profusion of rose quartz from the Bumpus Quarry became an extremely important source of income for Stan's mineral and gem business. Also about this time, Stan was interested in making his own gemstones

(*Lewiston Journal,* November 19, 1960): "To learn gem cutting, I had contacted Robert Bickford late of Norway. He had very kindly let me use his [gem cutting] shop, and he and Mr. [Elbridge] Woodworth [of West Paris] had been kind enough to give me instruction." Although Stan made very little jewelry himself, the instruction gave him the skills to appraise the value of gem rough for that purpose. Elbridge Woodworth [~1897-February 28, 1965 Portland] was a cousin of 1930s Maine State Geologist, Lucius Merrill.

In the mid 1920s, Raymond A. "Deanie" Dean [August 8, 1909 - March 5, 1963] became acquainted with Stan Perham and by the time Deanie graduated from West Paris High School in 1927, Stan had already kindled Deanie's interest in minerals and seeing the giant beryls from the Bumpus Quarry excited Deanie's interest in cutting and polishing gems (*Lewiston Journal,* February 18, 1956). Deanie would go mineral collecting occasionally with Stan and by the mid-Depression, when there was no employment available, Deanie would visit Stan at the mineral store for extended parts of the day. Frequently, Stan and Deanie would go collecting minerals for sale in the store. The Depression also meant that the tourist business was disastrously affected and Stan and Dean tried to think of ways to increase the Maine Mineral Store's sales. One of their resources was free time and during the Depression Stan and Deanie began making gems, jewelry, mineral collections, and souvenirs from local minerals to supplement the inventory of purely museum and collector's specimens.

Large Bumpus Quarry beryl crystal section (lower center) in front of Maine Mineral Store.
(Note similarity to Field Museum's beryl.)
Photo by Ben Shaub.

By about 1935, Deanie became an employee of the Store learning to wire wrap jewelry, to cut and polish minerals, etc. (Raymond Dean, taped interview, 1958). Norway naturalist and gem cutter, Bert Hamilton, would come into Stan's store to socialize and would watch Deanie cut gems, and would occasionally offer suggestions, but Bert was known to be temperamental. One time Bert offered a suggestion which Deanie, passively, did not adopt and Bert didn't return to the store for a long time. Stan told Deanie that the very next time Bert came in, that Deanie was to "drop

The Rose Quartz masterpiece in diffuse illumination.
Stones with asterism show the effect
only from point source illumination.
Perham's of West Paris collection.

everything" and return to the cutting bench and do exactly as Bert had previously suggested. Deanie did as Stan asked and Bert looked on without comment, but resumed as a frequent visitor after that and resumed offering help. [46]

During these hard times, Deanie was paid $5 a week, but he said that it was better than "going on the town" [welfare]. Deanie recalled that he had been approached by Dick Nevel in 1938 to work for him at "The Stone House", also on Route 26, just north of the intersection with the north end of the road to Paris Hill. Nevel had hired Priscilla Stearns to work in his mineral store summers while she was going to high school beginning in 1936 until 1938. Priscilla lived in the house formerly owned by miner, gem cutter, and mineral dealer, Luther Kimball "Kim" Stone, Loren Merrill's partner at Mount Mica. (That house, now known as the Dew Drop Inn, at the north end of the Paris Hill Road was next door to the Stone House and Nevel's store was probably named for "Kim".) When Priscilla finished work and went to Bliss College, Nevel looked for new help and approached Deanie with an offer. Deanie told Stan of the offer and Stan matched it and Deanie remained Stan's employee. Dick Nevel was killed in an accident in September 1938 and Priscilla had her own mineral business the next summer (1939) at her family home as she was offered consignment minerals, on the same basis that Dick Nevel had been offered minerals, by the Schortmann brothers Ray and Alvin, of Northampton, Massachusetts. (The well-known Schortmann Mineral Company had only just begun and needed outlets for its minerals be-

fore eventually becoming an important mail order company.) Priscilla married Stearns J. Bryant [March 30, 1918 – November 20, 1973] of Buckfield on May 5, 1942 and they eventually founded the Winthrop Mineral Store, after a short career in Norway selling insurance. The Winthrop Mineral Store is very active at this writing and is still operated by Priscilla and her husband, Levi Chavarie.

During this time, one of Stan Perham's most important gem materials was rose quartz, no matter from what location, although almost all of it came from the Bumpus Quarry. One of Deanie's early masterpieces was a 106 carat rose quartz cabochon from Scribner Ledge Quarry, Albany: "I worked off and on some two years on that stone," he said with a chuckle. 'In fact, I finished it once and still didn't think it quite suited me, so I took off a bit more and finally achieved this.' As he displayed the giant stone, one couldn't help but be impressed by its beauty. It literally seemed to glow with color, and in the center could be seen distinctly a 12-rayed star instead of the usual six-rayed star." (King, unpublished; and audio taped interview, Deanie with Ben Shaub, 1958). Deanie eventually taught more than a hundred Maine gem cutters the art of faceting and/or making cabochons.

While the best money was obtained from rose quartz that was faceted or fashioned into jewelry, as well as objects

Rose Quartz, Bumpus Quarry, about 35 x 25 cm.
Perham's of West Paris collection.

d'art, there was use for every piece of rose quartz. Deanie remembered how he used miscellaneous mineral pieces relating a story about local naturalist, George Howe of Norway (taped interview with Ben Shaub, 1958):

"Howe gave away a lot of minerals. He was a friend of all the youngsters. Back in those days, I remember, when I was wrapping wire around the stones for Stan, sometimes he'd come in with 30 or 40 camp girls with him. He'd been out on an excursion [mineral collecting]. He always would come up there and they'd have their little piece of

Priscilla Stearns, ~1938.
Courtesy Levi "Sonny" Chavarie.

stone for me to wrap up, that they'd picked up. I've done millions of those things.

In those days [1930s], it'd cost them 10 cents apiece. Then they'd have something [jewelry] that they'd use that they found themselves. George always took a handful of stones, not gems, in his pocket - rough stones that were good for something so that if some girl who was accidentally out of luck, who didn't find anything, she could have something, too. Everybody got one who needed one. That's the kind of man George Howe was."

The Maine Mineral Store also had specimens for sale and, over the years, Stan acquired literally thousands of tons of rose quartz from the Bumpus Quarry and other localities. The store buildings were surrounded by an outside mineral cache from which rose quartz could be selected by the pound by customers. From the store's stockpile a great many rose quartz pendants, and other items of small jewelry were fashioned by Deanie into wire-wrapped baubles. As a gem material, Bumpus rose quartz certainly holds the Maine record for number of jewelry items produced from her gem rough, although tourmaline certainly holds the record for number of Maine faceted gems. Many times, Stan mentioned that when he was 12, his first store in his home had only one faceted tourmaline for sale. He got a piece of rough gemmy material and had it faceted by Loren Merrill. It is certain that Stan's most famous piece of tourmaline, he ever had for sale, was Loren Merrill's 422 gem nodule found at Mount Mica Quarry, Paris c1904-1910. Stan bought some of Merrill's

collection after Merrill's death in 1930, although Merrill's son-in-law, Arthur Vallee, kept much of the mineral collection and gem rough. (The last of the Merrill-Vallee mineral collection was sold to Verlon G. Guyette [August 28, 1932-May 29, 1977] owner of the Rock-N-Stop Mineral Store in Dresden, Maine in the mid-1970s.)

On May 2, 1935, the Merrill gem nodule was displayed to members of the New York City Mineralogical Club at their regular meeting. Later, Stephen Varni, a famous New York City lapidary, gem, mineral, and equipment dealer, advertised an appeal, including direct mailings, to attract a benefactor who would preserve the specimen in its natural state. On March 7, 1936 (*Boston Transcript*), it was reported that the nodule would make a tour of Maine as "A $1,500 Maine tourmaline" and the tour was organized by Freeman Burr, Maine State Geologist. Burr's itinerary may have included the Skowhegan and Bangor State Fairs, and possibly the Eastern States Exposition in Springfield, Massachusetts. Varni still had the tourmaline nodule in early

George Howe discussing nature with Jean Herd and Pat Brewster (son of owner of Birch Rock Camp for Boys, Waterford), ~1929.
Courtesy David Sanderson.

1938 (*Avocations*, February, 1938).

One of the Depression era projects sponsored by the WPA was the Federal Writers' Project. Just as there were manual labor projects for those who could wield a shovel, there were projects for those who had formerly earned a living by the pen. One product was the book, *Maine. A Guide Down East.* This 1937 publication was "commercially" published, but was probably subsidized as well to keep every aspect of the economy afloat. The book told of Maine's history and culture, described her Nature and Industry, and featured proposed routes to travel on a tour or vacation. One suggested stop was "… the *Maine Mineral Store,* a museum of Maine gems, particularly those of the immediate area, and a souvenir salesroom." The age of the tourist had only begun and the guide served to alert the people on tour of the interesting experiences available along the route: "On display are beryl and tourmaline crystals in their original state…" When the *Guide Down East* was revised in 1970, the description was briefer, but still noted of the Perham store: "… a museum of Maine gems on display and for

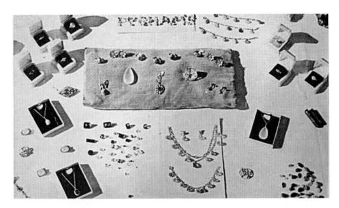

Perham's Maine Mineral jewelry selection featuring Bumpus Rose Quartz, 1940s postcard.

sale…"

What was Happening at the mill?

The Depression era crude and ground feldspar market is summarized in Appendix B. That section is a succession of reports relating to the prices of crude and ground feldspar, as well as tons produced, comparisons with other states, etc. and is suitable reading for those interested in "following the money". The numerical summary is important as it suggests motivations on the part of the various mill owners. Nonetheless, the mid-Depression section is interesting and that section is reproduced here.

In 1932 and 1933, demand for ground feldspar continued to diminish in the products used in the housing industry, although the manufacture of china tableware was not nearly as seriously affected (Rogers and Galiher, 1934). Feldspar mining in New England was severely affected with a 75% decrease in production, but Maine was still third in the nation during the year with an output of only 8345 tons of crude feldspar worth $5.02/ton. Ground feldspar continued to decline to $13.18. Part of the decline in value was because of competition, increased efficiency, and lower prices, from new grinding mills constructed in North Carolina and Virginia." (Rogers and Metcalf (1934) stated that ground feldspar from Maine averaged $16.79/ton in 1932, but the calculations used here have been based on total reported tons of ground feldspar produced and total dollar value reported.)

The passing of the Cullen Act of April 7, 1933, permitting the sale of beer, and, eventually, the repeal of the 18th Amendment to the Constitution on December 5, 1933, were very important to the feldspar industry and brought some minor relief to Oxford County. Rogers and Metcalf (1934) reported on the situation:

> "Overshadowing all other factors influencing the feldspar industry in 1933 was the legalization of beer and the consequent rise in demand for bottles, glass-lined tanks, enameled vessels, and allied specialties. The total output of crude feldspar, including both potash and lime-soda spars, was 150,633 tons in 1933, an increase of 43.9 percent over 1932…"

SDFWP (1938) noted feldspar processing activity in Keystone, South Dakota: "A FELDSPAR MILL (L) is at the fork of two roads, its large building covered with a coat of white dust. The plant has been operating day and night since repeal…" Maine's increase in 1933 was only 35%, below the national average, but it must have been welcome. Maine crude feldspar production was fourth in the nation, with 11,273 tons with a low value of $4.29/ton, as Virginia catapulted into second position between leading North Carolina and third place New Hampshire. Despite the desirability of Maine's feldspar, it was still more expensive than the "Southern" producers, and wonderfully-firing ground feldspar was expensive to use in bottles. Ground feldspar values were not specifically reported for all states for 1933.

The next year, 1934, Maine feldspar production rebounded to 14,685 tons worth $5.64/ton, a price that hadn't been seen for five years. Due to lowered production in Virginia and New Hampshire, Maine reoccupied second place behind megaproducer North Carolina. Maine's ground

Perham's Maine Mineral paperweight featuring Bumpus Rose Quartz and other Maine minerals, 1940s postcard.

feldspar brought $15.22/ton (Metcalf, 1935).

In 1935, there were big increases in the amount of housing construction and remodeling and there was a consequent large rise in demand for porcelain plumbing fixtures, etc. Maine feldspar mining rose to 17,103 tons worth $5.83/ton, but the increased production in Colorado and South Dakota dropped Maine to fourth. Maine ground feldspar was $15.01/ton (Metcalf, 1936).

In 1936, glass production, particularly for bottles, consumed half of the nation's feldspar production, whereas it was about a third several years before. The repeal of Prohibition brought prosperity back to feldspar mining, but Maine production slipped slightly to 16,392 tons, just good for fifth place, but given the reported weakness in the Sagadahoc County feldspar production, the West Paris mill, and therefore Oxford County, may have been "up" for 1936 (Metcalf, 1936): "The Trenton Flint & Spar Co., Trenton, N.J., with grinding plant at Cathance, Maine, is reported to be out of business." and "The Ceramic Feldspar Co., Bath, Maine took over the plant formerly known as the Cummings Feldspar Co., and started grinding operations early in 1936." From the notice, we might gather that the Topsham/Bath District suffered a collapse in feldspar grinding facilities and that there must have been a consequent collapse in Sagadahoc County mining as crude-feldspar purchasers disappeared. The demise of the Cathance River mill was probably the reason the Cummings mill was re-activated. It must be also imagined that the West Paris mill represented most of the State's ground feldspar production as the formerly major millers in Sagadahoc County were no longer among the strong players, although the Consolidated Feldspar Company's mill was still active in Topsham. Feldspar mills were built in the USA during 1936, but not in Maine. The Consolidated Feldspar Company had opened a new mill in Erwin, Tennessee and there were new mills, one each, in Colorado, New Hampshire and South Dakota and a mill was revitalized in New York. Sagadahoc County's fate may have

Deanie's home just south of Trap Corner on Route 26. 2007.

rested with a change of philosophy with the Consolidated Feldspar Company and its shifting focus in operations. Metcalf (1937) wrote: "Normally, the tonnage of ground feldspar produced from domestic crude is about 87 percent of the crude-spar output, the remaining 13 percent representing spar sold for purposes that do not require fine grinding and that lost or discarded during the grinding process." Maine ground feldspar sold for $14.65/ton. With the easing of the effects of the Great Depression, the mill made an improvement:

"… few changes were made until the year 1938 when a Granular Unit was installed. This is a magnetic separator used for the purpose of removing iron-bearing minerals from the feldspar, thus enabling the company to use a lower grade spar than formerly. Previous to its installation, all lower grade feldspar had to be discarded and was considered a waste product. This waste, called 'tailings'. Was removed by the magnetic separator and sold to department stores for use on floors to prevent slipping.

About this time the company began to produce a 'glass spar', a substance used in glass. In order to get the desired alumina content it was necessary to import nepheline syenite from [Bancroft District, Ontario] Canada to blend with the feldspar. However, this proved to be too costly an operation to be profitable and it was shortly abolished."

Working at the Bumpus Quarry – 1937

The Bumpus Quarry continued to be mined during the 1930s, although there were many idle periods. Earlon "Bud" Paine of Albany remembered working at age 15 for Harlan at the Bumpus Quarry during the summer of 1937:

"I was born on the Valley Road, July 28, 1922, not far from the Bumpus Quarry. My father Alton "Joe" Paine, who was a WWI veteran, was working for Harlan for at least a year previously, about 1936. There may have been five or six men working at the quarry and my job was pick and shovel work. I'd push my loaded wheelbarrow to the truck and then shovel in the feldspar. The truck was not very big and only held about a ton or ton and a half. There wasn't much mention of other minerals when I worked there. I got 35 cents an hour and did not work very long and that was true of many men in the crew. They were always trying to find better wages. Harlan treated me very well and I liked him. Harlan worked the equipment and was the dynamiter. However, Harlan was not always present while I was there and he would frequently work logging for my grandfather, Fred Littlefield. Harlan

Deanie at his workbench wirewrapping mineral jewelry. 1950s postcard.

had a small tractor he could use to haul logs out of the woods."

Matthews (1945) summarized that Maine beryl production was 25 tons in 1938, was 45 tons in 1942, and was only 2 tons in 1943. In 1937, 1939-1941, beryl production was grouped in the "other" category, in order to prevent releasing proprietary information, suggesting that there was only one producing locality in Maine during those years, but it could also mean that all beryl was purchased from one Maine distributor. These figures may be mostly attributed to the Bumpus Quarry, but may include production from Black Mountain, Rumford and other places.

With the passage of time, the first episode of the huge beryl discoveries have become clumped in people's memo-

Bottle from the Reymann bottling company. Wheeling, West Virginia, a city where Maine ground feldspar, particularly from the Consolidated Feldspar Company, was sold by the Franzheim Company to various bottle manufacturers.

of the mine is solid rose quartz [3] while in other parts perfect hexagonal cross-sections of green beryl are seen in white feldspar." The next year, apparently conditions changed and it was noted (*Boston Transcript,* September 12, 1940): "A few weeks ago I had the pleasure of meeting an enthusiastic group of mineralogists, all members of the Boston Mineral Club, returning from a 'treasure hunt' through the Bumpus mine in Albany. They had been touring the [Maine pegmatite] belt and were taking home extremely interesting specimens, including beryl, aquamarine and rose quartz crystals (sic) of good size from which they expected to get some rather fine cut gems." The columnist, Eleanor Stone, certainly meant "beryl and aquamarine crystals" and "massive rose quartz" and the subtle distinction of "crystallized" *versus* "massive" minerals seemed to have unrealized.

Maine Beer Revenue stamp, 1935, to be placed on brewing kegs, proof of paying the "Sin" tax.

ries along with subsequent events and many people have supposed that all of the Bumpus beryls were found only in 1928-1929 with a twenty year gap in time before new significant discoveries were made. The *Lewiston Evening Journal* (November 16, 1937) announced a new giant beryl find at the Bumpus Quarry complete with predictions of 200-500 tons of beryl probable, but the wording and the familiar vintage photograph of the former Giant Crystals suggests that it was merely a reprint of a very similar article of ten years before. However, Morrill (1939), then a recent mineral collector, seems to have known of the newspaper article and noted: "A new crystal was exposed about two years ago which is probably the largest in existence." [41] While it may be possible to dismiss the accuracy or dating of the newspaper article, something seems to have been found at the Bumpus Quarry. Morrill further noted that by 1938-1939: "Visitors, however, are not permitted to carry off samples from this locality." and "The colors are amazing as one end

Cameron et al. (1954) wrote that, during the 1930s, the Bumpus Quarry was "… worked … intermittently until 1940." Mineral collector Rudolf Bartsch had a number of articles in *Rocks and* Minerals magazine devoted to "New England Notes" and, in June, 1940, he warned collectors:

"The Bumpus mine or quarry is an exceedingly interesting place to visit. But do not expect to do any collecting. The locality is noted for its fine colored rose quartz and enormous beryls. If by any chance you should drop your 'uppers' [false teeth] with its glistening row of parallel 'crystals,' it is doubtful if you could get them back without first laying a five

Bumpus Quarry during idle period, 1940s. Courtesy Ava Bumpus.

spot on the old apple stump. And further, if you decide to purchase a rose quartz specimen, have them show you dry material from the house. Many of the specimens displayed out doors will be quite disappointing when you get them home and dried."

Collectors were not used to having to pay an entrance fee to go mineral collecting and, much more, were unused to visiting a locality and not having an opportunity to collect, even for a fee. The reputation of rose quartz fading was inflated because the specimens left out in the open had a minute film of water in fractures that increased the amount of light that could penetrate a specimen, thus revealing more color. As specimens dried inside a building, the specimens would show less intense color as light penetration diminished.

Governmental Interest in the Bumpus Quarry

An unidentified newspaper clipping from 1939-1940 based on references to Ralph Owen Brewster, Maine member of U. S. House of Representatives, who was on a committee on monopolies and beryl, reported: "At the present time there are 20 or more beryl crystals showing above the water in this [Bumpus] quarry, and there is one below the water that has two sides and an angle showing which, if it runs true to form should be six feet in diameter and many feet in length." The report suggests that the Bumpus Quarry was idle. The underwater beryl may have been the giant beryl reported in 1937. The *Lewiston Journal,* July 11, 1940 reported that Brewster had introduced a bill: "asking the government to make a survey to produce and store beryllium for war purposes". The article featured the Bumpus Quarry, with several factual errors, and mentioned other Oxford County beryl localities. A bill was passed on June 7, 1939 authorizing the stockpiling of strategic minerals and Brewster's bill made sure that beryl was considered to be strategic. The day before the Japanese attack on Pearl Harbor, Hazel Perham, Stan's wife, published an excellent historical summary of the discovery and uses of beryllium (December 6, 1941, *Lewiston Journal*).

The National Forest and the Demise of Albany as a Town

The year 1937 was an ominous one for Albany as it had had a town charter since June 20, 1803, soon after settlers arrived about 1800 and two years before Oxford County was formed. [13] There were 853 residents in 1860, but after that time, there was a continual decline in population. Folklore held that the loss in population during the Depression was the "final straw" and why the town lost its charter, but the reality was somewhat different. To be sure, there had been three Oxford County towns which had lost their town charters about the same time Albany did: Grafton in 1919, Mason in 1935, and Milton Plantation in 1944.

Some of these town charter losses were to be ex-

Rose Quartz, Bumpus Quarry. Addison Saunders Collection.

pected. In 1910, Grafton had only 64 residents and could not afford to support a town government. When Grafton was incorporated on March 19, 1852, with 108 residents, there may have been the expectation that small towns would naturally grow as agricultural and forestry businesses increased. Mason, incorporated February 3, 1843, was similarly small with approximately 61 residents when it lost its charter, down from its pre-Civil War high of 136 residents in 1860. By comparison, the other entity which lost its charter, Milton Plantation, organized in 1842, had 127 residents, down from its high of 271 residents by the 1860 Census.

The population ranking of towns and townships of the County were, according to the 1940 Census: Rumford (10,230), Mexico (4,431), Paris (4,094), Norway (3,649), Bethel (2,034), Dixfield (1,790), Fryeburg (1,726), Oxford

Bates Railroad station, West Paris looking northeast, left is Baptist Church steeple; from a postcard. Knoll in background is Berry Ledge. Feldspar mill was "left" along the railroad tracks. From 1930s postcard.

(1,316), Porter (892), Peru (965), Woodstock (913), Buckfield (903), Waterford (836), Hiram (787), Brownfield (741), Andover (757), Canton (706), Hebron (678), Lovell (647), Greenwood (564), Sumner (541), Denmark (532), Hartford (430), Roxbury (346), Stoneham (238), Sweden (225), Hanover (178), Upton (174), Newry (167), Gilead (160), Stow (153), and Byron (125). The unorganized townships in the County included Albany (288), Milton Plantation (127), Lincoln (89), Magalloway Plantation (84), Mason Plantation (43), Township 4 Range 2 (40), Grafton (26), Richardsontown (6), Lynchtown (4), C (3), Batchelders Grant (0). Scoville (1937) noted that the 1930s were not kind to Maine's towns, a total of 11 towns and three plantations were "disorganized". Almost all of the towns affected were demoted because of insupportable populations and hence inability to maintain roads or financially support a town

Rose Quartz, 25 x 20 cm, Bumpus Quarry. Addison Saunders collection.

government. In many small towns, public officials served with little or no salary. (It was this lack of any salary that nominations for candidates for town offices made orally during annual Spring Town Meetings would quickly be followed by oral responses: "I decline." Residents who showed up for Town Meeting late might find that they had been voted in as a new selectman, etc.) Interestingly, in 2007, Maine had an entity with town government, Frye Island, which had zero year-round residents in the 2000 Census. In the same year, Glenwood Plantation also had zero permanent residents. Plantations had a limited self-government while unorganized townships fell entirely under State supervision and regulations. Centerville and Beddington towns had populations of 26 and 29, respectively, by the 2000 Census, but Centerville surrendered its town charter in 2004. (http://en.citizendium.org/wiki/Maine#Organized_municipalities; accessed 2007).

Albany was the largest of the towns to loose their charter in the time period. In 1860, there were 853 residents, but in 1920 there were 360 residents with only 309 residents ten years later by 1930. After the 1940 Census, the number had dropped to 288. However, it wasn't size that mattered, because there were eight smaller towns than Albany in the County which survive with town governments to this day.

In an effort to avoid default, which already may have been coming, Albany ceased maintaining low traffic roads, but that measure was not enough as there were few services that could be cut by the small town. The survival of any town was based on its ability to assess and collect taxes. In 1935,

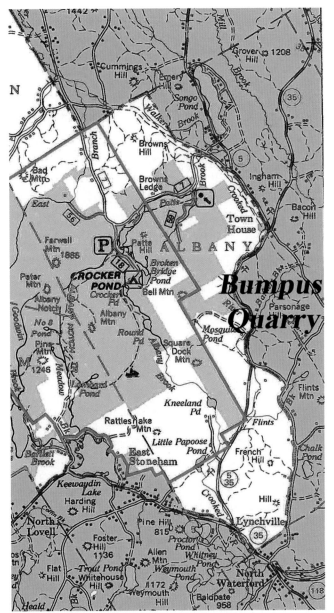

White Mountain National Forest boundary in red. Dashed line in center to upper left is the Albany/Stoneham line.

nearly 11%. Paris increased only slightly to 0.052 as did Rumford at 0.0525. Canton's rate increased to 0.086, while Upton's rate shot up 44% to 0.072. Greenwood's rate mercifully declined 13% to a still high rate of 0.075. The shrinkage of the tax rolls certainly meant that Albany could not pay for town expenses and Albany had to surrender its town charter on November 30, 1937 and since that time, as a township, it has been administered through Oxford County and State services. To be sure, a much more thorough examination of economics is necessary than merely examining tables of rates and other summaries, but the numbers all point toward insolvency.

House fires, frequently with subsequent abandonment of property with consequent migration to employment centers certainly contributed to the Albany's decline, but the population decline had been constant, even in the comparatively affluent 1920s. The loss of 21 in population during the Depression could not have been the straw that broke the camel's back. It was more than likely that Government "help" was involved. The formation of the White Mountain National Forest removed 3464.5 acres from the tax rolls of Albany, alone, or just over 10% of the town. Subsequently, more land was acquired by the National Forest and at this writing about 25% of Albany township is in the National Forest. The acquisition was by power of eminent domain for

Harlan's House and shed near Bumpus Quarry, 1940s. Courtesy Ava Bumpus.

the small towns usually had the highest taxes as the property values and, therefore, "wealth" was concentrated in bigger towns. A comparison of the 45th and 46th *Annual [Maine] Reports of the Bureau of Taxation* (1935 and 1936) shows that inland towns were frequently taxed at a higher rate than coastal towns. In Oxford County for 1935, Paris had a rate of 0.05 and Rumford had a rate of 0.0515. Most towns were near these values. It was a bargain to live in Lincoln Plantation with a rate of 0.0245, but there were essentially no services to provide in that settlement. Albany's rate was 0.065, which was 30% higher than Paris's rate. There were a few rates higher in the County that year: Hiram (0.065), Porter (0.066), Sumner (0.068), Canton (0.072), and Greenwood (0.086). As the Federal government began buying up land in 1936, Albany's efforts to maintain an income from taxes was redoubled. It's 1936 tax rate increased to 0.072, up

the stated purpose of "watershed control", although the tiny Crooked River was the only important waterway involved in the town's "condemned" land. Mason lost more than half of its area to the National Forest.

The State of Maine approved the proposed acquisition of western Oxford County Land for the purposes of establishing a National Forest on April 6, 1935. Subsequently, the National Forest had to search for the legal owners of all of the tracts of land to be affected and to determine if there were any potential claimants not known. Many tracts had not been registered under the names of new heirs and there were often family members in possession of lands that solely paid taxes on them for which there may have been other heirs, unaware of their inheritance. There were also adjacent land owners who had to be notified to permit them to verify that none of their land would be lost due to errors of land boundary claims, etc. The petition to the U.S. District Court, Southern Division was made at Portland on December 3, 1936 with notices to be published in the *Norway*

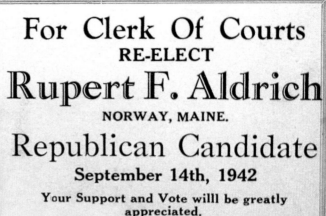

For Clerk Of Courts
RE-ELECT
Rupert F. Aldrich
NORWAY, MAINE.
Republican Candidate
September 14th, 1942
Your Support and Vote willl be greatly appreciated.

Advertiser Democrat of a hearing in the same court to be held February 16, 1937. The notice included as complete a list of potential claimants as could be discovered.

There is anecdotal evidence that the land acquisition process was not entirely hostile. The amethyst-bearing prospects on the Eastman Farm on Deer Hill in Stow were apparently excluded from the proposed National Forest by mutual consent of land owner and National Forest agents, although the main portion of Deer Hill, including the original 1880 Andrews amethyst location, was acquired. Chester C. Eastman of Fryeburg had sold 551.2 acres to the National Forest in 1937.

By February 4, 1939, Harlan Bumpus mortgaged his home for $83.42: including a 1927 Model T Ford, three rifles, a pair of snowshoes, a pair of skis, and a Silvertone battery radio all being kept "upon the premises leased by Harry E. Bumpus and Laura J. Bumpus and located in said Albany. [325] (From the mention of the battery radio, we might conclude that the Valley Road in Albany, had yet to receive electrical service from the Central Maine Power Company.)

Harry E. Bumpus
unidentified newspaper
c1930

Herbert Haven, a noted candy manufacturer from Portland, Maine visited Kimball Ledge (Songo Pond Quarry) and the Bumpus Quarry and wrote in his diary on October 29, 1939:

"Maine Mineralogical and Geological Society of Portland] trip to Albany, Me. We first went to Kimball's Ledge which is just east of the south end of Songo Pond which is the headwaters of Crooked River. Several specimens of a remarkably deep blue beryl and aquamarine were found. ...

We next went to Bumpus Quarry, Albany. Here we saw tons of rose quartz that had been mined and was piled up. Mr. Bumpus showed us a perfectly terminated crystal of golden beryl and some dendritic black tourmaline. I bought a large piece of rose quartz for $2 and Jessie Beach bought a large block of rose quartz for her father's grave for $35. Mr. Bumpus also had for sale pieces of fractured blue beryl, bottles filled with small pieces of rose quartz, golden beryl, blue beryl, green beryl. He also had some sheets of zoned mica for sale at 50¢ each. He had a large crystal of colorless beryl in the house that weighed several hundred pounds. In the ledge, we saw a small face of rose quartz with green beryl crystals. We ate lunch here. Mr. Bumpus permitted us to pick specimens only from a very large, poor, weathered dump." [170]

United Feldspar and Minerals Corporation
versus
Oxford Mining and Milling Corporation

On February 21, 1938, Francis E. and Winifred L. Haag of F. E. Haag & Company bought the Oxford Mining and Milling Company by obtaining a $40,000 mortgage from UFMC. The Oxford Mining and Milling Corporation had survived as an entity as it was being purchased partially by using the profits from the sale of its output. The Depression had intervened. Based on the behavior of the parent company, one recognizes the potential for desperate measures on UFC's part. The slow return of "prosperity" brought with it new fiscal attitudes. As the financial climate improved, there was an opportunity to form an even larger company and sometime in 1937, perhaps earlier, there began preparations for a new company to evolve and that new company was to be the similarly named, United Feldspar and Minerals Corporation, although under new leadership, Francis E. Haag. On January 9, 1940, United Feldspar and Minerals Company officially owned the Oxford Mining and Milling Company (OCRD v. 423 p. 418). OMMC also owned part interest in the Cummings Farm and its mineral rights, as well as all mineral rights to the Perham Quarry. UFMC also owned other Oxford County properties and

Jessie Beech with her mineral collection in Portland, c1960. Photo by Ben Shaub.

rights acquired during the Depression. Eventually, on June 28, 1940, Judge Guy H. Sturgis of the Maine Supreme Court issued a final decree in the case of Francis E. Haag vs. Oxford Mining and Milling Company dissolving the latter corporation. [361] The *Maine Register* continued to list OMMC for several more years.

One of the contemporary accounts requires comment. Neumann (1952) indicated a hiatus in mining at the Bumpus Quarry: "In 1936, upon the death of some members of the Cummings family, the United Feldspar & Minerals Corp.

obtained title to the Cummings farm at a court sale. The title was cloudy, and much litigation followed." Actually, United Feldspar and Minerals Company did not purchase the Bumpus Quarry until 1943 and there weren't any family deaths in the 1930s affecting new inheritance, etc. There seems to have been breaks in mining at the Bumpus Quarry during the 1930s. The mining breaks may have occurred in 1932, when Harold Perham went bankrupt, or as late as 1934, when Harlan Bumpus said his seventh year of mining at his parents' quarry was completed. The *Maine Register* indicated that Harlan was mining in 1933 and in 1934, and subsequently, Oxford Mining and Milling Company was responsible for mining, although it has been noted that the *Maine Register* was frequently slow on its updates. There seems to have been a cessation, perhaps beginning in 1934 and continuing through the next year, but there was a resumption of mining at the Bumpus Quarry in 1936. There are insufficient records for the time period, however. Despite the economics of the Depression, the West Paris mill still needed some feldspar and was actively seeking sources of it. On September 18, 1935, OMMC leased the Fred E. Scribner property in Albany in order to explore for minerals. [27]

The UFMC attempted to foreclose on Alfred C. Perham's and Harold Perham's various land mortgages, on August 16, 1937. [323] The mortgages had been made with the Casco Mercantile Trust Company and the mortgages were purchased by UFMC. It is probable that the Company had learned of the pending foreclosures by the Casco Mercantile Trust Company and may have purchased the mortgages to expand their control over the land, thus nullifying any royalty payments, etc. The foreclosure on the Alfred Perham lands had a one year expiration date of September 4, 1938 and Alfred obtained a temporary injunction of the foreclosure on September 3, 1938. [324] The Oxford County Clerk of Courts who attested the registration was Rupert F. Aldrich. There were similar temporary injunctions of foreclosure granted to Guy L. Ward and Nicolaus Harithas against United Feldspar registered on September 7, 1938 suggesting that United Feldspar was engaging in very aggressive business.

Search for Strategic Minerals in World War II

In WWII, the United States government made an analysis of essential, "strategic" ores and minerals, including beryl, needed for the war effort. Maine State Geologist, Freeman Burr, had already published a sketchy list of beryl occurrences in January, 1938. Of the Bumpus Quarry, he merely added: "More large crystals recently uncovered." Burr's Maine beryl inventory was used verbatim in the *Bangor Daily News,* May 13, 1939. The most important publications about pegmatites to come out of the Federal war explorations programs were **Internal Structure of Granitic Pegmatites** by Cameron et al. (1949) and the regional **Pegmatite Investigations** series. For Maine, the pertinent publication was: **Pegmatite Investigations: New England**

1942-1945 by Cameron et al. (1954). These publications made available the knowledge gained by the various U. S. Geological Survey teams throughout the war. The books explained that granite pegmatites, important sources of beryl, mica, and feldspar, had definite zonal patterns within a deposit of these important minerals. These publications were widely read by miners as well as industry scientists and the new information resulted in greatly improved exploration almost exclusively imported. In order to increase the domestic market supply of certain strategic minerals, the Federal government organized several companies whose job it was to stimulate production. [20] Cameron et al. (1954) wrote: "Therefore, in 1942 the War Production Board, through the Metals Reserve Company and the Colonial Mica Corporation, sponsored a program for intensive exploitation of domestic deposits." One of the methods of choice was to pay

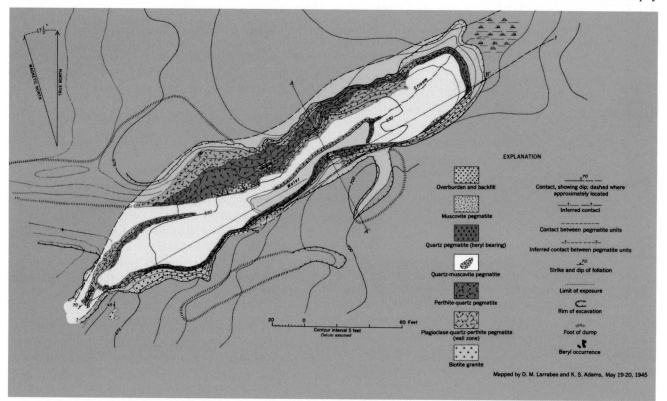

EXPLANATION

Map view, looking from overhead, of the Bumpus Quarry as of 1945. The thickness of the pegmatite was about 45 feet wide while the excavation was about 300 feet long. The map showed that most of the pegmatite had been removed along the contact and that most of the remaining rock was either deeper than the back fill of sandy rock debris in the quarry or extended beyond the limits of pit area along the strike of the exposed pegmatite.

methods and management of individual mineral deposits and quarries. Previously, miners had sensed that there were mineral distributions that could be used to predict success of a deposit, but there were also quarries which continued to be exploited in the vain hope that somehow the richness of a deposit might improve if one only dug deeper. Unfortunately for some miners, the reports were issued years after their struggle although some miners were given advice by the visiting geologists to help them be more successful. On the other hand, Government subsidies encouraged mineral production and helped make marginal mining efforts more worthwhile.

The New Products Committee, during WWII, was the Federal agency charged with stimulating U. S. industrial sources of supply of minerals that normally had been imported (*Portsmouth Herald,* New Hampshire, April 24, 1942). Some minerals were plentiful and could be acquired from domestic mines, while some strategic materials were

a subsidized price or premium for an item, which had formerly been obtained as a relatively inexpensive import: foreign sources frequently being richer, easier to mine, and/or having cheaper labor costs. Among the minerals sought of interest to Maine miners, mica, beryl, and columbite were bought at a high enough premium to entice the miners to explore and operate otherwise uneconomical sources of these minerals or to recover by-product minerals formerly ignored. (Most Maine pegmatites had been dedicated to recovering one mineral, feldspar, despite the presence of a variety of minerals.) Mica mining in the U.S.A. had been relatively stagnant, experiencing only sporadic growth in the early twentieth century, North Carolina being the exception. When the Federal government announced their premium payments for domestic mica, Oxford County miners responded with enthusiasm: "After WWII, the price of domestically mined mica dropped to less than twenty-five percent of their wartime highs." and:

"Beryl is commonly recovered as a byproduct of operations for feldspar, lithium minerals, and mica. Feldspar was somewhat of a strategic mineral, although in abundant supply, as land mines were manufactured with ceramic casings, rendering them invisible to metal detectors.

Under the low prices prevailing in former years ($35 per ton), only large beryl crystals were recovered, but under wartime prices ($90 to $180 per ton), small crystals have been recovered and sold."

Despite the subsidies, Trefethen (1945) noted: "It is unfortunately true that in spite of various forms of government assistance and high wartime prices (currently $8.00 per pound for sheet mica 1 ½" x 2" full trimmed, as against $1.00 per pound prewar) most New England operators have not been able to do more than break even." Nonetheless, the Great Depression had made Oxford County miners, as well as those elsewhere in New England, appreciate the joys of "breaking even".

Despite government efforts aimed at procuring strategic minerals, the needs of the West Paris Mill continued. Some of the new miners sold feldspar to the Mill in order to help meet expenses while hoping to make a bonanza in beryl or mica. The Perham Quarry, to be sure, was the Mill's mainstay, as were its other leased properties. Nonetheless, the Mill continued its search for new feldspar producers. For example, on July 23, 1941, United Feldspar and Minerals Co. leased the former Ralph Herrick farm on Patch Mountain in Greenwood from Edward W. Oman and Albert J. Stearns for the purposes of prospecting for minerals (OCRD v375 p432).

Wartime Search for Albany and Oxford County Minerals

Because of the specialized uses for beryllium, includ-ing the newly invented nuclear weapons, beryl had become a priority for mining. The very first place the U.S. Geological Survey visited, in New England, was the Bumpus Quarry. Lincoln R. Page and John B. Hanley visited the Bumpus Quarry in May, 1942 and made a geological investigation. Later, David M. Larrabee and Irving S. Fisher returned in August and mapped their findings. Larrabee and Karl S. Adams remapped the quarry in May 1945 after they learned that the quarry was drained and new work had been started. The resulting geological map contained a lot of information. The obvious feature was the size and shape of the excavation. The original map (Cameron, et al., 1954) was black and white and symbols were used to show the distribution of mineral exposed in the quarry. On casual inspection, the walls of the quarry, even on a freshly exposed surface, were composed of somewhat uniformly distributed mineral patches. Many rock specimens were taken from the Bumpus Quarry back to their laboratories to examine under the microscope so that subtle mineral associations could be verified. The map, especially when color coded, might suggest that it was easy to spot changes in the distribution of minerals, but the reality is that the map is used merely to illustrate the subtle differences so that there might be an understanding of the structure of the pattern. For example, the most interesting mineral assemblage, the places where beryl occurred, is shown as purple on the current map. Cameron, et al. (1954) wrote:

"Beryl occurs in the rose quartz of some parts of the southeast face in series of three or more over-lapping prisms; the crystals range from 2 to more than 10 feet long and from 6 to 48 inches in diameter. The longest crystal is exposed for ten feet and ranges from 8 inches to 2 feet in diameter. Thirty-four crystals have a total exposed surface of 80 square feet, or approximately 7 tons per foot of depth."

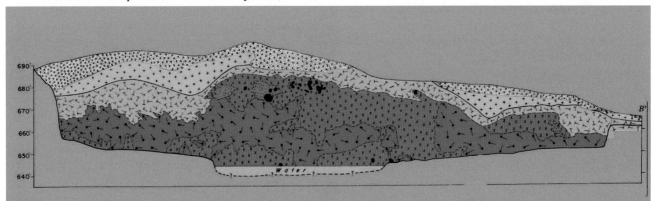

Wall view (southern), looking from within the Bumpus Quarry as of 1945. The depth of the pegmatite was over 45 feet while the deeper portions of the excavation were back-filled with rock debris, perhaps to an additional 10 feet. The map of the exposed rock on the wall might excite beryl miners. The area, of beryl-bearing rock (shown in purple with the black patches representing exposed beryl crystals to scale), was large, but the map view showed that the thickness of the mineralized rock was thin, probably five feet or so. The bright green areas were very rich in mica, while the tan and red zones, as well as the purple, had value for feldspar. The orange zone was soda feldspar pegmatite. The yellow area was the host rock, while the tan area on top was dirt. Redrawn from Cameron et al. (1954).

This description is very important and provides some information regarding future motivations regarding the Bumpus Quarry. It also begs for answers to questions. The primary question was, "Why was the Bumpus Quarry really idle?" There was obviously no shortage of beryl to mine and sell, according to surface indications. The only indications as to why the quarry was not being operated had to lie in the family feud. There were no obvious court actions among the family members in the late 1930s and earliest 1940s. More influentially, the U. S. Geological Survey report projected a bonanza of "7 tons [of beryl] per foot of depth". Extrapolations of subsurface reserves, especially in granite pegmatites, are highly speculative, even for the first foot of depth. Nonetheless, such statements frequently influence planners who are directed to find more beryl. The map should have been understood to mean that there would be "7 tons of beryl

Harry and Laura Bumpus residence, Vining Street, Auburn in 2007.

per foot of depth in the areas exposed and mapped, so long as that type of rock continued into the ground." (The 10 foot by 48 inch diameter crystal may have been the mystery beryl reported in 1937 and which was still in place.)

Miller and Wing (1945) visited pegmatites all over southern Maine as part of a strategic (war) minerals search, ostensibly for mica, in 1943 and 1944. Land owners throughout Oxford, Androscoggin, and Sagadahoc Counties were prospecting their land because of premium prices being offered for mica. Mica was particularly good in making capacitors for field radios and the Federal government adopted the philosophy that if there were money to be made in a venture, people would supply what you needed. Unfortunately, not all prospectors met with good news. Miller and Wing (1945) wrote of the Walter Conwell prospect, about 3 km (2 miles) SE of the Bumpus Quarry: "Neither the size nor quality of the mica warrants mining operation." and suggested that the Preston Flint Prospect, actually only two test pits made in 1941, 1.5 km S of Hunts Corner, did "not make mining operations advisable". Another prospect, that of Charles Holt, was a single test pit 6.5 km SE of the Bumpus Quarry, and was said to "merit no consideration", although it is now

shown as a quarry on the current U. S. Geological Survey maps. Scribner Ledge Quarry, and nearby prospects, had been initially worked for feldspar by about 1935 [27] and were being vigorously explored for mica and although the authors had some complementary things to say about the Scribner feldspar deposit, it did "not warrant further operation" for mica. However, they were also discouraging about mining the famous Mount Mica Quarry in Paris Hill. It was recommended that the Donahue Prospect, about 3 km (2 miles) NW of the Bumpus, be re-activated with a small crew.

When the USGS mapped the Bumpus Quarry in May and August, 1942, the quarry was idle and remained idle until 1945, but Miller and Wing (1945) did not report on the Bumpus Quarry. When mining resumed in 1945 at the Bumpus Quarry, the USGS made a new map of the quarry. In 1945, the Bumpus Quarry was 270' long x 40'-60' wide x 20-60' deep and 39,000 tons of rocks were estimated to have been removed.

As a measure of local mining activity, Harold Perham said in a lecture to the Rumford Rotary Club (*Rumford Times,* May 11, 1944):

County Court house and railroad station, South Paris. From a postcard.

"Eighteen mines have been opened in Maine during the past year as foreign sources of the stone [mica] were closed by the war and domestic production received encouragement from the Government. ... Mr Perham said that until recently Maine had had a mica 'rifting' mill, but that this has recently been closed and the nearest 'rifting' mills are now in New Hampshire."

More Feuding and Court Appearances

There was apparently very little court activity concerning the Bumpus Quarry and the Cummings family immediately after the 1931 decree. There was apparently a disposition of the feldspar royalty payments to the heirs. The mill owners, who owned slightly more than half interest in the farm ($^{19}/_{30}$) did not have to pay royalties to themselves. Laura and Harry didn't need to pay themselves for their interest in the farm, but in 1941 there began new court action.

The reason that mining did not continue into 1941 was a joint lawsuit brought by United Feldspar & Minerals

Corporation (UFMC) against Wallace E. Cummings, Adelia A. Cummings, Viola C. Ballard, Laura E. Pinkham, and Laura and Harry E. Bumpus. The timing of their initial notification of the complaint is not known, but the formal complaint was entered into the May 23 session of Oxford County Superior Court. It should be remembered that Adelia Waterhouse and Laura E. Pinkham already had sold their share of the half interest owned by their deceased aunt, Sybil. They apparently still had a share of the other half of the unsold portion of the Cummings Farm. During the Summer (June 27, 1941), Oxford County Clerk of Courts, Rupert Aldrich, forwarded the bill of complaint to Androscoggin County Supreme Court Judge, Harry Manser, as Androscoggin County was the jurisdiction affecting Harry and Laura Bumpus. Interestingly, lawyer George C. Wing, Jr. of Auburn represented Wallace, Viola, and Harry and Laura Bumpus with notice to Aldrich on July 1, 1941. UFMC apparently had two law firms involved in the case: Clifford and Clifford as well as Brann, Isaacson & Lessard, both of Lewiston. The

Harlan (right) and son, Kenney (left), cutting ice in Bumpus Quarry, 1940s. Courtesy Ava Bumpus.

UFMC complaint was slightly amended on July 18.

Harry and Laura replied to the complaint on July 29. The interesting features of the complaint included:

UFMC declared that the joint ownership of the land could not be "occupied in common by the plaintiff and defendants" Harry and Laura denied this was true.

UFMC declared that the Cummings Farm and Bumpus Quarry contained "feldspar, crude spar, quartz and beryl" Harry and Laura replied that they knew "not what the premises contain".

UFMC alleged: "That since the first day of June, A.D. 1927, the said Laura E [sic]. Bumpus and the said Harry E. Bumpus have received all of the income from said property and though requested has neither paid any of said income to the Petitioner or its predecessors in title nor rendered any account thereof to the said plaintiff or its predecessors in title." Harry and Laura, of course, denied this quit damning allegation.

UFMC claimed that its right to collect its "propor-

tionate" royalties derived from the previous purchase of Allen E. Cummings' and William B. Donahue's interest in the Cummings Farm obtained by Oxford Mining and Milling Corporation and all of which rights were purchased by UFMC on January 2, 1940. These rights were denied by Harry and Laura.

UFMC alleged: "That the defendant are operating said mine and will continue to so operate and unless the said defendants are restrained by injunction, destruction of the property of the complainant will take place and irrevocable injury result." Harry and Laura denied this section.

UFMC asked that the property be evaluated, that Harry and Laura Bumpus be "enjoined temporarily and permanently from boring, mining and operating for feldspar and other minerals on said land; that the defendants," Harry E. Bumpus and Laura E. [sic] Bumpus, may be ordered to account for the rents and profits of said premises, received by them and may be ordered to pay the plaintiff the amount found to be due it. It was further asked that the land be sold and the "proceeds thereof be divided among the plaintiff and the defendants".

In their denial of the complaints, Harry and Laura further added an 11th section: "And further answering the alle-

Feldspar Mill and Bates Railroad Station. From 1930s postcard.

gations in the plaintiff's representations and bill, he says that as to the rent reserved in said lease, payment therefore has been made by mutual understanding and agreement between the parties thereto and the plaintiff and its predecessors in title, if any, have by their conduct and representations waived any claim that they might have to said rent reserved and to which they might be entitled and have estopped themselves by their arrangements and agreements to claim any part of said rent and have received therefore the consideration for the same in another form." This section was probably explained in court.

1943 - The Second "Final Decree"

Oxford County Superior Court, appointed Judge Harry M. Shaw, to oversee the decree reported the resolution on October 14, 1942 (OCRD v 444 p339):

"...to sell at public auction at the OX-FORD COUNTY COURT HOUSE at South Paris, Maine, all the right, title and interest which the United Feldspar and Minerals Corporation, Wallace E. Cummings et als. had in the real estate, hereinafter described, ... subject to all unpaid taxes, and further subject to all rights of LAURA J. BUMPUS AND HARRY E. BUMPUS ...[and] in consideration of TWENTY EIGHT HUNDRED DOLLARS paid by Geneva B. Levinton of Forest Hills, County of Queens and State of New York, sell to said Geneva B. Levinton, she being the highest bidder therefore, and do

Dana Douglass and Barbara Anderson. 1939.
Courtesy Barbara Douglass.

hereby give, grant, bargain, sell and convey to her in accordance with the terms of said decree, the following described real estate: a certain lot or parcel of land with the buildings thereon situated in the Town of Albany ... To have and to hold said premises to the said Geneva B. Levinton, her heirs and assigns to their use and behoof *forever* [emphasis added]."

The deed of December 18, 1942 was received by the County Clerk, Rupert Aldrich, February 8, 1943. (Harry and Laura paid $2,361.68 in "overdue" royalties for minerals obtained from the Bumpus Quarry between October 1, 1934 and October 16, 1940. [358]) However, all during those six years, the mill was the only purchaser of the Bumpus Quarry feldspar and would have been aware of what feldspar royalties were due to themselves as $19/30$ (63%) owners of the land. The remaining $11/30$ of the royalties would have been due to Laura and Harry, Laura's siblings and nieces. Sales of beryl and rose quartz were probably also made to other buyers and of those sales, the mill might have had no direct knowledge. The absence of a significant new beryl find probably made the court decision more of a relief to the feuding family than a disappointment.) Court records [358] indicate the amount of

Bumpus Quarry, idle and water-filled, late 1940s.
Courtesy Ava Bumpus.

activity: "During the next six years something over 8,300 long tons of feldspar were recovered, along with 500 tons of quartz or thereabouts, a little less than *2,500 tons of beryl* [emphasis added] and mica scrap in negligible quantity. The machinery was borrowed or rented largely from the plaintiff or its corporate predecessor and the entire output sold to it." At no time, does there seem to be a valid inventory of production of Bumpus Quarry beryl or feldspar and the accounts, claims, and statements vary with whimsical imprecision. However, the 2,500 tons of beryl claimed to have been produced is astonishing. The court summary suggests a large beryl production in 1934-1940. If there were 2,500 tons mined through those efforts, with sales prices near $30-50 per ton, the royalty owed would have been "50% on by-product minerals" with a calculated sale of at least $75,000 and an owed royalty of at least $37,500, minus their estate share! The 2,500 ton figure is most likely a simple typographical error for 250 tons, a figure generally used in later times. The amount of "over due" royalty assessed in the decree might have been a compromise reduced amount or may have indicated that some royalties had been paid and that detailed accounting revealed a discrepancy. The number, otherwise, can not be reconciled against any inventory.

Kenny Bumpus, son of Harlan in idle and water-filled Bumpus Quarry, ~1949.
Courtesy Ava Bumpus.

The court summary of mining activity misreports that there was no mining at the Bumpus Quarry: "As a matter of fact they were not commenced until after a lapse of about seven years." Harlan's autobiographical story does indicate that he worked seven years at the Bumpus Quarry but his having been a miner during most years from 1927 through 1940 may mean that he was full-time during the first seven years and occasionally a miner in the latter years. After 1940, Harlan did not return to mining. He raised animals, worked in the woods, and was a hunter.

The $2,800.00 paid by Geneva B. Levinton was also split among the various claimants. As new owner, Geneva B. Levinton, exercised some clout: "On March 1, 1943 the purchaser at the judicial sale made demand on the defendants [Harry and Laura Bumpus] that mining operations be resumed promptly and conducted during the balance of the leasehold term with reasonable diligence and advised that if operations were not resumed proceedings would be taken for the cancellation of the lease." Of course, ownership of the land did not permit its being mined directly by her authority. Previously, UFMC had been the mining company of record. The new action by Levinton expected Harry and Laura to organize a new mining company and UFMC was not asked to resume as miner of record. Of course, there is no longer any memory of why Harlan didn't resume mining for his parents. This demand seems to have been unmet.

An agreement on March 19, 1943, between Levinton and the United Feldspar and Minerals Corporation to take

affect, less than a month after the court decree, may have intended to test what reaction Laura and Harry had, as Geneva let a two year lease to harvest trees from the Cummings Farm. The agreement was odd as UFMC had not ever dealt in wood (OCRD v456 p125). UFMC then leased the logging rights to the Philip H. Chadbourne Company, a wood products company in Bethel, for two years beginning July 1, 1943 (OCRD v409 p592-593).

Laura and Harry had the right to remove trees to facilitate having a quarry, but they did not have actual logging privileges. No reaction is recorded and Laura and Harry may have had no interest in the value of trees which were on the lot. The action may have simply been a test to see if any reactions would occur. The action was seemingly initiated to show that Levinton, as a land owner, expected her land to be productive. Five months to the day later on July 8, 1943, Geneva B. Levinton sold the Cummings Farm and Bumpus Quarry, but of course not its mineral lease, to the United Feldspar Mining and Minerals Company. (In a Supplemental Indenture to the earlier mortgage of the United Feldspar and Minerals Corporation, Geneva B. Levinton was a notary

Tydol gas station, Trap Corner, West Paris, 1930s postcard.

public from 62-65 Saunders Street, Forest Hills, Queens County, New York (October 5, 1942, OCRD v444 p244)! The association with the previous affidavit on the same page identifying Francis E. Haag and Winifred L. Haag, principal shareholders in UFMC, is compelling. Geneva B. Levinton seems to have acted as an agent on behalf of UFMC in the acquisition of the various Cummings Farm rights and land, when UFMC was directed to divest itself, along with the Cummings heirs, of rights or ownership in continuing litigation originally brought against the Sybil E. Cummings heirs by William B. Donahue and Adelia A. Waterhouse. In the same Supplemental Indenture, the clerk of Oxford County Superior Court, Rupert Aldrich, witnessed the documents of notary public, "Geneva S. Livingeston" (OCRD

Perham Grocery and Sunoco Gasoline Store with Maine Mineral Store on the far end. Trap Corner, Early 1950s .postcard.

v444 p245). The name was obviously a slip of the pen. It is interesting to note that Rupert Aldrich later owned the Bumpus Quarry!

There were minor communications preserved between Aldrich and several lawyers. On October 31, 1942, Judge Harry Manser issued his Final Decree: "This cause came on to be heard this day, and upon consideration thereof and by agreement of the said parties, IT IS ORDERED, ADJUDGED AND DECREED that the bill be dismissed without costs and without extended record."

"*Forever*", as it turned out, was somewhat shorter in Oxford County than in some other localities that may be mentioned. UFMC purchased the Cummings Farm and Bumpus Quarry nine months later, on July 8, 1943. [326] The unmet Levinton demand effectively expired when she sold the Cummings Farm. On October 16, 1943, UFMC started a new lawsuit to cancel Harry and Laura's lease. The final final decree of this action occurred in 1946. Meanwhile, Dana Douglass became a new player in the Bumpus Quarry saga.

Dana Carroll Douglass, Jr.
(August 21, 1914 – July 30, 1999)

Dana Douglass, originally of Portland, Maine, did not come from a mining or construction background. He went to the prestigious Dummer Academy in Massachusetts and graduated from the Ivy League, Dartmouth College, in 1937. He first worked in an investment company in Jersey City, New Jersey and lived in Forest Hills, New York before marrying Barbara Harte Anderson of Newton, Massachusetts in 1939. The couple soon moved to Portland where Dana established a business, Stanley Dana Corporation, selling "sound equipment" (motion picture cameras, movie rentals, etc.) The business continued into 1945.

L-R, Kim Stone and Loren Merrill at Mt. Mica, 1893 (Bastin, 1911).

Dana started the Douglass Mining Company and joined the mica boom begun by the Federal Government. His scrapbook has a few newspaper clippings that suggest that the news of subsidizing mica mining was an influence in his choosing an additional new business, although the clippings may have indirectly referred to Dana's new professial activities. One headline read, with details, probably from a 1943 Portland newspaper:

"Mica mining, a new Maine industry aimed at providing this essential material

Arthur Vallee demonstrating Loren Merrill's original machine. Photo by Ben Shaub, c1958. Machine now on loan to Paris Hill Historical Society, Dorothea Murphy, personal communication, 2003.

for America's war effort, was in full swing today with disclosure that five mines and a rifting plant were already in operation.

Seven other mines have been located, and by the end of this year at least 20 mines are expected to produce from 800 to 1,000 tons of mica for Government consumption in the war effort. ...

Mines now being operated in the State are located in West Stoneham, Rumford, Hale [Mexico], West Bath and Brunswick. Other sites where the mica deposits have been discovered and scheduled to be developed before the end of the year include, Fryeburg, Waterford, Lovell, Greenwood, Albany, Buckfield and Rangeley. ...

Demand is Unlimited

Rifting work is conducted at a Portland plant**."**

The Douglass Mining Company worked or leased the Donahue Mica Prospect, Albany (1943), Gogan Mica Prospect, Mexico (1943), Guy E. Johnson Mica Quarry, Albany, [346] John Lobikis Mica Quarry, Peru (1943-1944); [347], Matti Waisanen Quarry, Greenwood (1943-1944), and Willis War-

Early wooden patterns from which the Mercropon motor housing, lap, and splash pan were to be made. Courtesy Winsor Rippon.

ren Feldspar Quarry, Stoneham (1943-1944, the latter with Lawrence Anderson, of Stoneham, as a miner). Douglass prospected the Freeman Bennett Farm on the north side of Pattee Mills Road, Albany with a lease from Ralph H. Kimball on July 3, 1943 and is known to have tried to locate the Beryllonite locality by bulldozing on the Melrose Farm, Stoneham. [348] This is assuredly only a partial listing of his interests. Douglass did arrange for the traveling U.S. Geological Survey pegmatite teams to visit his properties as can be seen in Appendix F.

"Vaino Oja is seen here working at the Mercropon cutting machine, while Vernon Inman watches the operation. The cutters at Perham's frequently collaborate when cutting chore offers some unusual problem."
From Lewiston Journal Magazine, April 3, 1948

Dana's experience in western Oxford County, particularly in Albany, during 1943 and 1944 brought him into contact with various people interested in minerals and he certainly was known to UFMC staff, Stan Perham and probably, to Harry and Laura Bumpus. With the knowledge of the litigation, it is possible to suppose that Dana worked on behalf of the Bumpus's and at their invitation to satisfy the need to have the Bumpus Quarry operational. The record shows that their association was brief, with April 1945 the start date of Douglass' mining at the Bumpus Quarry. (April is about the earliest that a long idle quarry could be reopened in Maine.) Court records indicate: "The case lay dormant thereafter until April 21, 1945 when the plaintiff was authorized to inspect the leased premises on proper motion."

Brief Bumpus Mining Activity - 1945

Cameron et al. (1954) wrote: "In April, 1945, the Douglass Mining Co., Portland, made an operating agreement with Bumpus, and contracted with the C. C. Smith Co., Cambridge, Mass., to pump and clean the pit, and to remove the beryl, scrap mica, and feldspar. The work began in late April. The open pit is 270 feet long, 40 to 60 feet wide, and 20 to 60 feet deep. An estimated 39,000 tons of rock has been removed from it."

Neumann (1952) tabulated the total amount of minerals quarried by Douglass in 1945: 1,000 tons feldspar, 25 tons mica (mostly scrap), and 15 tons of beryl were sold and a stockpile of 800 tons of feldspar and 560 tons of quartz was created. In 1944, the entire state of Maine had produced a total less than 10 tons of beryl. [228]

Sampter (1945) provided more detail:

"Only two places [visited in Oxford County] had any blasting since 1944 – the famous Bumpus mine, which we visited six times and Bennett's [Buckfield]. When we first visited the Bumpus place, at Albany, Me., it was being worked for feldspar and beryl by Dana Douglas [sic], of Portland. The whole top of the East wall had been cleared by bulldozers and about 20 holes had been drilled for blasting. We anticipated great finds. Practically no blasting was done. The little bit that had been done, just broke up some marvelous large beryl crystals into little pieces, which they saved in a shed. We got some, as well as plenty of their famous rose quartz, for a rock garden, but no good crystals. Something happened – perhaps V-J Day – but all work there has ceased and the Bumpus quarry is again idle. The shed is still full of beryl and the pit full of water, with nice large green frogs. There are still some perfect and very large blue beryl crystals in the walls and under the water for some collectors to acquire in the future."

(F. Lawrence Sampter was an independently wealthy mineral collector who had his chauffer drive him to mineral localities throughout Oxford County and elsewhere during the late 1940s and early 1950s. Sampter kept an office on Wall Street in New York City where he could go everyday, just for the purpose of studying his mineral collection (Neil Wintringham, personal communication, 2007).) The Sampter observation has to be understood that he was probably only speaking of mineral locations that he had visited. The West Paris mill was fully operational and had to have constant feed of freshly mined, and therefore "blasted", feldspar.

Dana may have entered into the Bumpus agreement with the intention of long-term mining. Certainly, Harry and Laura were desperate to get anyone working at the quarry to satisfy the litigation. The above reported production figures suggest that Dana did not operate the quarry for very long. The property was unworked from the rest of 1945 through

1948. For example, Sampter (1947) wrote of the Bumpus Quarry: "Saw the Bumpus mine again [in 1946], which is still not being worked, but has good prospects of being operated soon."

1946 - United Feldspar in Court, again, and a Precedent is Set – Another Final Decree

Litigation plagued the Bumpus's almost since their acquisition of a lease. On October 18, 1943, the new official land owner, UFMC, brought a law suit. [358] The contention by UFMC was that there was no provision in the Harry and Laura Bumpus lease, with Allen and Sybil, for a minimum annual rent on the property. UFMC contended, therefore, that the lease implied that there should be active mining. The cessation of mining did not yield any royalty income for the land owner and, therefore, the lease should be cancelled for non-performance. The court's ruling indicated:

> "The owner of a reversion, subject to a mining lease, was not entitled to cancellation of the lease on the ground that a covenant to carry on mining operations with reasonable diligence is implicit in any lease providing for rental on a royalty basis where there is no provision for a minimum annual rental, and that there has been a breach of such implied covenant, where suit was brought within a short time after acquisition of title, and price paid for reversion at the sale must be considered as having been determined to some extent by the omission from the terms of the lease of any express covenant to carry on operations, and abandonment of mining right might be attributable to litigation initiated by the plaintiff."

The Court was not impressed that UFMC wanted to change an existing mining agreement after it purchased the Cummings Farm. In effect, the Court felt that the purchase of the land was made with UFMC's "eyes wide open", rather than their being "asleep" and that the lack of a performance clause was only discovered after the purchase. Further, it was probably because of the long duration remaining on the lease that made the purchase price as low as it was. A legal precedence was set in that the Maine Supreme Court felt that land ownership and various rights could not be separated by court decree. The various implications argued by UFMC were theoretical and not part of the original agreement. The court was particularly unimpressed that UFMC was _**both**_ plaintiff and defendant. This status was unacceptable. UFMC owned a portion of the land and as plaintiff, UFMC was suing itself to perform duties according to implications it saw in the lease it was bound to! During the proceedings, UFMC claimed an extra 1/30 interest in the Cummings Land. It does not seem to have come up that Harry and Laura did have provisions of their own for an annual payment from Harold Perham and other provisions regarding work performance in their leases. Nonetheless, Harry and

Laura won the case and UFMC appealed, only to lose to Harry and Laura again. UFMC continued with lawyers Brann, Isaacson & Lessard of Lewiston and added Thomas Delahanty and Raymond Burdick. Harry and Laura continued to be represented by George C. Wing. The brief Dana Douglass mining episode figured into the Maine Supreme Court's ruling. UFMC also tried to contend that there had been little or no mining after Harry and Laura were directed by Geneva Levinton to start mining by March 1, 1943, but the Court felt that the litigation itself may have impeded the mining effort. Even after this final decree, the Bumpus Quarry remained idle and full of water. On October 21, 1946, the Supreme Court of the State of Maine handed down their decision.

A New Location for the Maine Mineral Store

Stan and Hazel Perham had weathered the Second World War. Stan had been a beryl and mica miner and had bought and sold these minerals in order to make a living, along with his friend and employee, Deanie. The Bumpus Quarry had not been very productive of rose quartz and Stan had to make a choice of how to increase his income after war-time subsidies had ceased. The choice was to combine the Maine Mineral Store with a grocery and gasoline store. Collectors were still infrequent visitors, except during the tourist season, but everyone needed food and gasoline year round. Stan and Hazel bought the grocery store and Tydol gasoline station across the street from them at Trap Corner, in 1946. (Stan and Hazel had owned the lot immediately north of the Trap Corner store and sold it to John A. Gibbs on April 18, 1941, having purchased the land on May 15, 1936 from Lauri Immonen.) The store was eventually renovated. The family dedicated their efforts to making the combined businesses prosper. Unfortunately, family stores usually survive only because they are a communal effort. The individual workers receive no salary, but the proceeds keep the family in enough money to continue. Corner stores in the time period were usually marginal, and the same

Deanie demonstrating his Mercropon, ~1958.
Ben Shaub photo.

might be true today for many of them, especially in rural areas. Stan confided that there were many times he had to scrape even to be able to pay the gasoline distributor for the next shipment of gasoline. There were times that he feared he might go into bankruptcy.

The Perham family continued to live in the original building. Stan continued to look for opportunities to have additional ways to make the Maine Mineral Store thrive. One possibility involved providing equipment and supplies, in addition to gem materials and services to customers.

Harold Rippon and the Mercropon Gem Lap

The gem cutting industry had languished in Maine after Loren Merrill's death in 1930. Merrill had come from a family who worked in metals and castings and he had the knowledge and opportunity to manufacture faceting machines for gem cutters who wanted them. Merrill also had taught most of the local gem cutters the lapidary arts although some faceters taught students of their own. Merrill taught gem cutting to his son-in-law, Arthur Vallee, as well as the four Bickford brothers (William, Robert, Ross, and the very productive Knox) in Norway. It is noteworthy that Merrill had also taught Clarence Leslie Potter to facet gem tourmaline from Newry, although it was supposed to be a secret that Newry had produced gem tourmaline in 1902 and 1903. Merrill probably taught other local cutters including Martin Keith, Henry Cullinen, Bert Hamilton and others, but there were few new gem cutters to replace those who retired or stopped for one reason or another. Few of these men were very productive after WWII. Bert Hamilton had mentored Deanie for a while and Deanie also learned faceting from other local artisans.

The rejuvenation of feldspar, beryl, and mica mining in Oxford County after the Depression and during the war resulted in new finds of gem materials, but there were few new gem cutters entering the field as would otherwise be expected when a resource is "refreshed". A conspicuous reason for the lack of "new blood" in the region's gem cutting family was the unavailability of new gem cutting machines. The man who remedied the situation was Harold Richard Rippon [~1898 Lynn, MA – 1960s Randolph, VT].

Harold was born in Lynn, Massachusetts and was a veteran of WWI, in which he was a motorcycle messenger near the front lines. It was perhaps this experience which prompted him to bicycle all through the White Mountains of New Hampshire soon after the war. Harold taught "shop" or Industrial Arts/Manual Training for two years in Lynn Public Schools, a position that did not require a college degree at the time. "Dad could build anything. Our home was filled with every manner of tables, chairs, desks, chests of drawers, puzzles, knickknacks, etc. made by him." (Winsor Rippon and Dean Rippon, personal communication 2008). Harold mastered metal work as well as wood work and made a wide variety of metal objects ranging from lanterns to salt and pepper shakers. He loved to craft replicas of U.S. colonial age antiques.

Winsor (L) with Harold Rippon (R) holding recently found prize rose quartz boulder from Bumpus Quarry, Portland Sunday Telegram, October 13, 1940.

Harold was brought up in the Episcopal faith and married his pre-war sweetheart Laura Mildred Neal [November 6, 1897 – June 7, 1994 Burlington, VT]. Mildred was an artist and ceramicist. She frequently taught jewelry or art classes, especially in the summer. Harold was never known to have uttered a profanity and he and Mildred were never known to have said a cross word to each other. Their interests were harmonious and complementary.

Harold changed from teaching in Lynn to teaching at the Harvard School in Charlestown for two years and then he was attracted to the Rice Grammar School, in Boston's Back Bay area, where he remained until retirement, decades later. Despite his schools' locations, Harold always lived in Lynn and commuted by train into Boston.

Harold was a friend of George Howe's and frequently went on field trips with him. George was always providing mineral collecting advice for Harold's children: "Be sure to dig under that bush!" (Winsor Rippon, personal communication, 2008). One particular project was well-remembered by his sons, Dean and Winsor. Perhaps inspired by George Howe's 13.5 inch telescope, a gift of Rev. Alpheus Baker Hervey about 1930 and used to teach the youth of Norway about astronomy, Harold put his hand to grinding mirrors for an eight inch reflecting telescope as well as to making the housing, mounts, tripods, and complete accessories for it. The project took two years to be satisfactorily finished.

Harold was an immensely gifted craftsman, draftsman, designer, etc. In 1949, Harold devised a wood marking gauge and received U. S. Patent #2,579,205 which was purchased by the Stanley Tool Company in late 1951.

Harold was very inquisitive and had many interests, although some were short-lived: archery, photography, fly fishing, etc. About 1930, Harold became a mineral naturalist and traveled widely in Maine and New Hampshire in this pursuit and, of course, the family home was full of minerals. During the early 1930s, Harold taught woodworking at the O-At-Ka boys' camp on Southeast Pond in the town of Sebago for two summers. Harold and his family particularly enjoyed boating and had an eighteen foot Old Town brand canoe and they frequently traveled on the Saco River for explorations. Because he was a public school teacher, Rippon naturally had his summers free to go mineral collecting and he focused his energies on this interest for many years. Harold was a frequent visitor to the Maine Mineral Store in West Paris and a good friend of Stan Perham's.

By the mid 1930s, Harold began to develop an interest in gems and by 1936, Harold had become good friends with Edward Everett Oakes [1891 Boston, MA - 1960], who had a studio in Franconia, New Hampshire. (Edward Wigglesworth's 1936 pamphlet on *How Gems are Identified* was among Harold's library references.) Oakes was a well-known jewelry craftsman who had been named "master craftsman" by the Boston Society of Arts and Crafts in 1916 and won the Society's medal in 1923 for his designs and craftsmanship (http://chicagosilver.com/oakes.htm, accessed 2008). Oakes was the first living jewelry craftsman to have his work purchased by the Metropolitan Museum of Art. During the summers of 1936 and 1937, Oakes worked very closely with Harold frequently closely mentoring him in the design and techniques of making jewelry. A surviving ring of Harold's shows much of Oakes' influence on him and has a characteristic floral motif favored by Harold.

There are no written records of when Harold envisioned making a faceting machine. Nonetheless, Harold was what we would call a "doer". When he was inspired, he would focus his attention until a task was completed.

Merrill and Stone had influenced amateur and professional gem cutting with their machines, first developed about 1889. By about 1902 or 1903, their machines were being copied for Oxford and Androscoggin County lapidaries who also wanted to cut their own true gems. As already mentioned, later generations of would-be Oxford County artisans lamented the absence of a gem cutting machine that they could use. However, there was no efficient amateur faceting machine for Harold to copy, although he certainly had seen an example of the huge Loren Merrill and Kimball "Kim" Stone faceting machine.

The Merrill gem lap was heavy and lacked portability. Rippon, an inventive genius, designed a more compact unit. The miniaturized gem lap, Rippon eventually developed, provided the convenience required for practical home use. One must believe that Harold had discussed the design of faceting machines with Stan Perham and Robert Bick-

ford, with many other contacts possible and probable. One of the few, if not the first, contemporary mentions of the Mercropon, was in *Lewiston Journal Magazine*, April 3, 1948): "The past 14 years, [Perham] said, he [Perham] has carried on intensive research into cutting methods. A new cutting machine has resulted which is called the Mercropon, a name designed to give credit to every man in Maine who has advanced gem cutting. The machine built for Mr. Perham represents the composite thinking of the various cutters. Final refinements on this machine were worked out by Harold Rippon… The new machine saves between five and 20 per cent of gem stock cut, and a number of them have been sold to other cutters."

The precise date of the finished invention is uncertain, but by Spring 1948, Vaino Oja of West Paris, had learned the art of faceting and was using his Mercropon to cut gems for Stan's store (*Lewiston Journal Magazine*, April 3, 1948). After Vernon Inman got out of the Army, he too, became a gem cutter for the Maine Mineral Store, using a Mercropon, so that by Spring of 1948, Perham's was employing three gem cutters. The newspaper article suggested

William Cross, (Portland Press Herald, November 25, 2004)

Harold Rippon ~1918, punctilism drawing by Mildred Neal. Courtesy Winsor Rippon.

Winsor Rippon demonstrating one of his father's woodworking machines, 2008.

new gem laps. Rippon eventually made about a dozen compact, high quality faceting machines sold to Oxford County enthusiasts who wanted to cut their own gemstones. Although legend holds that there were 12 Mercropons made for Stan Perham's exclusive distribution, the variety of machine designs seen including patterns suggests that more than 12 "Mercropon" units were made. As Rippon had a steel cutting lathe, as well as a full compliment of wood and metal working machines, at home, evidently he was able to tweak his design until it was near perfection. No two Mercropons seemed to have the same design splash pans, but they units may have been altered to the tastes of the owners.

When it came time to name the new compact unit, it was called "Mercropon". The first part of the name, "Mer", paid homage to earlier gem cutting enthusiasts, Loren B. Merrill [February 14, 1853 Paris – March 29, 1930 Paris] of Paris Hill and the second part, "Cro", was for William M. Cross [September 2, 1877 Manchester, ME - October 17, 1931 Portland] jeweler of Portland. The unit was also obliquely named for Rippon in the last syllable, "Pon", of the synthetic word.

One feature the Mercropon machine had was its robust durability, having been made with the hardest metals that could be worked by Harold. Despite the small number of faceting machines made, the Mercropon was very influential on Oxford County gem cutting as a cottage industry, just as the original Merrill faceting machine of the 1890s to the early 1900s had been. It may be imagined that gemological tourists became interested in manufacturing their own

that Inman perhaps had begun learning to use his Mercropon by late autumn 1946: "The latest addition to the cutting crew is Vernon Inman, a former G.I., who has learned fast in his 1 ½ years with Mr. Perham. Already he has turned out a number of splendidly cut stones."

Family memory agrees that a faceting machine had been in the works by the latter part of WWII. Harold first made the components of the faceting machine out of wood and then reproduced the patterns into metal. During that time, Dean had enlisted in the Navy and was in the first graduating class of Navy Seals, while Winsor also enlisted in the Navy.

Stan Perham was the exclusive distributor for the

Gold ring designed and crafted by Harold Rippon. Dean Rippon collection.

gem cutting machines after seeing a Mercropon at the Maine Mineral Store. Before the production run of Mercropons was sold out in the late 1950s, there was a coincident rise in lapidary arts across the USA and also by the late 1950s faceting machines were being made available by several "national" companies. It would be difficult to assess the influence of a

few machines sold by a tiny gem store in a remote portion of the country. However, word of mouth was also a powerful agent in stimulating the amateur and semi-professional lapidary arts population and there were probably many who saw the Mercropon at Perham's, among the thousands of annual visitors, who carried away the idea of making their own gem cutting equipment. To be sure, the rise in the interest in amateur mineralogy and gemology by GI's just after WWII and the Korean War led to another gem interest: that of fashioning cabochons, a gemstone made from ornamental materials which were not transparent.

Landers (1955) wrote of her experiences with her Mercropon. Helen Landers of Dixfield had recently begun to cut and polish cabochons as the gem hobby was experiencing a rebirth in Oxford County and she was a member of the Oxford County Mineral and Gem Association, founded in 1948. The times mentioned should not be thought of as work at a "9 to 5" job. Sessions may have varied from one to several hours separated by periods of inactivity, especially as her earliest experiences involved commuting from home to her machine's location:

"I chose this machine especially, for it was said that it did the thinking for you and that you mainly just had to sit and guide it.

I practiced on my first stone and was pleased with the results. It only took me three

Mercropon faceting machine demonstrated by Harold Rippon. Courtesy Dean Rippon.

while my instructor plotted a different one, and

Upper Right: Dean Rippon demonstrating his Mercropon, 2008.
Upper Left: Mercropon detail.
Lower Right: Indexing head of Mercropon showing holes for each facet position.
Lower Left: Index heads for various numbers of facets.

months to cut it. It was a beautiful smoky quartz pendant, 15 carats. It was chosen for me to work on thinking the larger the stone the better I could see what I was doing.

Next, I started a light colored sparkling amethyst but it was jinxed from the start. My mathematics were rusty and I plotted one angle

as the two angles just would not co-operate, I had to start again. After three weeks of hard, and I mean <u>hard</u>, labor I succeeded in cutting a good crown. I then turned the stone over to cut the bottom or pavilion but when I went back to finish it I was told that some guests had been in the workshop and had knocked my stone off the dop stick, so I had to start over again. {The machine probably had been stored at a location where Helen could conveniently receive lessons from Deanie and may have been in a back room at Perham's.} …

I took my machine home, deciding to make my own blunders and I did just that. I thought I could tell when a stone was straight on a dop stick. I bought one of those three way gimmicks that are advertised to positively align the stone straight. I worked about one-half hour measuring the stone, the three ways it takes to align it, and the gimmick says it's O. K. and my eyes say it's O. K. but that ma-

Dean Rippon showing changeable abrasive pads (upper left) and examining faceting design (lower left). Shelves with brass dop sticks, grinding compounds (right), 2008.

chine has the last laugh. It comes out lopsided.

...

I sent for a coarser copper lap 600 grit. I bought more diamond dust for my other coarser lap and my fine one, and it was working better, but what a job, three copper laps and a leucite one for polishing, to cut one stone."

(Gem polishing is a time-consuming process and when a piece of gem material has completed one of the stages in the process from rough mineral to completed gemstone, there are periods of cleaning, disassembly, and reassembly so that the equipment can perform the next stage of tasks. In order to increase the efficiency of gem cutting, many artisans will have a selection of stones that go through the various stages sequentially, instead resetting the equipment for each stone.)

By the mid-1950s, Harold Rippon had ceased to cut gems, but he never abandoned his woodworking. Harold's sons also did some woodworking. In recent years. Winsor primarily kept up the woodworking family tradition, while Dean maintains his father's gem cutting. Harold died suddenly during a ping pong game at a friend's house in

Randolph, Vermont, due to undiagnosed rheumatic fever heart damage, although earlier that day he was told by his family doctor that he was in great health and "would live forever".

It has been through the spirit of Harold Rippon's faceting machine that gem cutting and faceting is still widely practiced in Maine. There are newer and better machines which are commercially available, but Harold's invention, inspired by Loren Merril and William Cross was an important part of Oxford County gem history.

Bi-color blue and green 2-carat tourmaline cut by Raymond "Deanie" Dean in 1955

Chapter Seven: Post-war Search for Beryl, Renewed Mining at the Bumpus Quarry, and Surrender of the Bumpus Lease - 1949

Northern Mining organized by
Dana Douglass

Due to the forces at work in the Cold War, the Bumpus Quarry took on a new importance as a beryl producing

Dana Douglass, ~1940.
Courtesy Selenda Giradin.

locality. What might seem as an abrupt turn around, on October 11, 1949, the United Feldspar and Minerals Corporation, the newly-formed Northern Mining Company, and Harry and Laura Bumpus, met to surrender and reassign the lease agreements relating to the Bumpus Quarry. [327] United feldspar and Mining Company also bought out the remaining interests in the Cummings Farm land.

"WHEREAS, United is now the record owner and true owner of the premises described in said lease and all of the right, title and interest of said Allen E. Cummings and Sybil E. Cummings as Lessors in and to said lease, together with all rights and remedies accruing to the lessors thereunder; and

WHEREAS, for good and sufficient con-

sideration tendered by said Northern Mining Corporation to said Lessees, said Lessees have heretofore agreed with said Northern Mining Corporation that all their rights and interests under the said Lease should be subordinated to said Northern Mining Corporation so as to be wholly controlled by it in whatever manner may be most efficacious for its purposes, said Lessees, however to be held harmless thereunder …"

On that day, Harry and Laura sold their right to mine minerals at the Bumpus Quarry, forever. Two days later, Northern Mining Company obtained a new five-year mining lease from UFMC to mine feldspar from the Bumpus Quarry, a logical activity while one was also looking for beryl. Feldspar royalties were the familiar $0.50/ton, while quartz was only $0.15/ton, but with a 15% royalty of the gross selling price "for all other minerals", including beryl. Interest-

Harry Bumpus, ~1950.
Courtesy Ava Bumpus.

Rose Quartz stockpile, 1951.
Courtesy Ava Bumpus.

ingly, the feldspar could be delivered and sold to the West Paris grinding mill for a premium: $8.25/ton. (In 1949, the State average price for crude feldspar was only $7.12/ton.) The minimum royalty payment for a year was $200. [328] Given the stress and heartache, not to mention tenacity, by which the lease was held, the sale of the lease must have been accompanied by a lot of money. (The author suspects that the money to purchase the lease was clandestinely subsidized by the Federal Government in order to renew mining at the Bumpus Quarry in anticipation of a bonanza of beryl. Given the previous Bumpus Quarry inactivity, it is also a wonder that UFMC, or even Geneva B. Levinton, didn't just try to purchase the lease years before? There would have been a lot less money spent for the benefit of lawyers and the grief would have been nil on both sides.)

Neumann (1952) wrote: "In 1949, … [Douglass] brought in machinery and started to mine pegmatite. During 1949 and 1950, Douglass and crew moved approximately 14,100 tons of pegmatite and 15,700 tons of overburden and hanging wall waste." Tabulated sales of minerals were 1,100 tons feldspar, 200 tons mica, and 100 tons beryl, while 7,200 tons of feldspar were stockpiled as were 3,000 tons of quartz, presumably rose quartz. Previously, about 36,700 tons of pegmatite and 19,000 tons of waste rock and overburden were removed from the Bumpus Quarry. Of the total pegmatite mined in 1949 and 1950, only 6,800 tons were put on the dump, suggesting a richness grade of 82.3% saleable minerals! Counting the removal of 15,700 tons of overburden and waste still meant that the richness of all moved materials was 38.9% commercially valuable. The reason why many Oxford County pegmatites closed was not solely due to declining mineral grade, but the amount of effort needed to expand into mineral-rich portions of a well-worked pegmatite. As has been mentioned, waste rock produces no income. The declining economics of mining, partly influenced by the County's only feldspar mill which failed to pay a reasonable rate, did not inspire local miners to do development work as they went along mining. Miners

were content to cut the heart out of a deposit and then try to find new easy ground to work. The Perham Quarry was fortuitously located. It also was a big pegmatite exposed at the surface and there was little need to continuously strip away worthless rock. (One imagines that the effort to mine the Bumpus Quarry's 15,700 tons of overburden and waste rock to rehabilitate the Bumpus Quarry was also subsidized by the Federal Government. Thee mill's special price of the Bumpus' crude feldspar was probably also subsidized.) The fact that Northern Mining Company was able to more than double the recorded historical mineral production of the Bumpus Quarry was an example of modern technology superseding old, not to mention financial incentives. Douglass also began exploring for beryl in Newry in 1949. [125]

The surrender of the lease had no provision about Harlan's residence on the property. One might imagine unwritten understandings with Harry and Laura that were placed in the subsequent October 13, 1949 lease between UFMC and Northern Mining Company: "… the rights of any other persons occupying parts of said lands with the permission of the Lessor, and will not damage any structures thereon. In the event that mining operations on said premises be extended beyond the limits of the present time so as to in any way threaten or impair the safety of any house, barn or other structure on the premises, the Lessee will at his own expense move said house, barn or other structure to another location on the premises to be designated by the occupant of such structure, but only with the written approval of the Lessor, provided that in such case the Lessee shall first ob-

Harlan's house being hauled by horses in snow,
April 21-22, 1951. Courtesy Ava Bumpus.

tain the written consent of the occupant of such structure to the said removal thereof."

Governmental demands on beryl production meant that mining had to be aggressively pushed. Just before a dynamite blast, a shrill steam whistle would warn of an imminent explosion. Bumpus family members in the immediate vicinity of the house or camp would seek shelter inside as small rocks would soon rain down on the area and roofs. [402] In light of the "safety" provision of the lease, after a year and a half, Northern Mining Company admitted that it was no longer safe to allow a residence so close to the quarry. Harlan Bumpus obtained a camp building from the "Beckler place", located next-door to the Cummings farm, and this was moved on the weekend, of his son's, Edwin's wedding to Ava Hutchinson, which occurred on Sunday, April 22, 1951. Harlan's new residence was transported by horse team and sled to its new location just below Hunts Corner on the Hunts Corner Road, Albany. The old Cummings family residence was probably demolished soon afterwards.

Harlan's house relocated, reassembled, and painted, 1950s. Courtesy Ava Bumpus.

A New Bumpus Quarry Beryl Record - 1949

In October, 1949, the United States Bureau of Mines (USBM) had "gotten serious" about finding beryl, undoubtedly because of the USA military's concern of the USSR's first nuclear bomb test on August 29, 1949. The USBM realized that the Bumpus Quarry had been one of the largest producers of beryl in New England.

There was a discovery at the Bumpus Quarry that validated the Government's attention:

"The aqua beryl occurs in clusters of crystals from macroscopic size up to 4 feet in diameter and up to 27 feet in length, usually in masses of rose quartz. The golden beryl occurs in the same manner but is associated with smoky quartz, and the crystals are much smaller. One beryl (aqua) crystal was extracted in October 1949. It was 27 feet long, 4-½ feet across the base, and 9 inches at the top. The crystal yielded 26 tons of beryl. This is probably the largest beryl crystal found at any deposit to date. [Actually much larger beryl crystals had been found in South Dakota during WWII.] ...

Beryl has been found throughout the quarry, but the greatest concentration appears to have been near the center of the pit, where the nests of larger crystals were found. It is not evenly distributed in the deposit but is concentrated in nests or clusters of crystals. From production figures, the average beryl content of the deposit appears to be 0.8 percent. In the sections where a nest of crystals was found, the beryl might average 10 percent of the total by volume. The beryl is good grade, averaging 13.0 percent beryllia content; and the large crystals, having a very distinctive color, can be readily hand-sorted.

Feldspar is segregated in some sections of the deposit and can be mined and shovel-loaded into trucks without hand sorting. On the fringe of the feldspar zone, hand sorting is necessary. It also occurs in graphic granite, which is waste." [176]

After the discovery of its biggest crystals to date at the Bumpus Quarry, there must have been jubilant feelings of satis-

Dynamite shed at Bumpus Quarry, ~1949. Courtesy Ava Bumpus.

faction, although they would have been short-lived. Unfortunately, it is not clear if the new giant beryl crystal was found before Northern Mining Company bought Harry and Laura's lease or after. More than likely, there were plans in the works to buy the lease even in September. Nonetheless, the discovery of the new giant beryl would have made Government agents anxious to control a quarry that yielded such large beryl crystals.

As part of the "serious" program to recover beryl, mineral collectors were discouraged to visit the quarry and Douglass instituted what was then a very high fee of $5 per person per day to collect, but he found that he was constantly being "chased down" by collectors willing to pay for the privilege of collecting and had to eventually close the location to all visitors [405] (Fee mineral collecting localities were rarely higher than a $1 per day per person, at least in New England into the 1960s. This was still the price at Mt. Mica Quarry and at the Tamminen Quarry in the early 1960s.) Before every blast, there had to be an accounting of the whereabouts of the collectors, who may have wandered away from the spot where they were advised as "in bounds". Blakemore (1952) wrote: "He fairly sees red if the word Gem is mentioned or a Rockhound appears on the horizon. He has gone to fantastic lengths to discourage it in his Miners." [18]

Clark (1951) supplied slightly more detailed, if not more accurate, measurements of the new giant beryl: "Northern Mining Corp. produced beryl in 1949 from the Bumpus quarry, Albany, and the Black Mountain deposit in Maine. ... A spectacular beryl crystal was uncovered in the Bumpus quarry measuring 27 feet 7 inches long, with end

Crane at Bumpus Quarry, 1950.
Courtesy Ava Bumpus.

diameters of 39 inches and 11 inches." The weight recovered was unstated, but if the conical mass of beryl were symmetrical these dimensions would weigh 33 metric tons using Clark's data. If Blakemore's numbers are used the result would be 55 metric tons. As both figures are too high, the reported value being only 26-27 tons, it must have been either that the thickness of the beryl was somewhat shallow or the figures were reported artificially low, intentionally.

Douglass was probably interested in the Black Mountain locality because of a recent, 1949, USBM report on the beryl potential of that locality, although eleven exploration drill holes showed that the pegmatites were relatively thin. [140] Douglass may have resumed explorations at Black Mountain in 1954 (E. L. Sampter, written communication to B. M. Shaub, September 15, 1954).

Though possibly apocryphal and with incorrect dimensions, the description of the new record beryl from the Bumpus Quarry was related by Blakemore (1976) who "quoted" Dana Douglass, and is interesting to read:

"'When the smoke of the dynamite charge cleared I nearly fainted. Imbedded in a cliff of rose quartz was an unbelievably large single beryl crystal.

It was twenty-seven feet, nine inches tall and weighed twenty-seven tons.

It looked for all the world like a giant ice cream cone, upside down, graduating from eleven inches at its rounded top to six and a half feet at the bottom, It was all shades of

blue and blue-green (aquamarine). Incorporated within the crystal and sprouting around its base were thousands of small crystals...

To complete the amazing picture, a scoop of 'vanilla ice cream' [quartz?] was there, too, right in its proper position. At the bottom of the aquamarine cone was a curved line of demarcation below which, scoop shaped, were six tons of heliodore [sic], the rare golden beryl.'

Did Dana preserve this marvel for future generations to admire in the Smithsonian Institute? He did not!

He ordered it broken into chunks. Fifty-two thousand pounds of it were loaded onto box cars and shipped off to the United States Government."

Bumpus Quarry truck and stockpiles, 1950.
Courtesy Ava Bumpus.

U. S. Bureau of Mines Explored the Bumpus Pegmatite - 1950

After the new giant beryl had been excavated, there may have been a desire to find all the "gold in the goose's egg" immediately, so great was the demand for bomb-grade beryllium. The USBM visited the Bumpus Quarry for the purpose of discovering the exact location of beryl concentrations in the Bumpus Pegmatite and, between September 27 and November 27, 1950, a field crew drilled seven exploration holes with recovery of the drill cores so that the minerals occurring at depth could be sampled and evaluated. Neumann (1952) reported that the pegmatite was at least 200 feet longer than had been exposed to the west and at least 350 feet longer to the east. The pegmatite was found to be about 40 feet in thickness with a dip rather consistently 50° to the southeast for another 50 feet below the quarry floor then known in 1950. The maximum length of their drill cored holes was 229.2 feet. The strike of the pegmatite is S 76°W and the rake is 11°W. The Bumpus quarry had received frequent attention previously, however, and unpublished surveys of the Bumpus Pegmatite had been made by other USBM geologists including Margaret Fuller Boos and H. O. Hammond and USGS geologists Lincoln R. Page, John B. Hanley, and David M. Larrabee (initials corrected). [64] In 1952, only two Maine locations were producing beryl on a regular basis, Bumpus Quarry and the Scotty Quarry, Newry (latter locality unmentioned) and the Neumann (1952) report was interpreted to have increased the proven "beryl reserves of New England. [70] The drilling actually did not reveal any beryl, as none was mentioned in the drill

Northern Minerals blasted rock, 1949.
Courtesy Ava Bumpus.

cores. The drilling proved that the Bumpus Pegmatite did not pinch out, but continued at a relatively constant thickness. The nature of the distribution of beryl in the Bumpus Pegmatite means that drill holes could easily miss even large beryls. The unspoken conclusion was that if the pegmatite continued at depth, so would the beryl. Because the pegmatite was active, the Neumann report made no remarks encouraging or discouraging the continuation of mining.

During the flurry of interest in the Bumpus Quarry and the proving of feldspar, if not beryl, reserves in the pegmatite, Harlan W. Childs [1909-November 12, 1960 West Paris], then superintendent of the UFMC West Paris mill, announced plans to double capacity of the West Paris Mill (unidentified newspaper clipping, c1950), however, those

plans did not reach fruition. The plant was then officially known as the Oxford Division of the UFMC and the then current production was 12,000 tons of ground feldspar a month. There was a particular need to stave off "old man winter", however: "'We have to store about 7,000 tons to get through the Winter,' Childs says, 'and we don't have bin storage to do it. Some of it has to stay outdoors and that makes it almost as hard to handle as it would be to mine it in the Winter.'" From the above figures, it is clear that the 7,000 ton stockpile was more of a back-up source, should a bad stretch of weather prevent mining, than as a steady source of feed for the mill. At the time, the feldspar mill had about 50 miners at various locations in the summer months:

USBM drill frame, 1950.
Courtesy Ava Bumpus.

"'We could use more,' Childs says, 'but it seems to be a pretty difficult job to find men who want to do that kind of hard work.'"

A Third Bumpus Beryl Record - 1950

Because of the richness of the feldspar and the strong encouragement by the USBM as well as the premium price subsidies to stimulate beryl production: "... by 1950, two blasts a day were usual; the entire rose quartz output was purchased by Stanley I. Perham (West Paris), the "rose quartz king" of the East." [247] "Rose quartz was the material that sustained the shop for many years. My father sold rose quartz by the pound and Deanie wire-wrapped rose quartz for earrings and pendants. When the State moved the road

[Route 26], we had to pick up and move 3,000 tons of rose quartz." [401]

It must have been gratifying that there was a third record beryl in that year. Perham Stevens (1972) wrote:

"Another outstanding beryl crystal was uncovered at the Bumpus Quarry in 1950. This single tapered crystal was thirty three feet long and about six feet on diameter at the base. The lower two or three feet of the crystal were composed of golden beryl while the remainder of the piece was blue-green in color.

As one would suspect, it is not unusual to find beryl crystals five to seven feet long and twelve to eighteen inches in diameter throughout the quarry."

The above record was also mentioned in the *Norway Advertiser* Democrat (October 27, 1950) along with a photograph of a 5000 pound block of rose quartz that was believed to be the largest block of rose quartz ever recovered up to that time from a location east of the Mississippi River.

The rose quartz was eventually destined to be used as a grave monument in Rochester, New York. It is the marker for Julius J. and Leona M. Andersen. Leona, was

Dana Douglass on bulldozer with miner, Bumpus Quarry, c1951. Courtesy Ben Shaub.

born in 1879, died February 1, 1938, while her husband, also born in 1879, died September 25, 1954. Julius was an immigrant from Germany in 1893 and a successful tailor in Rochester. Both he and Leona were active in the local church including its Missionary Society. Julius was a member of the Masons, Fireman's Benevolent Society, Chamber of Commerce, etc. Neither Julius nor Leona seem to have had any connection with West Paris or the feldspar industry (there were two feldspar grinding mills in Rochester) and the possible connection may have been a one time tourist visit to Maine with a consequent introduction to and appreciation of rose quartz. Scott's Rose Quartz Quarry near Custer, South Dakota was the "only game in town" as far as large pieces of ornamental rose quartz were concerned, so

the acquisition of a significantly large piece of rose quartz from Maine had to have been a very special "order" with some knowledge of the Maine Mineral Store. The monument stone is about 4.5 x 3 x 2 feet and would theoretically weigh over 4500 pounds in its present size. The original piece seems to have been reshaped a small amount with rough edges smoothed. Unfortunately, many fractures in the rose quartz have developed and they are now susceptible to winter cycles of freezing and thawing. Nonetheless, the fifty plus years that the monument has been continuously exposed to sunlight does not seem to have diminished the intensity of the rose quartz's color.

Stan Perham's Recapitulation of Record Bumpus Quarry Beryls

The inventory of Bumpus Quarry beryl records is frequently difficult to reconcile, particularly when the information got to be second or third hand. In order to set this progression straight, Professor Benjamin Martin Shaub, of Smith College, Northampton, Massachusetts wrote to his good friend Stan Perham and asked for a definitive accounting of the beryl records, as a first-hand observer of most of them:

"Between 1927 and 1930, we found several very large crystals at the Bumpus Quarry in Albany, Maine.

One was fourteen feet in length and a little over four feet in [its] greatest diameter; another was about twenty feet in length and something like four and one-half feet in [its] greatest diameter. One crystal was twenty-two feet in length and was standing almost vertically at the bottom of a great rosette of many

Maine Mineral Store across from Current Site, c1950 postcard

beryl crystals. This one measured six feet and two inches across in [its] greatest diameter. Largest one found there!

One or two small golden beryls set off at an angle from the side of the fourteen foot beryl, I think that is the right dimension of this crystal, about three inches through and about five inches long.

From this rosette of beryl was taken by actual weight, over one hundred tons of beryl.

A considerable tonnage was taken out at a later date from this same whorl.

While Mr. Bumpus, was operating, in subsequent years, Mr. Herbert Rowe of Bethel told me of seeing a crystal in the quarry floor about seventy-five feet from this original cluster that was between six feet and eight feet across it. No measurement was recorded as far as I can determine but his observations were usually reliable. It might well have been the largest [in diameter] but not the longest found there.

Stan Perham with Jane Perham (holding pry bar) and 5000 pound rose quartz block from Bumpus Quarry.

During the operation of the Northern Mining Corporation that followed they found a crystal standing quite straight in the quarry that was over six feet in [its] greatest diameter and appeared to me to be in overall length more than thirty feet. I always thought it was thirty-three but there is little to question of [its] being over thirty feet. The bottom of this crystal had about a ton of golden beryl in it according to one of the miners who worked on it the last day of that year.

One crystal that I saw about 200 ft E. of original finds appeared to be four feet through it and at [its] southerly end there may have been two hundred or three hundred smaller crystals in rosette groups. They were very beautiful.

Last fall while I was doing some work for Northern Mining we uncovered a crystal ten feet long and two feet in diameter. This one was deep blue aquamarine color. It laid in quartz and one end was embedded in mica, and the other in cream colored block feldspar. This crystal was laying horizontally at the present entrance to the quarry." (Stan Perham, written

communication to Ben Shaub, October 16, 1953).

Stan Perham was an outstanding mineralogist and while the beryl was not measured by a tape or ruler, it was certainly examined with care for detail and Stan's figures of 33 by 6+ x 6+ feet in diameter may safely be accepted as the record size for a Bumpus Quarry beryl crystal and appears to have bene a world record at the time.

Government Intervention at the Bumpus Quarry

Probably because of the secrecy associated with USA beryl production, the visiting USBM and other geologists did not inform Maine miners about discoveries of giant beryls in other parts of the country and probably did not want the record size of Bumpus beryl crystals publicized. Most of the U. S. Geological Survey reports about pegmatite investigations were made in the mid-1950s and concentrated on mineral production from 1942-1945, the subsequent Cold War production being largely ignored. The Federal Government undoubtedly did much to downplay the record size for beryl discoveries in various localities. Slowly and with little fanfare, some beryl production figures reached the public. Probably because of the McCarthy era's suspected invasion of enemy spies wanting to know the potential size of the USA's beryllium-dependent nuclear arsenal and other secrets, beryl production figures may have been often mixed with disinformation. [20] There may be some people who wonder if any of the stated beryl production figures are accurate given the secrecy of the times.

The Defense Production Act of 1950 was intended to stimulate the mining and production of strategic minerals by

Rose quartz from Bumpus Quarry, 2008.

private companies. On December 4, 1950, the Defense Minerals Administration was organized and it advertised to the public and mining companies of its needs. Uranium and beryllium were high on its list. In 1952 and 1953, the Beryllium Development Company of Temple, Pennsylvania, a long time refiner and fabricator of beryllium materials contracted with the Federal Government's Defense Minerals

Administration and the Defense Minerals Exploration Administration (DMEA) Programs, chiefly for prospecting on Plumbago Mountain, Newry and nearby areas for beryl (Frank et al., 2003). Leo Tessier also obtained a DMEA contract/grant to explore for beryl on Tessier Road in Jay, Franklin Co., Maine. Tessier had a series of holes drilled and 70 sticks of dynamite were used in the first blast in the first week of September, 1953. Beryl crystals were found up to 6 x 1.5 inches, some of which would cut into 1 carat pale aquamarine gems. The prospect yielded microcline crystals to over 1 meter, large flattened almandine crystals to 6 cm, as well as large masses of smoky quartz with coarse muscovite and biotite (Ben Shaub, unpublished notes). The Tessier Prospect did not produce marketable minerals, however. Under the DMA and DMEA programs, the applicant would receive between 50-90% of their exploration or development costs paid, depending on how important increased supplies of an ore was. The only other Maine beryl development grants were made to expand beryllium mining at the Scotty Quarry and Plumbago Mountain in Newry. Because the Bumpus Quarry already had been explored by the USBM and was producing beryl, there was no application

Detail of rose quartz monument from Bumpus Quarry showing concrete base, 2008.

made for financial assistance there. Nonetheless, the various actions in 1952 to 1953, especially relating to the Bumpus Quarry, suggest that some important agendas were being enacted, certainly with incentive "bonuses".

Wintringham (1955) noted: "Lawrence Anderson, mineral dealer of East Stoneham, operated the [Bumpus] quarry during the summer of 1952." Sampter (1953) used the same language: Anderson, who sold minerals from his home on Route 5, operated the Bumpus Quarry "all summer". More than likely, the report indicated that Anderson was the quarry foreman during 1952 as Douglass seems to have been still the official "operator". No beryl "of record" was sold to the National Stockpile from the Bumpus Quarry in 1952. However, Knapp (1952) reported that he had been

mineral collecting at the Bumpus Quarry in July of 1952 so the restrictive policy of collecting during mining seems to have been relaxed at that time.

Changes in the Beryllium World - 1952

The Bumpus mineral lease had been sublet to Northern Mining Corporation. On April 9, 1952 (unidentified newspaper clipping) a shift in mining controls seemed to be more than coincidental:

New Firm to Mine Scarce Materials in Albany Quarry. U. S. Bureau of Mines Announces Continuous Pegmatite Dike Found - Report Site to Change Hands and Government to Contract for Products – Douglas [sic] to Remain as Manager. Scarce materials discovered in large quantities in the Bumpus quarry area here will be mined by a new corporation in conjunction with the Federal government, it was learned Tuesday night. In Boston, earlier in the day, the U. S. Bureau of Mines informed the Associated Press that a pegmatite dike found in the quarry is continuous and extends at least 50 feet below the present quarry floor. To the new corporation, which will reportedly

Detail of rose quartz monument from Bumpus Quarry, 2008.

carry on large operations, the Northern Mining Corporation will turn over its holdings.
Auburn Man in Firm
Dana Douglass Jr. of Bethel, associated with the Northern Mining Corp., will remain on the scene as manager, it is understood. Other principal owners of Northern are Henry

M. Dingley Jr. of Auburn and William S. 'Pete' Newell of Bath, Bath Iron Works Corp. officer.

It is understood negotiations are underway at this time with those who will take over the quarry. One of those connected with Northern Mining Corp. said, "the egg is hatching now and we should not identify the prospective new operators by disturbing the heat."

Deal Within Two Weeks

Reportedly, the deal will be consummated in two weeks, possibly ten days.

The find in the quarry here was made as the Bureau of Mines conducted diamond-drilling studies of New England pegmatite deposits. ...

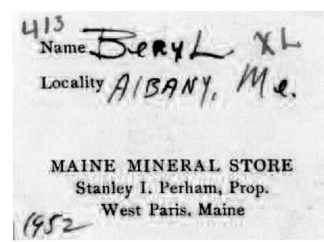

Drilling was done here in the fall of 1950. Some surprise was expressed that an announcement had not been made sooner in view of the fact this operation was about a year and a half ago.

Analysis of Pegmatite

The Bureau said the Bumpus pegmatite is estimated to contain 60% marketable feldspar, 20% quartz, 3% muscovite and biotite mica, .8 of 1% aqua and golden beryl, and 16% graphic granite, the latter a waste material.

Reportedly, the new corporation will contract with the General Services Administration of the government in getting out the scarce materials needed in the defense effort. ...

Beryllium obtained from beryl is used in copper-base alloys for the electrical and aircraft industries. It also has a strategic use in the field of atomic energy and is one of the Munitions Board's list of strategic and critical minerals.

Bumpus Records

The Bureau said available records show that from the Bumpus quarry had been marketed up to 1950 about 13,000 tons of feldspar,

450 tons of mica, 250 tons of beryl, and 260 tons of quartz. In addition, about 8,500 tons of feldspar and 7,300 tons of quartz were stockpiled.

Lest the reader believe all of the hyperbole. The report from the US Bureau of Mines was two years old and merely indicated that the pegmatite extended to much further depth. There was no actual discovery of new beryl, just a discovery that there was more pegmatite. It was left as a conclusion that more pegmatite equaled more beryl. The "new company" seems to have been Beryllium Development Corporation.

Later in 1952, beryl prices shot upward: "In August 1951, beryl was exempted from price control by the Office of Price Stabilization, and the price of beryl rose from $26. per unit of contained beryllium oxide to the September 1952 price of $38.50 per unit." [65] This meant that Bumpus beryl, with a strong 13%+ BeO content brought $500.50/ton instead of $338.00/ton. Coincidentally or not, on December 19, 1952, the feldspar grinding mill and some of the Maine holdings of United Feldspar & Minerals Company were purchased with mortgage by Richard Bell, president of Bell Minerals Company. Bell Minerals Company acquired mineral leases for fourteen properties and some land, including the Twitchell Farm and quarry (Paris) and the Alfred C. Per-

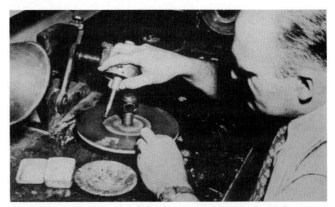

Stan Perham with Mercropon gem cutting machine, c1958 postcard

ham farm and quarry (West Paris). [329] The Cummings Farm (Albany) and the mineral lease to the Bumpus Quarry were not part of the Bell Minerals transaction.

The 1950s and 1960s, in Oxford and Androscoggin County mining, was the era of Bell Minerals. Bell Minerals seems to have been originally organized by Alphonso Edward Bell [September 29, 1875 – December 27, 1947] of California and was a feldspar mining company. Alphonso also organized the better known Bell Petroleum Company in 1922. Bel Air, Bell, and Bell Gardens, California were named for him (www.belaironline.org; accessed 2007). The Maine Bell Minerals Corporation was organized as a separate entity, not as a subsidiary. (Alphonso Bell won a silver medal in Men's Doubles tennis in the 1904 St. Louis, Missouri Olympics.)

Dana Douglass, ~1970.

Perham Quarry, expanded quarry with upper right derrick end and two full derricks in center. Early 1950s. Photo by Vi Akers. Courtesy Sid Gordon.

In late 1952, the Bumpus property rights were acquired by the American Encaustic Tiling Company when they bought control of the Bumpus Quarry from the United Feldspar & Minerals Corporation. By January 18, 1953, a photograph caption of the West Paris feldspar mill revealed: "American Encaustic Tiling Company, Lansdale, Pa., recently announced the sale of this feldspar grinding mill at West Paris. However, it retained title to the Bumpus Quarry at Albany because this quarry contains one of the largest known deposits of beryl crystals, which when processed, becomes beryllium. It is used in the manufacture of alloy metals." A recurring thread in all of the announcements is the use of beryllium in manufacturing alloys for aircraft, etc.

Tamminen Quarry, Greenwood, important feldspar source for the mill, 1952. Photo by Ben Shaub.

with minimal emphasis on nuclear weapons. It is known that the General Services Administration did not convert much of the country's mined beryl into beryllium. The authorization was the Strategic and Critical Materials Stock Piling Act of 1950 and the agency in charge of stockpiling strategic minerals was the Commodity Credit Corporation (later the Defense Logistics Agency) with its stockpile of raw beryl at Fort Belvoir, Virginia. Beryl and mica ceased to be added to the National Defense Stockpiles in 1962. There was also a "stockpile" of beryllium and beryllium alloys, but one must imagine that much of the refined beryllium quickly found its way into nuclear warhead production. [32] In 1992, the U.S. Congress voted to reduce the national stockpile inventory in a manner avoiding loss of original cost. Eventually, large lots of beryl from the stockpile were auctioned to companies, frequently in quantities of about 1,000-4,000 tons, but many years saw no offerings. A Beryllium exploration specialist formerly with the Cabot Company, Richard V. Gaines, said that chucks of faceting grade gem beryl were present on the stockpile amongst the thousands of tons of ordinary beryl. [40, 406] (Coincidentally, the rare beryllium cesium phosphate from the Nevel Quarry, Newry, Maine, gainesite, was named in his honor. Gaines was a perennial visitor to a resort at Small Point, Phippsburg, Maine.)

Chapter Eight: The Age of Bell Minerals and Afterwards

Dana Douglass Leaving the Bumpus Quarry and Other Transitions

In 1952, there was an expansion of beryl mining at the Scotty Quarry, Newry, while the Bumpus Quarry had not recently produced significant quantities of beryl, although it remained an excellent feldspar producer. The Bumpus Quarry and Newry both made the national news (*Christian Science Monitor* May 24, 1952:

> "Mining engineers for Beryllium Corporation of America are prospecting Plumbago Mountain in Newry, with Uncle Sam's backing. …
>
> The mining enterprise in tranquil Maine stems from restless politics in faraway Brazil. Here's how Plumbago diggings superintendent John I. ["Scotty"] Duffey explains the international quirk:

Stan and Hazel Perham home from 1957 onwards, West Paris 2007.

Some of the mineral shelves in Maine Mineral Store, mid-1950s, from a postcard.

> 'Most of the beryllium for many years has been produced in and imported from Brazil, but due to the political situation down there they have declared embargoes on it at various times. We hope to develop a domestic source.'
>
> There was no immediate estimate on how many hands will be employed at the Newry site, one of several being probed in this country for ore to feed Beryllium Corporation's Temple, Pa., processing plant.
>
> Beryllium Development, Inc., a Beryllium Corporation subsidiary, said it will explore and develop pegmatite ores at various sites in Oxford County. …
>
> In Albany, Maine Stanley Perham is managing the Bumpus Mine where a renewed demand for mica has revived dormant operations. Mr. Perham, who sells minerals as gems in West Paris, Maine, said Beryl will be mined at the Bumpus site along with mica for Northern Mining Corporation, lessee of the United States Feldspar Company [sic].

Wintringham (1955) noted that the Bumpus Quarry had been "inactive since 1952" and "mineral collecting prohibited". Because of the closing of the Bumpus Quarry, feldspar for the mill was needed from other localities and production had to be increased to make up for the deficit. The Tamminen Quarry, Greenwood and the Perham Quarry were two of the quarries that increased their feldspar production to compensate for the loss of Bumpus feldspar.

On June 2, 1953, Dana Douglass of Northern Mining Corporation surrendered the mining lease on the Bumpus Quarry to the United Feldspar and Minerals Company and the parent, American Encaustic Tiling Company, and turned his full attention to prospecting in the Newry/Andover area as a manager on behalf of Beryllium Development Corporation. [330] Beryllium Development Corporation was not listed as a mining company in Bethel in the *Maine Register* in 1953, although it was in 1954 and 1955. Beryllium Development Corporation was founded in 1929 and was merged with its parent Beryllium Corporation and the Kawecki Chemical Company to form Kawecki Berylco Industries in 1968. Ten years later, Cabot Industries purchased KBI and in 1986 Cabot was purchased by NGK Insulators Ltd of Nagoya, Japan. (Douglass later started a land surveying company, Dana C. Douglass Association, Inc., and in the 1960s worked on the Earth Satellite Station in Andover. He was a member of the Maine Soil and Water Conservation Board for 11 years. Douglass was also active in many charitable organizations, receiving many awards for public service, and was a clerk for 23 years at the West Parish Congregational Church.)

By September 28, 1953, Bell Minerals' financing paid-off its mortgage on the West Paris feldspar grinding

Perham Quarry tunnels, University of Maine at Orono geology students, 1968.

mill and was no longer beholding to United Feldspar and Minerals Corporation and by extension the parent company, American Encaustic Tiling Company. [331] Bell Minerals negotiated a five-year mineral lease on June 30, 1956 with Blanche Bennett for the Bennett Quarry, Buckfield. [332]

Other transitions were in the works as Carolina Pyrophyllite Company became the new name "by a certificate of ownership" on October 7, 1953, of the American Encaustic Tiling and the United Feldspar and Minerals Company. The president of United Feldspar, Albert P. Braid, became the president of Carolina Pyrophyllite. [333] Nonetheless, Carolina Pyrophyllite Company also had a short history as it was merged to form General Minerals Company on March 29, 1961. [28] General Minerals continued working the Bumpus Quarry.

Quinn (undated newspaper clipping) wrote about the retention of the Bumpus Quarry:

> "The transfer [of properties to Bell Minerals] did not include the Bumpus Quarry at Albany. The Pennsylvania firm was retaining that property, it pointed out, because it contained one of the largest known deposits of beryl. The Bell Minerals Company of Kentucky [sic] can have the mill and the feldspar but American Encaustic Tiling isn't giving up anything so valuable as a beryl deposit."

Although Douglass was still actively interested in beryllium exploration at Newry, an odd newspaper article revealed something happening in the background which has never been clarified (July 22, 1954, *Portland Evening Express*):

> "Maine mountains are yielding minerals for a new product that may be exported through the Port of Portland in great volume of use in the restoration of dikes in Holland.

Benny Benson West Paris feldspar mill manager, in baseball uniform. West Paris won the championship of the Pine Tree League at least once. photo date uncertain.
Courtesy West Paris Historical Society.

several reliable sources that export of this product developed by New England Ores, Inc. Bethel may reach 200,000 tons.

Dana Douglas Jr., firm president, denied the report that the new product is feldspar or quartz. He would only say that it is a new product developed by his firm. ...

The firm is building a silica mill in the lower part of New Hampshire, near Gilsum, and is operating a feldspar mill[sic] at Black Mountain, Rumford, and a mica mill in Vermont. One source described the new product as a 'crushed rock.' Other sources having a hand in shipping arrangements said they understood the product will be mixed with cement. They added that the product was selected because it makes, a stronger concrete than gravel does.

Holland still has extensive dike building and repair work because of the major floods of a year ago.

Tentative plans call for export of the product directly to Holland in 9,000-ton shipments. One of the two major steamship lines, operating between the U.S. and Holland, are 'nibbling at the possibility' of sending ships on regular runs to Portland to pick up the product in 2,000-ton lots, sufficient to fill one cargo hold.

Plans call for shipment of 3,000 tons to Portland every 10 days. State Pier, for one, offers exporting 10 days' free storage. Exporters asked the Maine Pier Authority if any new free storage time arrangement could be worked out because of the present plan to export in 9,000-ton lots. ...

When the shipments would start couldn't be officially learned. Estimate ranged from within the next 30 days to late fall. ...

A source said that the pier which handles these shipments 'probably would have to go out of business during its duration because it wouldn't be able to handle any other cargo.'

This expert is being considered carefully by various pier operators because Dow Chemical reportedly is going to import 3,000 tons of calcium carbonate monthly from overseas through the local port."

The proposed order seems not to have come through and so the preparations and predictions did not bear fruit. In all likelihood, the low density concrete mix may have involved a high percentage of ground scrap mica. Whatever the true material, the delivery of 3,000 tons of any mineral to Portland would have meant an enormous amount of mining activity and the magnitude suggested by the story suggests that it was either exaggerated rumor or a cover story to disguise some other effort project which was being planned. (The U.S. later perpetrated a similar hoax when one of the USSR's ships, carrying nuclear warheads, sunk in the deep ocean in April, 1968. Howard Hughes, the billionaire tool manufacturer, was enlisted to announce that he was intending to mine the sea floor for so-called manganese nodules. His mining recovery activities were the cover story for trying to recover the lost nuclear warheads and sensitive spy equipment.. The salvage operation, "Project Jennifer", in the mid-1970s, is believed to have been a partial success.)

By 1957, there may have been few restrictions on visiting the Bumpus Quarry and a fieldtrip report indicated the ability to make a casual visit. [137] The price of beryl was 20¢/pound or $400/ton "for beryl accepted on visual inspection", more than 11 times the pre-war price and more than double the WWII price subsidy, but beryl production was still below demand. [24, 180] In 1958, the USBM renewed its beryllium prospect evaluation program and it eventually studied New England beryl occurrences in 1961-1963. The only Albany location that was studied was the Wardwell Pegmatite as it was then selling beryl to the "depot" in Franklin, New Hampshire. [52] This USBM report particularly featured the Main Pegmatite at Newry because a mineral property evaluation report by a private engineering company in 1957, commissioned by International Paper Company, projected enormous beryl reserves. Unfortunately, the private report was based on faulty sampling procedures and faultier reasoning and the extrapolated figures suggested that the Newry pegmatite was one of the great beryllium deposits of the world. Several subsequently commissioned studies contradicted the initial evaluation report, but because the nearby Scotty Pegmatite, Newry, had been a more than satisfactory beryl producer, the Main Pegmatite was chosen for study by the USBM. However, after two seasons of core drilling in 1962 and 1963, the Barton and Goldsmith (1968) report revealed that the projected bonanza was really a bust. [122] In the summer of 1961, the Bumpus Quarry was water-filled and no new drilling was attempted. [407]

The West Paris mill management had the very popular Roland C. "Benny" Benson [April 24, 1897 Hebron – June 26, 1969 West Paris]. Benny had been a star pitcher for Leavitt Institute (high school in nearby Turner) and for several central Maine adult league baseball clubs, including West Paris. He was a WWII veteran. Benny was also an experienced miner.

During many years, the Bumpus Quarry was not mentioned in mineral commodity reports by the U. S. Bureau of Mines for feldspar, mica, or beryl. Oxford County mica miners had their own sales outlets in New Hampshire and the beryl miners who produced this ore as a by-product also sold their minerals in New Hampshire and that state got the credit for the production. The perennial A. C. Perham quarry continued to produce as long as the mill was grinding feldspar. The economics of a quarry virtually adjacent to the mill could not be denied. Contract miners eventually became more important to the mill as the Perham Quarry's feldspar diminished. Contract miners listed in the mid-1960s U. S. Bureau of Mines Annual Reports included Frank Perham

1967-1971, Carl Bonney 1967, Dave Buchanan 1967, Norman Jack 1967, James Ring 1967, Harold Thorne 1967, and Albert Herrick 1968. Russell "Rouse" Buck, then of Bethel was a miner for Bell Minerals and worked in the tunnels of the A. C. Perham Quarry. Wendell A. Pike was a contract miner who owned his own mining equipment. Pike worked mostly in the Stoneham-Albany-Waterford region and was good friends with Lester Wiley. Pike was a WWII veteran and had been a sulky horse racer in New Jersey. In Maine he had a career as a logger, while after mining he became a snow-maker at the Shawnee Peak Ski Area on the northern slope of Pleasant Mountain in Bridgton.

One of the innovations at the A. C. Perham Quarry in West Paris was the start of tunneling. Although the tunnels extended only about 100 feet, a great deal of feldspar was taken out from a very rich area. Unfortunately, when a quarry or mine begins to select only high-grade ore for removal, there is the implicit principle that the mining company is intending to phase out the quarry or mine because reserves are low. Selective mining, or high-grading, is the unmistakable first step before a shut down of activities.

Bell Minerals also worked Harold Perham's LaFlamme Quarry (Flamme rhymes with "wham") in Minot, Androscoggin County, until 1957. During the mid-1950s the Maine Geological Survey had been interested in granites and granite pegmatites as potential sources of uranium and thorium. There was a national exploration trend that grew out of the western states' uranium rush. The Survey identified the LaFlamme Quarry as a potential source of radioactive minerals, although few have ever been found there. The Western states had very high-grade deposits, but it was recognized that there were extensive low-grade deposits in granites. However, the economics of processing millions of tons of granite never became a reality. The Uranium Boom had made many millionaires in the Western States and some hoped to find their riches in Maine uranium.

About 1957, Armand LaFlamme [January 20, 1898 – July 16, 1963] let his contract with Bell Minerals lapse,

Priscilla Stearns Bryant Chavarie demonstrating the Geiger Counter to Levi Chavarie, 2008.

partly in anticipation of better proceeds from possible uranium discoveries. Uranium in Maine was being discussed far out of reason of its true potential. LaFlamme said that he had found pollucite at his quarry and hoped that he could make a good business on that mineral and uranium minerals, as by-products. Renewed feldspar mining in Minot was discussed. If there were a suitable local processing mill for these minerals (unspecified Lewiston newspaper, ~1958): "LaFlamme said the processing plant in South Paris [sic] utilizes the old method of crushing and processing feldspar, and with this method, the end product isn't suitable for the latest use of pure feldspar." At the time, the Littlefield Corner mill was in ruins. It is not generally believed that pollucite was actually found because the LaFlamme pegmatite was a simple one with no complex late-stage rare-element

LaFlamme Quarry, Minot, deep end on left, eventually 70 feet deep, Unidentified Lewiston newspaper, ~1958.

View of LaFlamme Quarry from the black-topped road, 2008.

LaFlamme Quarry, Minot, drained, looking toward the deep end, Ladder access on right to first level. Arrows indicate two miners on lower level. Unidentified Lewiston newspaper, ~1958.

enrichment or replacement units, although the newspaper said some pollucite had been sold for $3 per pound, but that price was an unreasonably low figure for true pollucite. A specimen of supposed "tested" pollucite was obtained by Raymond Woodman from Armand LaFlamme in 1958 and the specimen appears to be "misidentified" (R. Woodman, personal communication, 2008). About 1961, the Pechnik Brothers, Stanley and William, leased the LaFlamme Quarry as contract miners selling feldspar to Bell Minerals.

The LaFlamme Quarry was not a complex pegmatite and was not particularly known for its mineral specimens. The quarry did produced interesting black tourmaline crystals in long unterminated prisms to 30 cm. There were some large masses of silvery metallic arsenopyrite to 20 cm and some undistinguished apatite crystals. There were, however, outstanding free-standing vesuvianite crystals to 7 cm found in the pegmatite's contact zone with a calc-silicate pod near the black-topped road.

The LaFlamme Quarry was the scene of a bizarre accident in the 1950s. The feldspar truck had been driven near the lip of the quarry, above the working miners, so that the feldspar in the pit could be hoisted out by Reg Hollard and loaded onto the waiting truck. Hanness Hakala [July 22, 1891 – January 4, 1969 Norway] and Ed Herrick were on

the parked truck's load bed when Ed, who was pulling the boom, realized that it was swinging too fast and the heavily loaded bucket would cause injury if it hit them. Ed pushed Hanness off the truck and they landed mostly without injury in a heap on the ground. Coincidentally, the idling truck lost its parking brake and started to roll over the edge. Miners, including Armand LaFlamme and his son, looked up from the bottom of the pit and saw the truck coming over the edge and fled to a side wall, not knowing if they would be crushed or not. Roy Coffin was drilling and only saw the commotion in the pit, but could not hear the miners' shouts over the sound his drill was making. He realized that something was wrong, abandoned his working drill, and moved to an overhang not far from where he was drilling. The truck somersaulted into the pit and landed on its four wheels, with its engine still running and ostensibly undamaged, although the feldspar that was in the truck rained into the pit during the somersault. No one was hurt, but the still new truck had to be dismantled before it could be hoisted up to ground level. In the final stages of dismantling, the truck had to be cut into pieces using a cutting torch (Mickey Liimata, Milton Inman, personal communication, 2007).

The Wages of Feldspar

During the early 1930s, a feldspar miner with some responsibility was earning about $16.00-20.00 per five-day work week equivalent to $800-$1,000 per year, if they could get a full year's work. In 1938 the Federal minimum wage was $0.25 per hour and that rose to $0.30 from 1939 into 1944. In 1956 through 1960, the Federal minimum wage was $1.00. By 1957 and 1958, Reg Ross was a senior man working for Bell Minerals. He was a mine supervisor, but worked according to the needs of the mill. His pay checks for the two years were almost identical: $4412 and $4456 or about $2.20 per hour, the difference possibly related to overtime. As was common in the Bumpus District, Wayne Ross, Reg's son, worked for the mill, gaining two year's experience there before pursuing a career as a house builder and finish carpenter.

Microcline crystal (1.5 cm) with glassy apatite (left and left center), gray quartz crystal. LaFlamme Quarry, Minot. Specimen courtesy Richard Hauck.

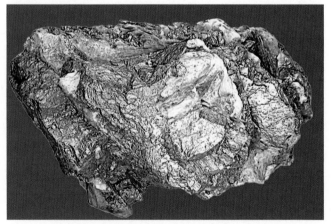

Arsenopyrite and muscovite, 15 cm. LaFlamme Quarry, Minot. Specimen courtesy Richard Hauck.

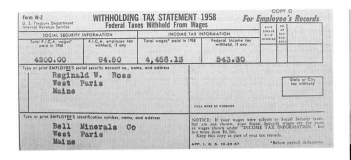

West Paris Achieves Independence - 1957

West Paris achieved independence from Paris in August 28, 1957. The new town status focused attention on local factories. It should be mentioned, as memories fade, that there were numerous small manufacturing businesses throughout Oxford County. Many provided support services in the form of machined parts, wood products, food processing, etc. When transportation was slow, local factories were necessary in the nineteenth century. Due to the improved transportation infrastructure of the State and the nation, larger, more competitive, distant companies either forced those companies out of business or prevented their continuation when key personnel retired or died. It remains to be seen if there will be a return to more localized factories as costs of transportation increase. The presence of so many factories in West Paris was unusual and represented a remarkable financial influx for the town.

Trap Corner Mining Company

Stan Perham's influence on Maine mining continued into the 1950s. Trap Corner Mining Company (TC Mining) was organized December 27, 1950 by Stan Perham, Philip H. Chadbourne, and Albert J. Stearns (OCRD v431 p267):

> "… To undertake, do, engage in, transact on any and all kinds of manufacturing, mechanical, mercantile, trading, contracting, commercial, building, agricultural, logging, lumbering, mining, quarrying, and real estate business; and any and all other kinds of business incidental, ancillary, related, pertaining, necessary or proper to or connected with any one or all of the purposes and kinds of business in this clause mentioned."

Share holders included Raymond Dean, Maine State Senator Benjamin Butler [August 7, 1905-December 13, 1980 Farmington], and others. TC Mining initially was seeking opportunities to mine beryl, mica, and other minerals that were subsidized as strategic minerals.

By July 13, 1959, TC Mining Company had New Haven, CT engineer, Willis F. Thompson [December 11, 1895-December 24, 1990 CT], as its president. TC Mining Corporation also had an association with a chemist who claimed to have developed a formula for a cesium-based rocket fuel, but the chemist died without giving notice. A search of the chemist's laboratory and notes did not produce

Reg Ross, Evelyn Hibbs Ross, Wayne Ross, ~1958. Courtesy Irene Card.

a recipe of the secret formula. In 1959, TC Mining Company repossessed the BB #7 Quarry, in Norway, along with the Noyes Mountain Quarry, and the Willie Heikkinen Quarry, both Greenwood, from the Miles Channing Mines Inc. [349] That company was going to be a refiner of pollucite to produce cesium that had been found at the BB #7, as well as any other sources that might be discovered in nearby deposits. (In 1961, Thompson, an influential engineer of Westcott and Mapes Inc., New Haven, Connecticut, was made an honorary member of the American Society of Mechanical Engineers.) The B.B. #7 Quarry in Norway, near the Greenwood boundary, also had a famous tourmaline strike on Friday, August 13, 1954.

The early fifties were relatively hard times for Stan and his family, and "many times, we were worried" (Stan Perham, personal communication, 1968), but after the Korean War tourism in Oxford County was on the rise, as well as nationally. There was also a great increase in naturalists interested in mineralogy. Tourism and mineralogy frequently went together and hardly a customer left without a Bumpus Quarry rose quartz specimen, pendant, or charm. In view of increased demand on the gem and mineral business, Stan sold his interest in the grocery and gasoline business and moved back to his former store on the east side of Trap Corner. The grocery and gasoline business had kept Stan away from the business he loved and became a drain on time of the family as everyone had to pitch in to operate these businesses. The effort just wasn't worth it, with low financial benefit. The 3,000 ton Bumpus Quarry rose quartz stockpile was moved as well.

In the 1950s and 1960s, there were other mineral stores in Oxford County, but none were so extensive as the Maine Mineral Store. One of the nicest competitors was Addison Saunders' "The Gem Shop" on Route 2 in Bethel where many extremely fine Maine minerals were on display including Charlie Marble's legendary Newry watermelon tourmaline "log". Amy Merrill had her store in a barn attached to her home on Route 2 just west of Dixfield Village. Donald "Stone Rock" Briggs and Charlie Holman (Cedar Valley Minerals) also had mineral specimen stores on Route

2 in Wilton and Dixfield, respectively. Thurston Cole sold a few minerals and gems out of his home in Rumford Point, also on Route 2. Stanley Paul sold tumbled minerals from his home: "Stan's Gem Tumbling Shop". Dean McCrillis had his store at Gum Corner in Roxbury, on the road to Rangeley Lakes Region, while further north, Joe White had a roadside store from his home and catered particularly to the gold panning tourist. Lawrence Anderson sold a few specimens from his home in East Stoneham as did Turner's Rock Shop on Granite Street in Mexico and Mike's Maine Mineral Stand west of Roxbury Notch on the road to Andover. Orman McAllister sold minerals on consignment at the grocery store in Lynchville near the famous road sign. Charlie Marble sold minerals and gems from his home and barn on the eastern edge of Buckfield village, while Charlie Bragg sold minerals from his home. Also in Buckfield, Blanche Bennett sold minerals, originating from her quarry, from her porch near the Bennett Quarry road entrance. Retired Colonel Joseph Pollack had a barn and outbuilding store near his "Brick House" in nearby Harrison, Cumberland County, but his store was important to Oxford County as he bought from many of the miners, such as Bill Liimata. Before the days of "garage sales" there were sometimes tables and porches, laden with minerals, inviting the mineral tourist to stop and make a purchase, but most of these informal roadside businesses had no name and didn't persist for more than a season or two.

Perham's Maine Mineral Store and the "Gem Man at Trap Corner"

In the mid-1950s, during the height of the vacation season, perhaps several hundred people a day were visiting the Maine Mineral Store. Many were international tourists, some were dignitaries, ambassadors, etc. An article had appeared in Maine's *Down East* magazine, April 1957, by one of Maine's prolific authors, David Oakes Woodbury [July 24, 1896 South Berwick - December 26, 1981 Ogunquit], featuring Stan and his store. While the *Down East* article was great publicity, the hugely popular national magazine, *Reader's Digest,* with it's 20 million or so subscribers, brought advertising you just could not buy. The May, 1957 issue featured a summary of the *Down East* article entitled: "Gem Man at Trap Corner" and told of Stan's life and of the enchantment a country mineral store could have. The article was not chosen by chance. Woodbury was a frequent contributor to *Reader's Digest* and the "Gem Man" article was certainly in the print queue for both magazines simultaneously.

The *Down East* article was, of course, slightly different than the one from *Reader's Digest*. Additionally, the first article had only a black and white photograph and featured the mining crew at the B. B. #7 Quarry, in Norway. The second article had only a color picture featuring the wonderful 106 carat rose quartz gem Deanie had cut. The color photograph also had a selection of gems from Maine and worldwide sources. There were, of course, several mis-reports in the article with several instances where one incident supposedly occurred before another, in contradiction to what is known. The two articles greatly increased business at the Maine Mineral Store, although it had already had become an attraction, especially for naturalists. The store was a mu-

Frank Perham, then West Paris selectman, at Trap Corner, c1964, holding a bear trap he would use to decorate the sign post. Caption on sign says: "Trap Corner at West Paris where the Indian princess Mollyockett buried gold and hung a trap in a tree to mark the spot." Unidentified newspaper clipping.

Perham's Maine Mineral Store, West Paris, Bumpus Quarry rose quartz stockpile on right, 1960s postcard.

A Topaz and quartz B Amethyst C Rose Quartz from Scribner Ledge Quarry, Albany D Aquamarine E Tourmaline, From D. Forbert photograph, May 1957 Reader's Digest.

Frank Perham (top) next to his gem tumbler, c1960. Stan (middle) and Hazel (below) Perham posing assembling jewelry, 1961.
Photos by Ben Shaub.

seum with many outstanding specimens on exhibit and most were for sale. Free maps were provided for self-guided nature tours to local quarries, mostly owned or controlled by Stan, but also including locations whose owners welcomed or at least tolerated the visits. The article suggested that 50,000 store visitors a year "crowd through those rock-cluttered rooms", but just three years later, Stan was quoted: "There will probably be around 20,000 visiting person in this region this year (*Country Correspondence,* Haydn S. Pearson, nationally syndicated column, September 2, 1960)."

The Maine Mineral Store became a significant local employer. Besides family, there were several counter persons who frequently doubled as jewelry assemblers, in addition to Deanie. The need for naturalist guides arose. However, Deanie's specialized work of gem cutting meant that he was more valuable producing gem inventory than any other chore. In 1959, mineral collector, Dale Waterhouse, native of Rumford Point and protégé of Thurston Cole's, was hired to be behind the counter and serve customers. However, when schools were booked to go on field-trips, Dale was usually their leader, a function that Deanie previously filled. Frequently, Dale would guide them to the Slattery Quarry on the east side of Paris Hill as it was a place that a school bus could drive to. The locality was also a sure success as there was so much rose quartz available. Dale's commute from Rumford Point was not unreasonable, but it was a benefit when Stan permitted Dale to use the upstairs as an apartment.

The one characteristic which many visitors remember is the amount of personal time Stan would provide to his customers, frequently to the consternation of the store's workers, including the family. The staff of the store increased as did sales. Stan and family had lived in the Store since 1930. In 1957, the increased trade enabled him to buy a separate home in West Paris village and devote more floor space in the Maine Mineral Store building to sales and displays. In fact, the increased need for floor space and storage was also a reason for moving.

*Bumpus Quarry, lower tunnel with
Janet Nemetz 2007.*

The rose quartz stockpile outside of the store eventually dwindled in size and, as the deeper colored specimens were selected by customers, the stockpile appeared to fade in color. In the late 1950s, Frank Perham constructed a tumble polisher made of a bank of large tires and in which rose quartz and other minerals could be inserted with appropriate grinding compounds. The bank of tires would be rotated around an axle and the action of the tires would tumble the contents and eventually yield polished stones including rose quartz pieces from the Bumpus Quarry. The polished stones were made into jewelry, paper weights, souvenirs, collections, advertising and promotional gifts. When Route 26 was widened at Trap Corner by the Maine Department of Transportation in the mid-1960s, the thousands of tons of rose quartz had to be moved yet again. The rose quartz stockpile was eventually discontinued about 1965.

Frank Croydon Perham
Contract Miner 1965-1966

It was Frank Perham who started the tunneling at the Bumpus Pegmatite. He was a contract miner for Bell Minerals for about a year and a half at the Bumpus Quarry, just after having finished mining for two seasons at Mount Mica Quarry. [35] Earlier, in 1962, Frank operated the Orchard Quarry adjacent to the Bennett Quarry in Buckfield and in 1963 he had worked at the Waisanen quarry in Greenwood. Frank also did occasional mining at the Pulsifer Quarry in 1964 and 1965 for Irving "Dudy" Groves and in 1966 and 1967 for Terrance "Skip" Szenics. Frank mined for the Plumbago Mining Company in 1972 to 1977, first at Newry later in a return to Mount Mica Quarry. Although many of Frank's various quarrying operations involved feldspar, Frank is currently known as the most significant gem miner in Oxford County.

Frank entered Bates College in Lewiston in September, 1952, with the Class of 1956, but:

"He joined the U. S. Army in January, 1955, for three years; 6 months basic and Advance Engineering training, 18 months in

Korea on 38th Parallel, assigned to the Wheeled Vehicle Maintenance Unit, Service Company, 34th Infantry, 24th Division.

He served 12 months at Fort Meade, Maryland, assigned to the 69th Signal Engineers." [187]

Frank married his high school sweetheart and classmate, Mary Tamminen, on April 18, 1955. He was a member of the Rifle Club in high school and the Jordan Ramsdell Scientific Society of Bates College 1954-1955 and 1958-1959. In public service, he was the RADEF officer of West Paris Civil Defense beginning in 1961 and he joined the West Paris Volunteer Fire Department beginning in 1962. In 1964, he was elected Selectman and Overseer of the Poor of West Paris and was made a board member of Stephens Memorial Hospital.

Before graduation as a geologist from Bates College in June 7, 1959. Frank spent a summer mining in the underground silver mines of Cobalt, Ontario, Canada. Although he had not been a driller in Cobalt, he had learned some of the techniques necessary for drilling and blasting tunnels, a

*Route from West Paris to Ruggles Mine.
Modified from National Geographic Atlas.*

Ice and snow in front of upper tunnel, Bumpus Quarry, 1970s. Courtesy John Kimball.

technique rarely used in Maine pegmatites. Frank asked the West Paris mill manager, "Bunny" Ring, if he could drive some tunnels into the Bumpus Pegmatite, and with this permission he was able to have an "all weather" operation. The tunnels did not yield a significant amount of either beryl or rose quartz, however.

During the 1960s, The West Paris feldspar grinding mill began to import feldspar from the Ruggles Quarry in Grafton, Grafton County, New Hampshire. This importation was necessary, because the feldspar coming into the mill was no longer high enough purity. The one-way driving distance between the Ruggles Quarry in Grafton, New Hampshire and West Paris is about 125 miles (202 km). The passenger car driving time was probably 4 hours in the time period, but the truck time was over 6 hours. Additionally, the usual load size was probably only about 10-12 tons. The trucks rarely got as much as 5 miles per gallon of fuel when gasoline was selling for about $0.25/gallon. The round trip between West Paris and the Ruggles mine, without counting the wear and tear and hidden costs of having a truck would have added almost $4.00 per long ton to the cost of the feldspar. In the ordinary market, when the contract miners were selling their local feldspar for $6.09 per long ton [1960], this was an outrageous disparity. One of the drivers, Bill Liimata of West Paris, would get up before dawn, drive from the mill to the Ruggles Quarry and arrive there about noon. The routine was similar for Wayne Lawrence and other drivers. The trucks would be loaded from a storage

hopper fairly quickly and would return by 6 or 7 in the evening. There were frequently two or three trucks per day picking up and delivering their loads. A particularly active week of trucking would mean that up to 180 tons of nearly pure microcline would be available to the mill. With this influx of high purity microcline feldspar, the West Paris mill was able to mix the imported feldspar with various sources of local feldspar. The tunneling that started at the Perham Quarry suggested that the Perham Quarry was running out of acceptable grade feldspar.

The importation of New Hampshire feldspar is not absolute evidence that Oxford County quarries were unable to supply #1 Grade feldspar suitable for porcelain and high temperature glaze pottery, but the mill was having trouble attracting the highest grade feldspar because of it's offering low prices. The mill was not able to supply the quantity of quartz-free feldspar that was on order with the mill. Frank Perham told the mill manager that the contract miners were not unaware of the expense the mill was laying out in order to improve the feed for the mill. He suggested that if the local miners were able to get more money for their feldspar, the extra price would attract more and better mineral. The suggestion fell on deaf ears.

The Bumpus Quarry had been a very important producer of highest grade microcline and there were batches of similar grade feldspar from other quarries, but the amount of locally produced pure feldspar was still insufficient to maintain the specifications of one of their most fastidious

Bumpus Quarry, lower tunnel interior, 2007.

Icicles near the entrance of the Lower Tunnel, Bumpus Quarry.

customers: Bon Ami.

Frank had a 1949 Ford F-600 truck with a capacity of about 10 tons ("when grossly overloaded"): "The truck could smell a hill half a mile away and took two miles to stop." His earth moving equipment at the quarry was made by John Deere. Frank would try to leave the Bumpus Quarry early enough to bring the day's production to the feldspar mill in West Paris so that there would be no extended delay going through downtown Bethel and then through to Locke Mills when the Stowell Spool factory got out at 5 PM. The tangle of Locke Mills traffic would result in a considerable time waiting on Route 26 whenever they left the quarry late, as the truck could not get moving fast enough to take advantage of any holes in the factory traffic. (The truck was also a "war torn" veteran and special positions had to be devised by the driver to keep one foot on the gas peddle and one foot on the gear shifting lever through an entire trip.)

The eastern wall of the Bumpus Quarry had histori-

cally been wet and enormous icicle walls always would form there during the cold months. After tunneling began, icicles built up quickly in cold months and after a particularly frosty weekend, icicles accumulated and they had to be knocked down with the front-end loader. There was also another problem. The "stalactites" (ice on the ground) were an issue, although not a serious one. Ice "stalagmites" would build up to three or four feet on a weekend and whenever these formed the tunnel entrance was effectively blocked equipment from being driven in. [401] Usually, the easiest way to remove the blockage was by drilling into the ice and loading the drill hole with a partial stick of dynamite: "Ice blasts awfully easily."

Tunneling is a specialized art, one that few Oxford County miners attempted. In ordinary blasting, the outside wall is the direction in which rocks are forced by explosions.

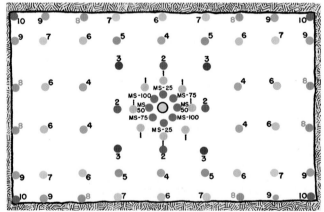

Typical arrangement of drill holes. Central holes (yellow and red) empty, uncharged, holes served as a pathway of collapse. The first charges to ignite (#1) are nearest the empty drill holes. Succeeding holes (#2-#10) explode toward the central cavity in sequence toward the edges of the desired tunnel size. Pattern shown is for a large tunnel. Blasters' Handbook, 1954.

Miner holding wooden tamping rod for inserting dynamite into a drill hole. Yellow tags illustrate drill hole pattern and are numbered to illustrate the firing sequence. In practice, markers were not placed in the drill holes by the miners and the blaster usually would hand-wire the sequence from memory. Actual working face at Zero-level, Sterling Mine, Sterling Mining Museum, Ogdensburg, N. J. 2007.

Upper Tunnel, Bumpus Quarry,
Note slope of tunnel ceiling and floor paralleling
the pegmatite's upper contact with the Songo
Granodiorite. 2007.

A simple line of holes is sufficient to move many tons of rock. The blasting caps are ignited initiating the explosive properties of the dynamite. The wall of rock in "front" of the explosives has only one direction to travel, although there are always some upwardly flying pieces. Modern dynamiters or blasters attempt to maximize the efficiency of their explosives by starting the chain of detonations at one point and letting the rock at that point begin to respond to the forces. By using microsecond delays, the caps can be ignited in sequence along the line of charged drill holes, helping to propagate the "crack" along the working face of the quarry.

In tunneling, the driller drills a pattern of holes for the explosives. He has to make an initial cavity towards which he wants the rock to move. The explosive charge is not behind the rock, but is essentially beside it. The central dynamite charges have to be detonated first. The center of

the face to be blasted is usually an empty drill hole or a set of empty drill holes. The empty central holes are collectively called a "burn hole". The charges immediately adjacent to the burn hole are detonated first. The charges next to the central ones are then detonated. The succession of delayed timing of charges expands the size of the burn hole. The "crater" exposing the central part of the working face is, effectively, a slight cavity and is a path of lower resistance toward which the force of subsequent explosives may act. Even into the 1940s, some major underground mines still used lighted fuses on blasting caps, but it was difficult to time the blasting sequence precisely. With modern detonating techniques, using electronic and radio-controlled detonators, it is possible to have explosions timed small fractions of a second apart. The short delay is still enough to begin to move rock away from a central part of the working face.

There obviously must be blasting to the edge of the planned tunnel otherwise the work area becomes too small. The back wall of the Upper Tunnel of the Bumpus Quarry shows the artifacts of the ends of drilled dynamite holes. (See illustrations with numbers or letters immediately above the holes.) The drilling pattern used by Frank was a standard

Arden Andrews driving earth mover at Emmons Quarry. Walden Rider (right) and miner (left) working in freshly blasted minerals. From Perham (1987).

Center view: Frank Perham and skiff, at Orchard Quarry 1962. Holes at back of skiff allow fine pieces to sift out. From Perham (1987), Lower view: similar skiff at Bumpus Quarry, 2007.

metal template then available from the Atlas Copco Corporation.

At the Bumpus upper tunnel, the roof of the tunnel was mined to points just below where the granodiorite was exposed. The granodiorite had no value for its feldspar and so the tunnel was made parallel to the contact with the pegmatite. Because the tunnel is in the upper portions of the pegmatite, where the roof is composed of the Songo Granodiorite, it became apparent that the concentration of giant beryls discovered in the 1920s had occurred under a "roll" in the contact. In everyday terms, a roll is where the upper contact of a pegmatite changes slope to become more horizontal to whatever was "up" during the crystallization of the pegmatite. Fluids might collect under the roll and form a concentrated zone of interesting minerals, such as beryl or a host of other minerals.

In order to facilitate the movement of rock away from the burn hole, a driller would frequently angle the central drill holes so the charges would be slightly behind the burn hole and thereby push the rock more directly into the tunnel.

Lester Wiley – 1965-1966 Bumpus Miner

The General Minerals Company, the most recent incarnation of the former United Feldspar and Minerals Company sold the land by warranty deed on March 31, 1965, to Lester and Lucy Wiley, indicating that at no time had the mineral rights been separated from the land, but had always been leased. [334] On March 16, 1967, Lester E. and Lucy Wiley of Waterford, sold a warranty deed to the former Cummings Farm to Bell Minerals Company. [335] Noted Oxford County miner, William "Bill" Liimata was a witness. (Liimata was particularly known for his being the miner at the Lord Hill Pegmatite, Stoneham in the mid-1960s.)

Also in the mid-1960s, Wiley mined and sold "feldspar and other minerals" from several quarries to the Bell Minerals Company. Wiley also had a mica trimming shop on the Waterford Road in Norway, during at least 1957-1959. (See Ashton, this volume.

Wiley's mining crew at the Bumpus Quarry, at various times, included Arden L. Andrews, Oliver Frechette, Stanley A. Pechnik [March 4, 1926 – June 19, 1983 Norway], and John Hatstat. "His daughter, Laura Wiley, was manager and clerk of the mineral specimen shop and "earned enough money from collecting fees [$1 per day per person] and the sale of Bumpus Quarry mineral specimens to start her in college at the University of Arizona. (Gregory, 1968)" (Naturalists were welcome to visit the quarry and *Rocks and Minerals* magazine (January 1966) featured a front cover photograph of two Bumpus Quarry beryls.)

Disaster at the Bumpus Quarry

Tragically, a Bumpus Quarry miner, Oliver J. Frechette, was killed by a falling rock wall while he was attempting to start a water pump near its base at the beginning of the mining day on July 21, 1966. A rescue crew made up of miners and volunteers under the supervision of Oxford County Sheriff's Department came to recover the body. At first there was hope that Oliver was merely trapped, but:

"About nine hours after the accident, the Oxford County sheriff's office said, Frechette's body was spotted under an eighteen ton boulder. Deputy Alton Howe of Bethel was able to get close enough to determine that Frechette was dead.

Sheriff Lester Horton told The Sun, 'There isn't any equipment in the state heavy enough to remove the boulder.' Preparations were under way to use dynamite in an effort to recover the miner's body from the rocky grave.

The cave-in left another section of wall hanging precariously. Operations to recover the body could not get under way until the rock ledge was blown loose. After the danger was eliminated, several miners and volunteers began efforts to locate the dead man, hopelessly buried at the bottom of an 18-foot pit, which was filled nearly to ground level with debris from the huge granite ledge. ...

Two of the miner's five children, Edward and Donald, were in a group standing by helplessly during the recovery operations.

At first it was thought John Hatstat [October 9, 1931- January 7, 1999] of Albany was with Frechette in the mine. Hatstat was late for work, however, arriving after the accident.

An unidentified miner had been in the mine with Frechette only moments before the cave-in. Officials said the worker had climbed out of the pit to go for coffee.

Mine foreman Lester E. Wiley said he

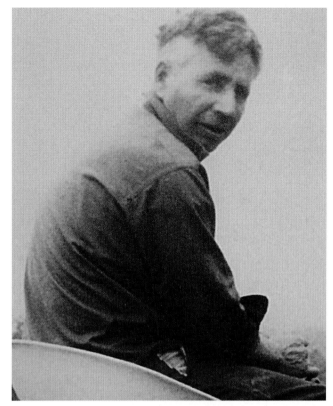

Oliver J. Frechette
(September 27, 1910 Canada – July 21, 1966)

was at a shack some 200 yards from the spot when the wall let go at about 10 a.m. No one else was at the mine, which is owned by the General Mineral[s] Co. of Greensboro, N.C.

According to the Maine foreman; Frechette had gone down to start a water pump and was just starting to ascend when the 10,000 tons of granite and rock cascaded into the hole. ...

A four man crew from the mine and volunteers worked feverishly throughout the day and night to locate the trapped miner's body.

Long idle, the mine was re-opened about 18 months ago with two men working days and one drilling and blasting at night.

Wiley said the mine employs from two to 30 men at times but at the time of the accident only two men were working. The two miners, who usually work on the day shift collect feldspar, mica and quartz, which is sold for various industrial uses. These minerals were loosened by dynamite blasts the night before the day workers collect them. ...

Frechette had been employed as a miner for the past 17 years. (*Lewiston Sun*).

Another unidentified newspaper added:

"Rescue workers awaited the arrival of a heavy earth moving machine equipped with a powerful winch before trying to remove the

Aquamarine beryl, Aldrich Quarry (Sugar Hill), Stoneham. Largest crystal about 15 cm tall. Perhams of West Paris Collection.

Bumpus Quarry, arrow shows debris collapse area. Photo from Norway Advertiser Democrat.

body. Machines at the mine weren't big enough for the job. ...

The pit was filled by the rock when the section of the 100 to 150 yard high wall gave way. 'The whole mountain must have come in on him,' Wiley said. ...

The cave-in was believed to have been caused by rain-loosened rock. It occurred about 9:10 a.m. ...

Thirty rescue workers manned electric drills, cranes, power shovels and bulldozers in the attempt to reach the body.

Hatstat helped in the rescue operations, but refused to enter the pit to try to move rocks. He said the shaft area was still too dangerous to enter.

Stan Pechnick [sic] of Paris, another miner, told newsmen he had quit last weekend because he felt the mine no longer was safe.

He was quoted as saying he had advised Frechette to follow his example."

The newspaper accounts differ on the time of the accident and the height of the overhang has the inflated claim of being 100 to 150 yards high, probably a misunderstanding of units where probably 100 to 150 cubic yards of rock collapsed. There is also a slight confusion of how many workers were at the quarry and their whereabouts. Photographs of the collapse scene show blasted rock fragments with little sandy soil or vegetation. The active rescue efforts did include Donald and Edward Frechette, who were both experienced construction workers, along with Arthur Farrar, Deputy Sheriff Robert Wood, Stan Pechnik, Floyd Carrier, Charles Millett, George Hammond, Reggie Swett, Gene Pride, Larry Thurston, Steven Brown, Randall Brown, and Roscoe Swain. "The Norway Rescue Crew provided lighting for the operation." (*Norway Advertiser Democrat,* July 18, 1966).

Mining-related deaths were known before in Maine. Several miners died in the underground silver mines in Hancock and Washington Counties in the early 1880s. Gold panner, Napoleon Bonaparte Jackson committed suicide by dynamite in Bethel in the early 1900s. Dick Nevel was killed in a dynamite blast in Rumford, on the road to the quarries in Newry in 1938. A zinc and copper miner, John Montgomery, died in an underground accident at the Callahan mine March 18, 1970. Charlie Bragg died of a heart attack, September 30, 1990, while strolling in the Songo Pond Quarry, although he was not mining that day. Despite five Maine deaths being directly attributable to mining over 125 years, the profession was certainly safer than many other lines of work such as logging, house painting, railroading, firefighting, or working as an electrician.

The Final Years of Mining

The Bumpus Quarry was shut down for about a year after the fatal accident, but Bell Minerals Company resumed the mining operations there in Spring, 1968. However, production was light. Swan (1969), Howard Irish's grandson, wrote of an Oxford County Mineral and Gem Association fieldtrip to the Bumpus Quarry: "The reason for going there

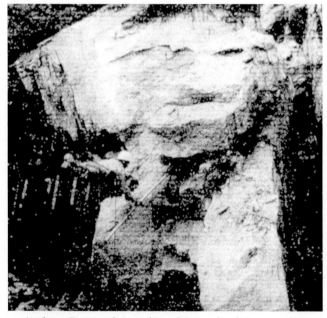

Arthur Farrar inserting dynamite into crevice. Photo from Norway Advertiser Democrat.

Perham Quarry feldspar being loaded onto a truck, ~1966. Courtesy John Kimball.

Remnants of dynamite drill holes, Upper tunnel, Bumpus Quarry, Letters placed immediately above the ends of the drill holes, 2007.

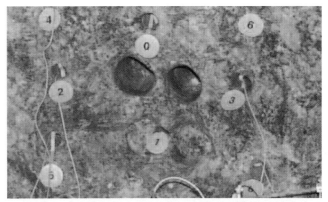

Close-up of the actual working face at the Zero-level of the Sterling Mine, Ogdensburg, New Jersey, 2007. There are two central large drill holes (about 10 cm) which would contain no charges. The yellow tags, numbered 0-6 are in the central part of the tunnel face and indicate the sequence in which the smaller (about 5 cm) drill holes, would be ignited. One or several sequential numbers might be ignited at a time, for example 0 and 1, followed by 2 and 3, etc.

was because a geologist, Mr. Frank Perham, had blasted a good find of beryl all over the mine, and he said that the beryl was of good quality so Mr. [Nestor] Tamminen arranged for our group to go and look around." Nestor was Frank's father-in-law. Some new tunnels were attempted during this time. A short middle tunnel was made as were a few blasts which could have been followed although the hanging wall was still close to the current walls of the quarry.

In one of the fate makes history coincidences, Edwin Gedney returned to the Bumpus Quarry, one last time in 1968. Dr. Gedney had written the first significant article about the Bumpus Quarry as a result of his visit there in one of its first years of operation. Gedney had been a professional geologist and science educator. He had maintained a mineral collection, partly as he had taught geology courses. When he returned, there was still rose quartz exposed in the working face of the quarry.

Wood products mill in Locke Mills. From a 1920s postcard.

Remnants of angled and parallel dynamite drill holes in large blocks or rock (about 2 x 2 meters), Sterling Mine, Ogdensburg, New Jersey, 2007.

Wall of nearly pure feldspar near the entrance of the Upper Tunnel, Bumpus Quarry, Note reflective cleavage about 1 meter across.

Bumpus Quarry ticket office and gift shop, 1966.

Disaster in West Paris - 1969

The accidental burning of the feldspar grinding mill on May 1, 1969 was a severe blow to all feldspar producing activities n Oxford County. The fire started before 3 AM and was apparently caused by a rock which jammed a conveyor. The pulley system continued to rotate and the rubber belts heated up and caught fire to the wooden cased chutes. Four units of West Paris firemen responded and both South Paris and Norway sent firefighting squads to battle the blaze (*Advertiser-Democrat,* May 8, 1969). The fire was out of control for four hours and one fireman was quoted when the flames were out: "There's not very much left to burn." The article continued: "Orders for replacement of the machinery destroyed have been issued, and the demolition of the building shell will be started this week to speed the construction of new building."

Ela (1971) reported that feldspar mining in Maine was suffering in 1969:

"Production of marketable crude feldspar was reported only from mines in Oxford County and was 17 percent lower than in 1968. All the output, together with a substantial quantity of high potash feldspar from the Ruggles Mines in Grafton, N.H., was processed at

Bumpus Quarry upper and middle tunnel, 1968. Note dark basalt dike to left. Courtesy Bob Gedney.

Bumpus Quarry debris collapse area, 2008.

West Paris, Oxford County, by Bell Minerals Co. Ground feldspar was nearly one third less than the previous year because a fire destroyed a portion of Bell Minerals Co.'s grinding mill near West Paris. The ground feldspar was sold primarily for ceramic applications to consumers chiefly in Pennsylvania, New York, Ohio, New Jersey, and Massachusetts. ... Scrap mica, the first [produced in Maine] since 1962, was recovered at the Bumpus Mine by Bell Minerals Co. The mica was stockpiled at a plant nearby for grinding at a later date. "

It was also reported that "Sales of small quantities of beryl concentrates, were reported for the first time since 1962. Feldspar from the Ruggles Quarry in Grafton, New

Edwin Gedney in lower tunnel, 1968.
Courtesy Bob Gedney.

Feldspar mill next to Little Androscoggin River, 2007.

Hampshire was trucked to West Paris for processing. The beryl concentrates were recovered by Bell Minerals Co. from the Bumpus Mine in Oxford County." [85] This was the last report of beryl ore sales from Maine and the Bumpus Quarry. The feldspar mill was partially renovated, but the Maine feldspar business had become marginally economic and by September 1, 1970 the mill closed. At the time of closing, the mill's crew and the variable number of miners amounted to about 30 to 35 employees. Al Wallen was the last feldspar mill manager.

Reviving the West Paris Mill

Initially, Bell Minerals Company announced within a week of the fire that it had "already taken steps to re-activate its mill", but the company was discouraged from continuing the mill's operations, even though it eventually re-invested

Bumpus Quarry, idle, again, c1969.
Courtesy John Kimball.

Edwin Gedney examining rose quartz pod at
Bumpus Quarry, 1968. Courtesy Bob Gedney.

West Paris Feldspar Mill and Office.
c1958 Courtesy Ben Shaub.

$200,000 in the restoration of the mill. Ground feldspar production had been declining, expensive New Hampshire imports were cutting into profits, and the fire was almost the "last straw". Bell Minerals decided to shut down its "Oxford Division" and concentrated on its southern holdings in North Carolina. Local newspapers contained articles about the mill's future:

"Civic minded citizens of this community have banded together, are pooling their own money to get the business [a Local Development Company] in operation.

Local people must put up their share of the cost for purchasing the closed plant, for operating capital and for some needed improvements. The balance of the funds needed will be sought through a Small Business Administration [SBA] loan, and a mortgage through a local lending institution.

Besides throwing 15 residents out of work, the plant shutdown meant a $5,600 loss in property taxes to the town – approximately six percent of the total revenue. In addition, the water bill for the feldspar operation averaged nearly $1,500 per year. …

Estimated figures show that the West Paris group would need a total of $175,000 to purchase the Bell plant and resume operations.

Alfred F. Wellen [sic], former general manager for the local branch of Bell Industries, already has said that to the best of his knowledge the plant never again will operate under Bell ownership. …

Of the $175,000 required for a takeover of the operation, $100,000 would be required to purchase the plant, its operating equipment and various parcels of mining property. Another $50,000 would be used as working capital and the remaining $25,000 used for plant improvements.

The improvements would involve primarily the installation of a floatation processing stage in the refining operation. This would enable the revitalized company to produce the high grade spar in demand by major customers. (undated, unidentified clipping).

An additional 20 or so miners also were unemployed due to the mill's closure.

The SBA required that at least 25 local people pledged to loan the recipient $25,000 and that a local lending institution had to underwrite some of the loan. The organization which would run the mill would be a Local Development Company, with investors and loans to insure success. Stan Perham mentioned that the loss of the mill would also close local quarries and that about 12,000 tourists a year came to Trap Corner and West Paris, all particularly interested in collecting minerals at local quarries, although he described the attraction of these tourists in more general terms. The loss of the mill would have a ripple effect through a broad range of the economy. (Another unidentified clipping indicated that only $5,000 were pledged.)

Frank Perham, who formerly supplied feldspar to the mill, had been a West Paris selectman. etc. volunteered to advise on engineering concerns for the mill and researched market conditions for the ground feldspar. When a customer is no longer able to buy from a former supplier, they find a new supplier and rarely return when the old supplier returns to the marketplace. Frank was quoted:

"'We were tailoring our product on the premise we could rely on ... two major consumers,' he explained, 'But the hard facts of the matter are that these two users … are now demanding certain specifications instead of preferring the as[-]manufactured product, and frankly, I rather doubt that we'll be able to get them back as customers.'

There were particular concerns about the amount of soda feldspar, mixed in the otherwise potash-rich feldspar. The two kinds of feldspar, albite and microcline were being produced by Oxford County's quarries and the two feldspars were not easily separable. (Soda-rich feldspar, albite, had very detrimental firing characteristics and could not be used in some porcelain and other ceramics but was suitable in abrasives and other products.)

December 1, 1973

The closing of the feldspar mill was not the only transition to affect Oxford County mining. One of the County's most important mineralogists died of a heart attack while taking a morning stroll: Stan Perham. Stan's life has been summarized in parallel with the mill and Bumpus Quarry story, but much more can be said about him as a person. One characteristic which will be remembered by those who knew him was his measured bass voice which freely flowed without stammer or fault when discussing minerals or history of mining. In many ways, his life was a struggle to adapt to changes that occurred in his own lifetime: the Great Depres-

Stan Perham at Black Mountain Quarry, Rumford. c1940s. Courtesy Ben Shaub.

sion, WWII, changes of businesses to and from groceries and gasoline and booms and busts of mining. Each adaptation underscores his resourcefulness. He had a life full of public service and he shared his time freely with friends and visitors. His last visit to a quarry was to one of his favorites: the Dunton Gem Quarry in Newry. In 1972, an extraordinary gem tourmaline discovery was made there and Frank Perham had been hired as the principal geologist and miner by the Plumbago Mining Company that had been organized to work at Newry. Stan was carried into the quarry on a large chair so he could see first hand the marvelous discoveries being made – but that will be another story. His daughter Jane, a graduate gemologist, and son Frank, a graduate geologist, succeeded him.

Rupert Fremont Aldrich
[April 1, 1908 Paris – August 2, 1987 Norway] Maine

Perhaps because of all the mineral actions Rupert Aldrich observed while he was Oxford County Clerk, he developed an interest in gems and minerals, although there had

Rupert Aldrich looking at a mineral. Note gemstone faceting machine in background.
Ben Shaub photograph, 1961.

Repaired West Paris Feldspar Mill 1971.

been mining in his family. Today, many current mineral naturalists owe their Nature interests due to the mentoring he provided. He was particularly influential on Ron and Dennis Holden who discovered the famous gem morganite beryl at the Bennett Quarry, Buckfield in 1989. In the 1970s, Rupert Aldrich discovered several blue gem beryls at what is now known as the Aldrich Quarry, or Sugar Hill Quarry, Stoneham. The crystals had flawless aquamarine areas and from one of these pieces was cut a 128 carat pear-shaped gemstone, the second largest faceted aquamarine gemstone from North America. The current North American record aquamarine gemstone is only slightly larger, but with small flaws, 137.16 carats, and was cut from a piece found on the nearby Melrose Farm, in 1882, also in Stoneham.

Rupert was also a lawyer, notary public, and an elected politician. His name appears as a notary or Clerk of Courts on many mineral leases filed in the Oxford County Courthouse Registry of Deeds Office. Rupert's grandfather, Charles Aldrich, was the first to discover feldspar in West Paris, at the future site of the Perham Quarry, in 1915. Charles trucked his feldspar to Littlefield Corners in Auburn (unidentified newspaper interview, c.1971).

Rupert had been Senator Margaret Chase Smith's suggested appointee to fill the vacancy of the Maine State Attorney General: "Her first choice, Rupert Aldrich of Norway, was rejected because the Justice Department felt he lacked sufficient training and experience. That he 'read law' in Lincolnesque fashion, instead of earning a degree from an accredited law school, appeared to be a reason why he was not chosen." (*Sun Journal*, June 27, 1953). It can be added that the time period was during that of Senator Joseph McCarthy's reign of terror. Smith's second choice was deemed to be too old, and, while her third choice, Peter Mills of Farmington, was accepted only after the Federal Bureau of Investigation had checked Mills' library records to see what books he had read and that they were politically acceptable.

Rupert was the son of Elmer L. Aldrich and Mineola Buck, also given as Minda Buck in some sources. He married Mary Swan on September 13, 1928. He was also a member of the Methodist Church and Kiwanis, was a Freemason and an Odd Fellow. (Noted Oxford County mineral naturalist, Henry Swan married Joyce Spencer whose step-mother, Sally Irish Spencer, owned Mount Mica.) On March 9, 1971, the Maine holdings of Bell Minerals Company, the feldspar mill, its mineral leases, and properties were purchased by Rupert F. Aldrich of Aldrich Realty Company, excepting: "This deed does not convey any right, title or interest in and to a certain lot or parcel of land now or for-

Scott Hartness holding coarsely ground
Rangeley garnet.

Wing Hill Quarry, Rangeley. Very large quadrangle in center with yellow dot is the Rangeley airport. Courtesy Google Earth, 2008.

Wing Hill Quarry, 2008.

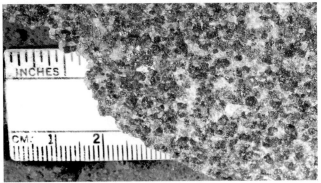

Garnet ore, Wing Hill Quarry.

Wing Hill Quarry, Rangeley.
Courtesy Google Earth, 2008.

merly known as the homestead of Joseph W. Cummings and Abbie A. Cummings, located in the County of Oxford, State of Maine, described in a deed to United Feldspar & Minerals Corporation from Geneva B. Levinton...” [336] The sale did include about 600 acres of land distributed in West Paris, Paris, Albany, Minot, Hebron, Stoneham, Sumner, and Buckfield.

Rupert made an abortive attempt to renew the West Paris mill as the Oxford Feldspar Corporation, in 1975, but only 200 tons of ground feldspar were produced and sold. [252] Feldspar was obtained mostly from contract miners. Unfortunately, the Aldrich Realty Company, which sought to revive the mill went bankrupt in the attempt to revive the feldspar mill and the mill fell idle.

After Life of the West Paris Mill – Garnet

There was a discovery in the 1950s of a huge garnet-rich deposit (1 km long east-west by 700 m deep), containing over 50% garnet, north of the Rangeley airport. The importance of the find reached a high level in the 1970s. The Industrial Garnet Extractives Corporation (IGE) was organized in 1978 by Michael O'Connor and Scott Hartness, to mine it to make garnet sandpaper, sandblasting compounds, filtering agents, and non-skid grit. The Oxford Hills Development Corporation, with Harvey Wiley as executive director, was important in the revitalization of the West Paris mill and facilitated a $50,000 loan provided through the Androscoggin Valley Regional Planning Board.

The Wing Hill garnet quarry was located about 2 km almost due north of Rangeley village, Franklin County. There was a pre-crushing stage mill erected on the Wing Hill quarry property to reduce the rock to 3 cm size fractions. The West Paris mill had not been dismantled and was purchased, also in 1978, and renovated to accommodate the new mineral for grinding. The mill used two large magnetic Kellogg separators that emanated a loud hum through the mill. Iron-bearing minerals such as most garnets are very slightly responsive to magnets and crushed garnet passing through a magnetic separator can either be removed as waste, as in a feldspar mill, or be recovered as saleable product, as in a garnet mill. The red waste pile that was rich in garnet, but was not pure enough for specialty users, was sold to sandpaper manufacturers. The recovery rate of high-grade garnet was about 35%.

Garnet crystals, largest = 2.5 mm, Wing Hill Quarry.

Wing Hill Garnet Quarry buildings, 2008.

The mill attempted to improve the recovery rate using water separation, in addition to the magnetic separation, but they were not successful (Frank Perham, personal communication, 2007).

Several Garnet Mill workers remembered that the huge motor, used to rotate the ball mill and inherited from the feldspar operation, sounded and looked much like the proverbial cinematic mad scientist's equipment operated during lightening storms. A long hand-operated shifting lever jutted up from the floor and was used to increase the rotation speed, with the shifting process taking four or five minutes to accelerate to the full speed of 10 rpms. As the 100 horsepower motor revved up, ozone and evil-looking electrical sparks would emanate from the motor. The motor had behaved the same way when it was operated by Bell Minerals in the 1950s. Formerly, the ball mill was operated "dry" when grinding feldspar. Garnet was ground "wet". The change became a problem in winter as outside temperatures would frequently reach -25° to -35° F and the water in the ball mill could freeze as the building was essentially uninsulated. For this reason, the mill employed a person to go into the mill to fire up the heater at 5 A.M. every day in the winter.

The garnet was lightly to thoroughly crushed and sieved to a variety of sizes (ranging down to microns), unlike the feldspar which all had to be pulverized "finer than flour". Some specialty garnet product was reduced to 3 micron particles. (A bill of lading dated February 18, 1986 indicated that #34 garnet sold for $0.24/pound, while #12 grade sold for $0.22/pound (Robyn Greene, personal communication, 2008). This particular shipment of just over 1000 pounds was sent by truck via APA Transport Corporation of New Jersey [founded 1947 and disorganized 2002].)

By 1982, IGE was the second largest producer of garnet in the USA. The West Paris mill was expanded in 1985 and increased its production by 60%. IGE had quite a few major customers and was looking forward to increased prosperity. Additionally, there were relatively few competing domestic sources of industrial garnet, the principal ones including Gore Mountain, Warren County, and the NYCO quarry, Willsboro, Essex County, both in New York and Emerald Creek, Benewah County, Idaho. [253] The IGE work force, presumably including the quarry, the trucking crew,

and the mill staff, numbered 32 and had been rapidly increasing. The mining season was only about six weeks as the blasting yielded up to 12,000 tons of raw material in just three large blasting events. A newspaper clipping (possibly Portland Sunday Herald, 1985) quoted: " 'On the overall picture,' Wiley said, 'I feel very sure that they have a very good future in front of them.' "

Austin (1993) wrote: "The Wing Hill garnet deposit, near Rangeley in western Maine, is one of the largest and highest grade garnet deposits in the world. The host rock is a homogeneous garnet granofels that consists of 50% to 70% almandite garnet. It is medium grained and consists essentially of garnet and andesine plagioclase along with biotite and minor quartz. The garnet is present as discrete, well-formed crystals from less than 1 millimeter to about 10 millimeters in diameter but averaging 1.5 to 2 millimeters. The granofels is an east- west trending, tabular body more than 1,000 meters in length and as much as 700 meters thick. … When crushed, the garnet breaks into blocky grains, a shape that persists down to the finest fragment."

Despite the size of the garnet deposit (1.8 million tons of reserves), the very high grade, the low percentage of free quartz, and the seemingly limitless size of the USA garnet demand, three years later the IGE president, Ralph A. Dyer announced that problems of "maintaining a reliable supply of high quality raw material, foreign competition, and inefficiencies in the engineering of the old mill" were reasons for shutting down the mill (unidentified May, 1988 newspaper clipping). The clipping further revealed that, by 1988, IGE was then getting much of its garnet from New York, not Rangeley, and that garnet ore, of only 10% grade more or less, was declining in quality. The turn-around from local Rangeley garnet to imported New York garnet, was attributed to: "The Company has also been involved in a legal dispute concerning its right to mine garnet ore from a site in Rangeley." It was further revealed that IGE could buy Chinese garnet delivered to Portland Harbor more cheaply than it could produce ground garnet form New York garnet ore. IGE had tripled its production and reduced its workforce to only 12, but still could not make a profit. Evans and Moyle (2006) added that between 1991 and 1993, Pittston Mineral Ventures of Greenwich, Connecticut, and its subsidiary, Rangeley Minerals Resources, could not get the town of Rangeley to re-zone the Wing Hill garnet area and they then abandoned their lease on the land.

Albert B. Kimball and a Naturalist's Location

Perham (1971) reported: "Some of the darkest rose quartz seen in some time came from the Bumpus Quarry dumps down near the roadside." It may have been that Stan Perham's report indicated that the Bumpus Quarry had begun to receive visitors in 1971, but certainly, by 1972, Albert "Cannonball" Kimball [February 22, 1908 –March 26, 1984] was hosting mineral naturalists' visits to the Bumpus Quarry for a fee, usually $2 per person per day, although after several years, the entrance fee was raised to $2.50. Kantz (1977) quoted Kimball:

"The most I ever took in here last summer, let's see, one day I think I took in $88. But, see, that's selling rocks and everythin'. But, then, I have a couple of men here that I have to pay out of it, see, my grandson, you might say, oh, I have a policeman out of Portland that comes up here weekends and helps me. …

When Rupert Aldrich took it over … and knowin' me and that I owned two sides of this land – I owned it, good God, let's see, I owned on three sides – and so he called me right up and I went down and he gave me this mine to take care of. I made it right out in my deed … he's got the minin' rights to come in here any-

time he wanted to and pay me so much a ton." The wooden mineral shop building was still there from the Laura Wiley time period. Tickets were sold and better quality specimens than could be easily found on the dump, could be purchased.

Because the quarry was so close to the paved road collectors did not always pay attention to the "No Trespassing" signs. When caught, some visitors, would refuse to pay the minimal entrance fee and they frequently had to be asked to leave, sometimes after considerable argument.

When the quarry was open to the public, Albert and his grandson John worked most days in the summer and, in the beginning, would arrive about 9 a.m. Every day, they would begin by patrolling the dumps to collect the entrance fee from local visitors who had arrive early, although they sometimes found evidence suggesting that a visitor had already come and gone, either on the previous evening or between dawn and the "official" opening time. Eventually the

Albert Kimball and grandchildren in front of revised sign, "If not willing to pay, Stay out". Courtesy John Kimball.

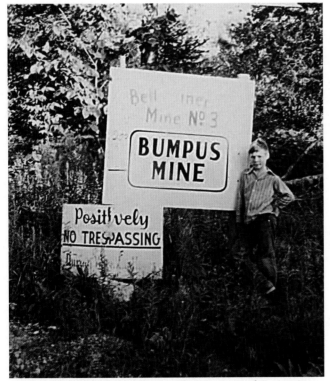

John Kimball at Bumpus Quarry road entrance, ~1972. Sign from earlier times indicating that the quarry was also known as the "Bell Minerals Number 3 mine". Courtesy John Kimball.

John Kimball in front of beryl in ledge. "For Display Only Keep Out!!! No digging-chipping or picking here!" Courtesy John Kimball.

signs specified that the collecting day was 8 a.m. to 5 p.m. John and his grandfather also developed an exhibit of beryl in the ledge, using hand tools to expose some crystals, to show visitors what the mineral looked like in its natural place in the ledge. Despite having a wood enclosure, the meager beryl displayed was eventually taken by a trespasser. In the 1970s, the Bumpus Quarry was one of the few fee localities in the State. Over the decades, there have been a number of fee-based localities with Newry one of the first when a 10 cent daily entrance fee was charged in the 1930s. The Bennett Quarry, Buckfield; Tamminen Quarry, Greenwood, Black Mountain, Rumford, Howe Amethyst locality, Denmark, and Mount Mica Quarry, Paris were also fee localities for many years. Generally with a $1 per day fee.

John Kimball had a good educational opportunity working with his grandfather. Albert taught his grandson the lessons of surviving in an area that had few large industries. Albert's generation lived through the Great Depression and had developed many skills simply because, if you did not personally learn to do every job relating to home, farm, vehicles, and woodsmanship, you could not make a living. Albert impressed on his grandson, "You always have to have a dollar in your pocket." Advice from an era when constables could detain people simply because they had no money in their pockets and thus were unable to support themselves that day. The "no visible means of support" charge was akin to vagrancy charge and was generally aimed at strangers in small towns, but the object lesson was still valuable. When John entered High School in Bethel, grandfather Albert entered a new business.

The Bumpus Quarry was closed to mineral naturalists due to Albert Kimball's and Robert Lowe's unusual new business: lining bees. Maine law provided that anyone locating a nest of wild bees in a tree may write their name on the tree and claim the bees as their own, even if the land was owned by someone else. Albert and Robert would watch honey bees gather pollen, catch them, mark them with a dot of paint, release them, and then would follow the bees' lines to locate their home colonies. They would then "domesticate" the bees for use by farmers who needed them for pollinating their crops. Albert, who had been a registered Maine Hunting Guide as well as a cattle dealer, used the Bumpus Quarry premises for keeping his bees in transition to local farmers. [245]

On June 1, 1977, the original mineral lease between Allen and Sybil Cummings and Harry and Laura Bumpus on the Bumpus Quarry, officially expired; although the lease had been surrendered and purchased on October 11, 1949, many years before. The drilling and blasting to recover tons of feldspar, beryl, or rose quartz were over. The Bumpus Quarry was variously opened or closed to mineral collectors and casual visitors in subsequent years, but after a while, was only closed. The quarry area began to be reclaimed by vegetation, gradually hiding and overwhelming a location that had an influential half-century run.

Pegmatites continued to be prospected in Oxford County, mostly by mineral naturalists. An internationally im-

Albert Kimball at the Bumpus Quarry,
Upper view from Kantz (1977).
Lower view courtesy John Kimball.

Albert holding a bee box for holding marked bees. Lewiston Evening Journal photograph.

Rodney "Sonny" Kimball, John Kimball, and Albert Kimball. c1972. Courtesy John Kimball.

Albert Kimball displaying symbols of some of his trades, hunting and apple growing. Courtesy John Kimball.

John Kimball in front of revised sign. Courtesy John Kimball.

West Paris mill buildings behind Penley Brothers Wood Products Company, 2007.

portant tourmaline discovery was made at the Dunton Gem Quarry in 1972-1975. The internationally renowned Mount Mica Quarry, the oldest gem locality in the U.S.A. was reopened in 1964-1965, 1977-1980, and again in 1991 to the present. There are other pegmatites in the County which have been purchased or leased by individuals and small groups who enjoy mineralogy and gemology and who have made important finds.

Lawrence Stifler - Mary McFadden
Recreate the Naturalist's Location

After the Kimball's had opened the Bumpus Quarry to visitors, the land had some intermittent legal activity. Albert Kimball's son, Rodney "Sonny" Kimball, had inherited the Bumpus Quarry lot in 1984. In the early 2000s, Lawrence A. Stifler and his wife, Mary McFadden, recognized that the heritage represented by the Bumpus Quarry was deserving of preservation. They formed the Bumpus Historic Mine, LLC organized under Maine law, on July 29, 2004 to preserve the location and to develop a program whereby the location would serve as an educational and historical resource for everyone. The purchase of the former Cummings Farm was made October 8, 2004.

The lot immediately north of the Cummings Farm was also available. Albert Kimball had previously sold that lot to Vernon I. Turner and Mildred E. Turner on November 18, 1971. The Turner's were interested in having a larger piece of land and Albert Kimball sold an additional lot, to the north of the first lot, to them on November 16, 1973, but they soon sold the lots back to Albert and his son, Rodney E. Kimball, on August 28, 1974. Albert conveyed his share in the land to John A. Kimball and Rodney E. Kimball on June 23, 1976. Rodney purchased John's portion, as tenants in common, on "Leap Day", February 29, 1996. The land was eventually offered for sale as a "Camp Lot", but was also sold to Bumpus Historic Mine LLC on October 8, 2004.

Naturalist Tours Resume at the
Bumpus Quarry - 2005

More and more locations are now generally off-limits to naturalists. In 2005, the Bumpus Quarry was reopened to public view with tickets and appointments available for guided tours and mineral collecting (*Sun Journal,* October 15, 2005). The first quarry guide and manager was Seabury Lyon of Bethel working for Maine Mineralogy Expeditions in association with Bethel Outdoor Adventures.

There has been a transition in catering to people interested in the out-of-doors and naturalists' pursuits. In 1820, fish and game clubs were first organized in Maine. Eventually, there were clubs in many locations, which by the 1870s were incorporated. Excursion boats became popular not only along the coast, but there were steamships on many Maine Lakes, such as Sebago and Moosehead Lakes. Even Lake Pennesseewassee in Norway had steamship cruise boats by the 1880s. The mid-nineteenth century also saw the dramatic

rise in Natural History Societies and Lycea in the Nation and every major American city established a Natural History museum. The Portland Society of Natural History and its museum were founded in 1843. The Victorian Age saw the rise of collectors, evidenced by the antique glass cabinets in "grandma's" house, now thought to have been only for the fine china rather than for curio collections. Perhaps during much of the early nineteenth century and Maine's gem and feldspar mining era, visiting naturalists were guided by

Nature kiosk at Bumpus Quarry. Beryl and rose quartz display specimens. Photo by Kreigh Tomaszewski, 2007.

Nature group at Bumpus Quarry. Courtesy Seabury Lyon, 2007.

West Paris mill building, 2007.

Top left. Ernest Yap, owner of the West Paris "mill", 2007.
Left center: South Wall of Mill adjacent to railroad tracks.
Lower left: Details of machinery drive belts and
arms of powdered-feldspar bag fillers

Upper right showing north side of mill.
(second down right) East side of West Paris mill building,
note new addition, 2007
Third down right view is the former east wall of the 1925 structure..
(Bottom right)   separation spirals

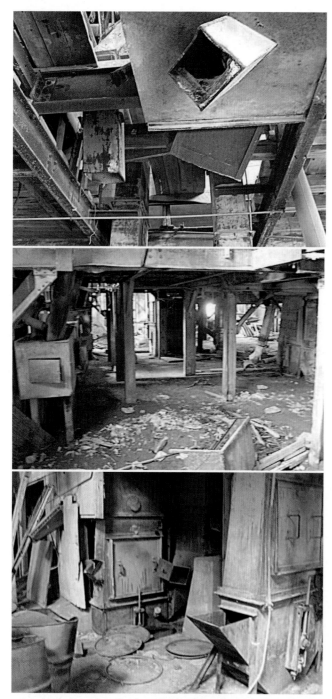

West Paris mill building, 2007. Upper left view showing cut off wooden ground-mineral chutes. Upper right showing ground garnet bin for filling bags. Center and lower left showing sub-structure of the 1925 section. Lower right view showing ground-feldspar elevators.

quarry operators or land owners to interesting mineral locations as professional courtesies. Both Ezekiel Holmes and Elijah Hamlin guided mineralogists to Mount Mica during the 1820s. Augustus Hamlin, Nathan Perry, Edmund Bailey, Edgar Andrews and others guided visiting mineralogists during the late nineteenth century. Sometime during this period, there must have been visitors who merely needed to be shown the way and hired local naturalists to assist them. Some may have been botanists or ornithologists, but the natural history guide may have been an occasional occupation in Oxford County by the beginning of the twentieth century. By the 1950s, Charlie Marble was often employed as a guide to Oxford County's mineral locations, although he might leave the visiting party to go off berry picking on his own

while collectors dug into the dumps for treasures. Stan Perham began offering mimeographed maps for self-guided excursions, and there also evolved full fledged guidebooks showing visitors where they might examine active and abandoned quarries. By the 1960s, naturalist-guiding became better organized and whale and seal watches proliferated along coastal Maine. There were even tours to watch lobstermen haul their traps. About the same time, the outdoor adventure concept became very popular inland and by the mid-1990s, there were probably dozens of opportunities for the public to discover nature alongside well-informed guides. One of the advantages to guiding for the visitor was the problem of locality access. Rising population in Oxford County meant that residents began to live closer to mining properties and "No Trespassing" signs began to be erected.

The West Paris Mill After the Feldspar/Garnet Era

What happened to the grinding mill? Shortly after Industrial Garnet Extractives ceased operations, the mill was the site of several specialty recycling companies, including cardboard and plastics, but each had difficulty in obtaining materials to feed their business needs and each quickly closed. The mill was no longer in use as a grinding facility, but merely a large volume enclosure in which the inventory was protected from the elements of nature.

In the 1990s, Ernest Yap became the owner of the mill buildings and established a recycling business for automobiles and parts. The mill building is now used for storage of automobiles and various metal stockpiles which may be recycled as "ore" and see a new existence. The site also is used as a resource of used automobile and truck parts that also will have an extension of useful life.

Epilog

Mining in Oxford County is no longer a commercial venture. Hobbyists and tourists dominate current mineral production. Mineral collecting is a popular pastime with the renewed hope of finding a crystal treasure once discarded

West Paris in 2007.

Railroad tracks and Bates sign looking west toward Bryant Pond, 2007.

and unnoticed by the frequently vigilant miners. A few robust naturalists still work in old quarries, generally turning over less rock in a year than former miner crews turned over in a few days, despite the new technology and equipment.

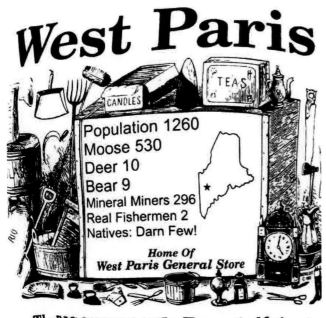

Popular tee-shirt of the West Paris General Store

West Bethel railroad station looking west. 1920s postcard.

Chapter Nine: Work at the Songo Pond Quarry, Albany
By Jan Neal Brownstein, Albany, Maine

The Songo Pond Quarry, also formerly known as Kimball Ledge, in Albany has been a mystical locality for beautiful blue beryl and aquamarine. It was known probably in the late 1920s and was prospected in 1930 (See King, this volume). The *Bethel Citizen* (April 14, 1993) quoted: "'When I was in the pit in 1974 it was very small.' [Jim] Mann said. 'I worked the mine for 30 days. I realized then if someone had the equipment and the technical capability to work the pit, it had the potential to produce some very

Songo Pond Quarry, looking east, Mid-November, 1992. Photo by the author.

fine blue beryl (aquamarine).'" Many years intervened and there was sporadic activity there and, in the 1980s, Charlie Bragg and Ron Larravee, began to pursue the locality in earnest, but Charlie's untimely passing brought that activity to a standstill. After a brief hiatus, I, too, became interested in this locality.

My reasons lay in my own distant past. I can't remember how old I was the first time that my mother brought me to see the Smithsonian Institution museum in Washington, D.C., but I'll never forget how I ran to each display case to see the gems and mineral specimens: to see colors and shapes that still amaze me. I might have been 6 or 8 years old and I was already looking for minerals any time I went into the woods around my home in suburban Maryland. By the time I was 12, I became the youngest member of a Washington area mineral club. Later, my friend, "Mike", gave me a geode for my 13th birthday. It must have cost all of $3.00 at a mineral show in October, 1962. My interest stayed with me and by the summer of 1992, I had decided to pursue my interest more seriously. I obtained a "lease", actually a handshake agreement, on the Red Hill Quarries in Rumford and prospected for rare minerals, mostly phosphates. My partner was Al Gestaut. We needed a bunkhouse so I moved an old church bus, painted sky blue, up onto the hill and lived in the bus with my dog, Topaz, a large Rottweiler. (I told people

she was an AKC registered "rockhound".) Topaz's duties extended to chasing off the harassing coy-dogs of the hill and even chased an angry bull moose away, early one morning. The bus had an LP gas furnace, stove, lights, etc. and we spent 4 or 5 days a week in the bus, prospecting during the day, and mining on a shoestring budget. That summer's experience was a dream as I learned new techniques to drill and blast rock for the first time. Mark Duhamel, a stone mason from Gray, Maine, taught me many techniques of mining including feather and wedging the rock into steps, blocks, posts, and walls. Mark and I first met when I was volunteering at the Maine Handicapped Ski Camp held at Sunday River Ski Area near Bethel, Maine. During this time, our blaster was Jim Mann, who mentored us drilling and blasting. It was under Jim that we began to learn the tools of the trade. Although we occasionally felt edgy, safety was always the watch word. We kept all of our fingers and thumbs with no incidents while we were there. Still, the rare rose quartz crystals of that place eluded us.

In August, 1992, I heard that the Kimball Ledge, even then known as the Songo Pond Quarry, was for sale. A little over 30 days later, Alan Obler, of Cumberland, and I purchased the more than 68 acres of land which included the quarry and in October, 1992, we began mining at the Songo Pond Quarry in Albany township. Mark, Alan, and I incorporated the Double Diamond Mining Company to mine the old Kimball Ledge. Jim Mann was again hired as mine manager and we began to work at the quarry. He also emphasized the use of calling the locality Songo Pond Quarry. We hired some local "diggers" to clean out the trench of loose rubble, rock, and dirt, not to mention old leaves. After this "housecleaning," we found that there was a baker's dozen of beryl crystals exposed in the surface of the ledge. They ranged from dark green to beautiful sky blue! Some were 2-3 inches in diameter,. These all were truly amazing. I was looking at

Songo Pond Quarry, looking west, 1993. Photo by the author.

Beryl, Songo Pond Quarry. Photo and specimen by Tom Klinepeter.
Photo by Nathan King.

more gem minerals in one place than I had found in all my previous years of collecting. That discovery was before I began doing "the blue dance". Nowadays, when I "hit" blue beryl, I get excited and end up jumping around doing a one-foot hop improvisational "dance". I love to do that.

One of our early chores in May of 1993 was to remove some of the overburden "granite", actually granodiorite. The quarry had been left with very little room as it pinched in about 15-20 feet below the original ledge; so Tom widened this area to allow room to work. We sold the granodiorite which is actually a beautiful stone for landscaping, etc. and in 2004 we purchased specialty tools, including diamond-bladed rock saws, and began cutting the granodiorite into blocks specifically for posts, steps, and walls and there is a local awareness among local stone masons about our quarry. Most western Oxford County pegmatites are enclosed in gneiss or schist: neither a soft rock, but much easier to remove than granodiorite. We hired Tom Ryan of Ryan Drilling and Blasting from Oxford, Maine, to "shoot" the overburden granite in a series of "development" blasts.

We began to work a series of small benches into new material and we were soon rewarded with more blue beryl and aquamarine of the wonderful color this locality is famous for. Because of the brittleness of gem crystals, we did as much of the mining we could using hand tools, feathers and wedges, etc. We also found some pockets containing a variety of crystallized minerals, including quartz crystals. Wonderful apatite crystals, muscovite, schorl, the rare rutile, and other species – all in well formed crystals. As is the case in most Maine pegmatites, only a very small percentage of the beryl was gem quality. Less than 1% by weight of the beryl was translucent or transparent (aquamarine), but the color was wonderful! Some of the beryl was true "Sky Blue", while others turned to a deep green or even a shade of blue-green that looked like crystallized mint julep liquor. I had never seen a more saturated aquamarine color without heat treatment. I have never considered using any treatments to enhance the color of this or other Maine aquas, although there are many techniques available which can be used to enhance the color of otherwise natural gem minerals.

Almost all of the gem beryl is locked tightly in quartz/feldspar matrix, but on rare occasions, pocket beryls were found that were deeply etched. These gem pieces could cut exceptional gem stones weighing 1-3 carats, but most were kept in their natural state. Pockets as large as 2 x 2 x 8 feet were opened and an exceptionally large parallel growth smoky quartz crystal was found in the largest pocket opened in 1993. It weighed approximately 125 pounds and showed double terminations. The entire piece had a beautiful gemmy exterior about 2 cm thick of parallel overgrowth. Many pockets had a cap of smoky quartz with other crystallized minerals lower in the pocket. Some pockets were filled with nothing but albite feldspar crystals or sandy albite fragments. However, I quickly learned to watch out for dark mud that was often an indicator of richer mineralization.

In our first full season, we began to find some exceptional color-zoned apatite crystals with deep blue exteriors and green to gray cores. Several specimens were found in fragments in a pocket, disassembled by the naturally violent forces present during the crystallization process, but the pieces could be reconstructed to their original combined length of 3 inches or more. These fluorine-rich apatites ranged through a spectrum of colors from white, gray, yellow, green, and even deep blue. To date, no pink or purple apatites have been found at the Songo Pond Quarry, undoubtedly due to the low manganese concentrations in the pocket zone, although there is enough of this element in the

Quartz (9x8x8 cm). Songo Pond Quarry.
Photo by Nathan King.

early-formed variety of apatite called manganapatite.

During July of 1993, three pockets were opened and together, they laid cross-ways to the thickness of the pegmatite vein, spanning about 5 feet. Each pocket had a distinctly different grouping of crystallized minerals. One pocket had mostly smoky quartz crystals, while another was almost entirely made up of albite crystals. The third contained many apatite crystals on matrix, as well as "floater" crystals. Even within the same pocket, the apatites showed a variety of crystal habits, elongations, and/or crystal forms of varying proportions. These pockets were left intact for several weeks so people visiting the quarry could reach in to feel the pocket interior lined with crystal points, as yet unseen by any of us. When we finally took out the pockets, the quartz crystals were somewhat disappointing to see, as we had imagined them with absolutely clear gem transparency and magnificently aesthetic shapes, but, as is normal, they

Beryl (~15 x 6 cm). Songo Pond Quarry.
Jan Brownstein collection.
Photo by Nathan King.

were not.

Even in out first season, we allowed visiting naturalist and collectors to enter the premises and collect minerals for their own benefit. We have felt strongly that we had an important educational experience available and the best way for the student or amateur to experience this rare aspect of nature study was to allow them access. We have continued this policy whenever conditions at the quarry permitted. On drilling and blasting days, of course, there are no visitors. Similarly, when pockets are precariously exposed or there is considerable loose rock, access is also limited.

In 1996, Doug Smith of South Paris joined us and he was ready to become a serious miner. We have frequently worked together since then and he has been a real asset to my work at the Songo Pond Quarry. During this mining season, I trained Doug to drill and blast. At first he was a spectator, but he grew into being a true apprentice. During this time, we were working in the middle of the original excavation about 20 feet below ground level when we hit an area rich in smaller blue beryl in matrix. The crystals were small (finger size or smaller) and none were gem grade, but we re-

covered fine matrix specimens and over 1000 crystal pieces from 1-3 inches in length and a rich blue color!

During 1997 and 1998, Doug and I, along with several occasional helpers, worked the remaining exposed core area of the pegmatite. As we worked deeper in the original pit, we began to find pockets on the just below on the edge of the core. All sign of large aquamarine crystals disappeared as we began finding pockets with more unusual minerals Apatites were again found with colors ranging through colorless, yellow, green, and an intense neon-blue. Other interesting minerals included black rutile, amber yellow hydroxylherderite, small colorless bertrandite, smoky quartz, white hyalite opal, and brassy chalcopyrite. Some of these were the first-time finds for these species at this location. Dark red feldspar was colored by the radioactive action of, and its close association with, blackish brown zircon and black uraninite. The very first uraninite crystal I found was locked in a solid matrix of black quartz.

Despite what seems like a lot of attention, the footprint of the quarry is barely the area of a small house foundation. Gem mining is slow and methodical unlike the former quarries in the area where volume was the key to success. Drill holes might be only 4 feet deep and dynamite charges would be measured in fractions of a stick, rather than cases. Big blasts would pulverize gem minerals and delicate crystals. Also in the mid- to late-1990s, I began to explore the ledge area west of the original pit whenever I had a bulldozer working at the quarry. Several other pegmatite veins were exposed as was an extension of the Songo pegmatite,

Apatite color-zoned crystal, 1.5 x 1.5 cm.
Songo Pond Quarry..
Photo by Nathan King.

itself. Doug and I shoveled off the areas to see the mineralization in these new veins. Some of them looked like they had good potential for aquamarine and we made several exploratory blasts. The complete veins were typically from 100 to 150 feet in length, ran parallel to the original vein, and dipped to the south at close to at 45°. There were some large areas of milky quartz, showing at the widest part of the veins. Only one of the veins had any depth to the quartz. The one next to the Songo vein began to widen as we blasted deeper and milky and smoky quartz became more abundant promising that were would find a thick core with beryl. We were constantly teased by some fine quality associated minerals in the new areas. Where the vein pinched off near the old quarry, we found pockets with large rhombohedral siderite crystals (up to 5 x 3 x 1 cm) together with light green apatite crystals to 2 x 1 cm in with several crystal forms. We also encountered large schorl crystals and masses, with parts of the massive schorl peppered with inclusions of small white apatite crystals. These and most of the Songo Pond Quarry apatite fluoresced a bright yellow to orange-yellow when exposed to shortwave ultraviolet light. Some gemmy smoky quartz was recovered in this new pegmatite and I faceted some beautiful coffee-colored gemstones from it. The only pockets found in the new area were in the thin, pinched out area where the pegmatite was approximately three feet wide. The maximum width of the veins has been about 12 feet wide. The new pit is, at this writing, about 20 feet deep. We can see two other veins which may connect to the original Songo vein. These did not show on the glaciated surface and there is but a few feet separating these veins at this depth. On the western end of the new pit, large salmon-pink feldspar blocks may indicate a core pod and the pegmatite shows more masses of quartz the deeper we blast: good signs when looking for gem crystals. Formerly, the pits at the Songo Pond Quarry were all "dry" over the winters, but now there may be six feet of water present requiring the pumping of 100,000 gallons of water before we can start a new blast sequence.

I continued mining the core and intermediate zones of the Songo pegmatite and that is where I have found numerous pockets. We have had operations every year except for 2000. In May, 2008 pockets were found which were filled with blocky albite crystals that fluoresce a strong deep magenta color. Additionally, colored zoned apatite crystals (light blue rind or cap on clear green centers) were encountered, but we did find clear, lighter colored floater apatite crystals in a subsequent pocket. The associated mineralization has resumed it varied nature with marvelous white albite crystals and black rutile crystals to several cm. Some of these crystals were coated with small mica and cookeite crystals. One doubly terminated smoky quartz crystal (9 x 8 x 8 cm) showed parallel growth layers and skeletal over-

Pockets exposed in the pegmatite (light colored), dark rock on either side is granodiorite. Jan Brownstein at left (2008). Photo by Nathan King.

growths on several edges with a glassy bright crystal surface and gemmy interior.

Over the years, many discoveries were made. The first lithium mineralization in the Bumpus Mining District was made when cookeite and green elbaite were found. Only several elbaite specimens have been found to date, but it is interesting that no colored tourmaline was found on the edges of the pockets, only in the interior. Yet another tourmaline was found in the Songo Pond Quarry, foitite.

Nighttime collecting with short wave ultraviolet lights is an exciting and dazzling experience as most of the pegmatite minerals show at least some, if not spectacular, response. The pegmatite in the pocket area shows strong fluorescence, presumably from rare earth elements (REE) and there is some coloration around tiny radioactive zircon crystals. The coloration or "burns" give the otherwise white to salmon pink albite a deep reddish brown coloration. Hyalite opal varies from a white response, when pure, to bright lime green when autunite is also present. Albites fluoresce intense deep red and apatite glow brightly in yellows and orange yellows. The minerals encountered were present in a variety of sizes. Of interest to many naturalists is the variety of minute, perfectly formed crystals in association with the museum grade specimens: gemmy red garnets, shiny black bladed columbite, crystals, etc. In total, over 30 different minerals have been found in what was once thought of as a "simple" pegmatite.

In the earliest experiences, our mining was of the "let's dig here" approach. Through the years, we have learned more of the science of the rocks. Minerals may be "where you find them", but there are underlying principles which dictate their occurrence. In order to be successful, one must be able to "read" the story the rocks are trying to reveal. When we have visitors, we try to share that knowledge we've gained so that the natural history of the minerals excites the mind as the beauty of the crystals excites the eyes.

As a postscript, it may be mentioned that the eastern

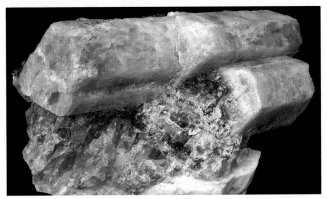

Blue beryl crystals in quartz and muscovite (7.2 x 6 cm), Wheeler-Saunders Prospect. Photo by Van King.

extension of the Songo pegmatite was prospected in the late 1950s by Roger "Stubby" Wheeler and Addison Saunders. The pit was only a few feet deep and discontinuous over about 15 meters, but it did produce some nice beryl, naturally of the beautiful blue shade characteristic of the main workings. OCMGAY (1962) noted that the prospect was reworked that year: "Charlie Gorman and Stubby Wheeler have been working on their prospect of unique blue beryl crystals in Albany township."

Wheeler-Saunders Prospect on Songo Pond Pegmatite. Photo by Addison Saunders.

Charles "Charlie" Bragg, c1960. Courtesy Ben Shaub

Emily Brownstein sitting on the contact of the Songo Pond Pegmatite with Songo Granodiorite.

Charlie Bragg carefully uncovering a blue green beryl crystal at the Songo Pond Quarry

East Pit of the Songo Pond Quarry. Note darker colored Songo Granodiorite above light Songo Pond Pegmatite. 2008

Blocks of the wall rock around the
Songo Pegmatite.
The rock type is granodiorite and the parent
bedrock has been named by geologists the
Songo Granodiorite Pluton.

L-> R Van King and Jan Brownstein.
Exploring pocket-bearing area.
Note elongated extension of pockets seen
in other photos. 2008

Close-up of crystal pocket
showing connected chambers. 2008

Addison Saunders at the
Mount Mica Quarry, Paris, 2007
Note Multicolored tourmaline fan in center right.

Wheeler-Saunders Prospect on
Songo Pond Pegmatite.
Photo by Addison Saunders.

Chapter Ten: Albany Rose Quarry, Albany, Oxford County, Maine

By Barry Heath, Waterford, Maine and Frank Perham, West Paris, Maine

Introduction

The first prospecting at the Albany Rose Quarry was in the summer of 1968, by the senior author on his own land. At the time, the granite pegmatite ledge was barely a tip of rock exposed in a field overgrown by scrub alders, but the ledge revealed some rose quartz. The prospect remained inactive until May 2008 when the senior author and long-time mining partner, Frank Perham, decided to reinvestigate the ledge for the purposes of recovering mineral specimens. The mining crew also included: Carlton Holt and Ken Budlong.

Previously, the senior author has been a miner at …. The junior author has been a miner working at the Waisanen, Tamminen, Noyes Mountain, Morgan, and Nubble Quarries in Greenwood, General Electric Quarries, Buckfield, Mount Mica, Whispering Pines and Slattery Quarries in Paris, BB #7 Quarry in Norway, Perham Quarry in West Paris, Perham Prospect in Rumford, Dunton Gem, Bell, and Crooker Quarries in Newry, and at the Bumpus Quarry in Albany.

The Albany Rose Quarry is located near the break in slope of a small hill and was ideally situated for further prospecting. As a rose quartz-bearing pegmatite located in the Bumpus Mining District, the mineralogy of the deposit was expected to be relatively simple and the plan was to expose the ledge and mine the deposit by creating benches As the depth of the granite pegmatite was revealed, exceptional minerals began to be found almost immediately: beautiful rose quartz, giant microcline feldspar crystals (at least one weighing over a half metric ton [calculated 1028 kg]), well-formed crystals of fluorescent apatite, bright black tourmaline crystals, and even modest but well-formed muscovite mica crystals! Most of the reports of rose quartz in Oxford County are for barely pink mineral and we were very encouraged by the quality and quantity of rose quartz that we were finding. The list of different minerals found is surprisingly short: annite, albite, apatite-(CaF), autunite, goethite, hematite, microcline, muscovite, pyrite, and quartz. There are also black "manganese oxides" which probably include at least cryptomelane and there is a tan to olive clay which is probably nontronite. The autunite is only visible using an ultraviolet light and the microcline may fluoresce faint red or faint blue white. The apatite-(CaF) is, of course, very fluorescent orange yellow.

When a miner is expecting to locate mineral specimens, blasting must be judiciously planned as mineral specimens are very fragile. Small charges were generally set and as the deposit was exposed, the internal structure, or distribution of minerals could be seen. The rose quartz was located in large coarse masses to several meters across and was within about 1-5 meters of the hanging wall edge of the deposit in contact with mica schist. Immediately above the rose quartz masses were some of the largest terminated feldspar crystals ever found in the State of Maine, probably

in all of North America, with at least a half dozen crystals each weighing nearly, or more than, a quarter to a half a metric ton. Some of the previous large microcline feldspar crystals from Maine which have been preserved, included a crystal from the now "lost" Horse Hill Prospect near the Norway/Oxford town line which crystal weighs about 150 kilograms (about 330 pounds) and crystals from the Wardwell Quarry which sometimes reached 50 kilograms. Edson Bastin reported that a "crystal" of microcline that was "20 feet long" and had been found at Mount Apatite, Auburn,

Albany Rose Quarry. Top: Frank Perham tamping dynamite with wooden pole, June, 2008; Middle: Working face of quarry, June, 2008; Bottom: Giant microcline crystal (~1 meter) in ledge.

but he did not given any description of the "crystal" and did not report whether it was a crystalline mass or if it actually had any crystal faces.

The rose quartz at the Albany Rose occurred in "pods" as it does in other Oxford County granite pegmatites, although the tonnage of rose quartz which has been produced from several quarries in Albany has been prodigious. Rose quartz pods are generally color-zoned with white to barely pink swaths of quartz on the edge of bright gorgeous pink quartz.

Rose quartz in place. Albany Rose Quarry.

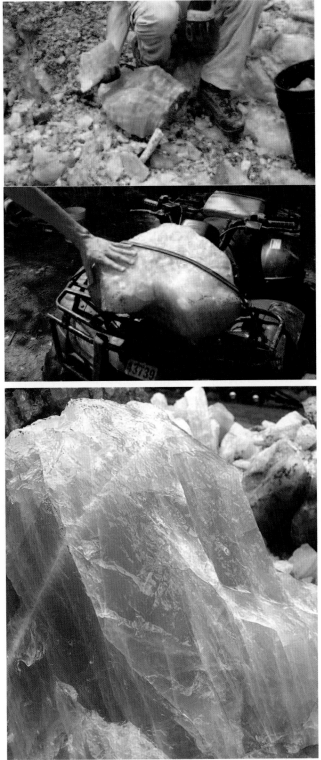

Albany Rose Quarry, 2008. Top-Bottom: Carlton Holt, Frank Perham, Ken Budlong. .

Rose quartz and white quartz at quarry (top); rose quartz block about to be transported to off-site stockpile (middle); rose quartz block [12 x 20 cm] on stockpile (bottom).

Rose quartz on off-site stockpile. Orange block is a hematite coated quartz.

Almandine crystals (5 x 5 cm).
Photo by Van King.

Derek Katzenbach of Farmington faceted some of the paler grades of the rose quartz have been faceted in order to determine their suitability as faceted gems. The initial trials have yielded good results and better quality rose quartz will be faceted. The largest rose quartz gem cut to date from this locality is 41 carats.

Of course, there are several excellent locations for rose quartz in Oxford County. The historical record holder for all of Maine in terms of quantity, has been the Bumpus Quarry, but, the Whispering Pine Quarry, Paris has had world class facetable rose quartz. The deepest and most beautiful shades of rose quartz came out of the Hibbs Quarry, Hebron. The Scribner Ledge Quarries have also produced some interesting rose quartz as has the nearby Stearns Prospect. Nonetheless, the beauty of the Albany Rose quarry rose quartz is very exciting.

Ken Budlong with giant feldspar crystals.

Tricia Perham with record size microcline crystal (54 x 61 x 122 cm).
Photo by Frank Perham.

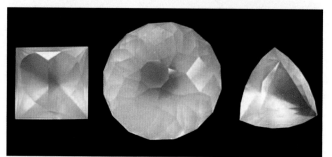

Rose quartz gems; L = 15 carats, M = 36 carats, R = 9 carats from Albany Rose Quarry.

Microcline crystals. Bottom right: FOV ~70 cm (VTK), Bottom Left (11 x 17 cm) (VTK)

Schorl crystals (5 x 8 cm) with diamond-shaped muscovite. Photo by Van King.

Apatite-(CaF) (4 x 3.5 cm) in matrix with dark color muscovite prisms. Photo by Van King.

*Microcline, top left, 10 x 16 cm.
Lower left, Muscovite crystals, showing rhombic "diamond" crystals (largest is 7 x 11 cm).
Top right, Multiple growth "fishtail twin" of muscovite. Photos by Van King.*

Chapter Eleven: Albany in the Twentieth Century – Glimpses of Rural Life by Lorraine Leighton Greig Parsons, Hartford, Maine

Our family came to Albany in 1913 from Stark/Dummer area, New Hampshire. Our farm, then known as the 1837-built "Old Parsonage" (Oxford County Records and Deeds v. 349 p. 253), was purchased by Alfred E. and Lottie Purington Leighton. Formerly. Alfred was schooled in horse breaking, treating animal diseases and animal injuries. He received a diploma from Professor Jesse Berry, Pleasant Hill, Ohio, USA. He and Lottie prospered with the farm along with trades of a blacksmith, butcher, selectman, road agent and sexton of Albany. Mary, his mother, lived with them after his father's death. Lottie died; both were buried at Hunt's Corner Cemetery along with A. E. Leighton.

I grew up "next door" to the Bumpus Quarry. It was just across the meadow behind our house, up the hill, across the ledges, down the hill, across a meadow, up the ledges and down to the Cummings farm house. There were some light woods and bushes in between, as well. For all that we knew, our lives were more or less the same, although we didn't have a quarry.

Any family story starts with a family, in this case two: Winslows and Leightons. Our "begats" follow:

The back ledges on the Leighton farm were often playground and picnic areas for the Leighton family. In the 1940s and 1950s many interested mining entrepreneurs visited A. E. Leighton trying to obtain permission to blast and look for important minerals such as beryl, feldspar, mica, etc. As these ledges are located at the back side of the Bumpus Quarry where the huge beryl crystals were discovered in 1928, many discussions were held and many offers were made, but no blasting was ever done. I visited the ledges October 2006 to assure myself that they were still intact.

Remembering the old days

As I revisit my past, I recall the foot paths we had connecting us to the neighboring properties. In the 1940s, the Bumpus and Hall families lived on the Valley Road. We thought nothing of walking this little trek to make fudge, play cards, gossip or perhaps join a working bee - "neighbors helping neighbors." In rural areas, possibly even today, children invented their own entertainments: hanging "May baskets " - the fun of trimming a box with crepe paper, making sure we had a handle to hang on the doorknob, fill it with fudge and sneak to the door, hang it and holler "May Basket". We all ran and the receiver had to catch (tag) us. What fun!! Often we walked miles just to participate in this event. After being caught, one helped find the others. It was usually a neighborhood gathering. We also played "Alli alli in free . This name was the yell for everyone to come inside, eat fudge and play "post office ". Boys and girls numbered off and when the opposite sex number was claimed, a kiss was sneaked behind the closed door. With a belly full of fudge

Leighton-Winslow Genealogy

Thomas Layton 1604-1672
Thomas Layton 1642-1677
John Leighton 1673-1718
Thomas Leighton *1696-1763*

William Leighton 1776-1868
Thomas Leighton 1809
Thomas Leighton 1833-1900
Thomas Leighton 1853(d. ~1913-1918)
Alfred E. Leighton 1877-1963
A.E. Leighton married
    Arlene Winslow of East Bethel in 1938.
Albany, Maine children
Lorraine Leighton Feb. 7, 1939
Lucy Leighton Nov. 8, 1941 d. Aug. 2, 1980
Alfreda Leighton Apr. 21, 1944
Alfred Thomas Leighton July 18, 1950

William Wyncelowe 1300
William Wyncelowe 1350
Thomas Wyncelowe 1400 c.
William Wyncelowe 1440 c.
Edward Winslow Oct. 17, 1560
Kenelm Winslow d. 1607
Kenelm Winslow *(child with 2"d w.-Magdalene*
    b. April 29, 1599/Mayflower
Nathaniel Winslow 1639
Gilbert Winslow July 11, 1673
Barnabas Winslow Feb. 24, 1701
    (child 2"d w. Deborah Bradford)
Philip Winslow Oct. 7, 1740
    (child with 2"d w. Ann Kimball)
Lyman Winslow Apr. 27, 1838
Lyman Winslow July 26, 1880
Arlene Winslow Dec. 2, 1915
Urbian (Leighton) Winslow May 9, 1938

Leighton Farm off Hunts Corner Road east of Cummings Farm on Valley Road, Albany. Distance about 1.6 km. Note straight line trace of the Portland Pipeline right of way just east of Leighton Farm. Courtesy Google Earth.

and the kissing done, we all walked home to our respective residences. In the 1940s and 1950s most neighbors in Albany were at least 3 to 5 miles away from each other. At the Bumpus Quarry, as darkness closed in, we were all called "in free". Fear terrorized our parents that we might fall into the quarry filled with water.

Activities were centered around the home. Clothes were home made or we might have "pass-me-downs" given to us by neighbors and friends. Dresses, shirts and blouses were made from the grain bags. Knitting, rug making, sewing, tatting lace, embroidery and spinning were always on the agenda. The old Singer Sewing Machine had a foot treadle – "Ya had to have rhythm!" Reading or being read to were enjoyments: Merry Hearts had sad stories, but *Uncle "Wriggle"* and the *Burgess Stories* were dear to our hearts.

Fall season brought slaughter of a beef (all this meat needed canning: bones were boiled and mincemeat was made) and killing of a pig (pork was salted, hams and shoulders were soaked in brine and cured at the farm) pigs' feet were pickled to make hog's head meat and hog's head cheese. There was also sausage and that was hung in the attic. In raising a pig, all the swill needed to be cooked to avoid hoof and mouth disease. Apples were picked and stored in barrels in the cellar. Potatoes dug and sluiced to the cellar into a bin. Vegetables were picked all summer and canned. Wild berries were similarly picked and canned.

Typical haying gang. Old Maine postcard.

Horses in full regalia at work with an ice cutting gang. Oxford County postcard.

Milking was done twice a day. The milk and cream separator hummed twice a day. Oh, what a job to take that apparatus apart, wash it, and put back together! Butter was made weekly and often sold. Cottage cheese was always wheying on the back of the old cook stove. The wood stove was always in use. Biscuits made daily and we ate three square-meals a day. Rainy days were for baking. Doughnuts by the dozens, six or eight pies, homemade yeast bread and molasses cookies were always snacks or desserts at our house. We also made fudge practically every day.

Hot water was heated in the side tank attached to the stove and teakettles perched atop the stove. Water was pumped from the hand pump installed at the old iron kitchen sink. Dishes were washed and wiped and kept in the cupboards. Lamp chimneys and lantern glass needed washing often. Lampblack sure did accumulate fast when the lights were in use. Laundry was a "headache" day. Water was heated in copper boilers on top of the stove. There was a gasoline motor washing machine with the exhaust hose poking out the window and two rinsing tubs, all set up in the kitchen for the entire day. Baskets of clothes were hung out on the clothesline and, in winter months, the clothes were brought back inside half frozen. Oh, My! This was just a dreadful day!

Farming means cutting wood (firewood for home use and pulp wood to sell), haying, and gardening as well. Horses were the power for all labors. First mowing, raking, making haycocks, pitching on the hay wagon, hauling and backing a loaded hayrack into the barn enabled the use of the hayfork. (I remember it striking the diamond and carried on a track the length of the barn via the ridgepole). Hunting was a necessity as well. Meat on the table kept everyone healthy. We ate bear, venison, raccoons, hedgehog, and rabbit meat. No job was done without a gun handy.

Harnessing a two horse team and hitching up to a pole and whiffletrees was no small task, either. All well-kept farms had big rocks that had to be trimmed around and road sides to be cut. Most of this cutting was done with a hand scythe. This makes another job of turning the grindstone or emery wheel to sharpen axes, scythes, chisels and the like. A weathervane is a must for a farm. No farmer can work against Mother Nature. The farmer must know which way the wind is blowing The Leighton farm had a rooster weathervane. With real hens clucking and roosters crowing, eggs had to be collected and the hens had to be fed. We were always looking for broken dishes to smash up to feed the hens to keep their egg shells hard.

During the school days, the bus had a turn around at the Gorman place and that is where the Leightons, Ruggs, and Stearns children were taken aboard. When we had extra milk, we were able to tote the milk on the school bus. The bus driver would let us off to carry milk into the Bumpus family house. After he picked up the Hatstat children and turned in their driveway; he would pick us and the Bumpus kids up to continue our trip to the Bethel schools. As we were the first students on the bus, we learned the lay of the land well. During mud season the roads were impassable and most snow storms kept us at the "Old Parsonage" for a couple of days. The plowmen always planned to arrive at noon. They always came to our living room, put their feet upon the stove hearth, settled back for their lunch and had hot biscuits and pies that my mother had made. When the snowplow crested Picnic Hill, we knew it was time to prepare lunch and bake off biscuits. We always watched them twist and turn until it reached the Leighton homestead. They had to turn at this dead end; therefore most of our yard was plowed.

Singing was a family trend at church, circle suppers, youth groups, skating and sledding parties throughout the entire town. We owned an 1888 Estes Organ that we pumped and played for entertainment. We had one radio that ran on a battery. To save the battery we, as a family, listened to the news and the "Jack Benny Show. Is it any wonder most of us are star gazers, mineral oriented, or day dreamers?

In Sap Season, we had a stone arch with a 3 x 5 foot tin pan to boil down the sap after we had collected it from many, many trees. We filled 25 gallon milk cans on a wooden hand drawn sled to haul to the arch area. The final

Albany school. c1885. L-C-R, front to back, L: Harold McNally, Ethel Skinner, Archie Bass, Gladys Grover; C: Laura Cummings, Lillian Skinner, Marjorie Barker, Bessie Skinner, Herman Cummings; R: Ella Skinner, Annie Cummings, Etta Cummings, Raymond Cummings. Photo courtesy Barbara Inman.

boiling down process and pouring into preserve jars took place in the kitchen on the old wood *"Home Comfort"* stove. We usually stored in our cellar about 40 gallons of maple syrup yearly.

In the winter farm chores needed to be done at least twice a day. Cows and horses were led individually to the wooden watering tub where a hand pump was installed above the well. Often it was my duty to pump while an animal drank. The pump had to be letdown or it would freeze and, of course, before pumping again it had to be primed. Hay had to be pitched down from the lofts to the barn floor. Turnips were chopped for cattle and they were "grained". Father milked while we busied with the chores, although the barn kerosene lanterns only gave a faint glow.

The Leighton Family grew and scattered. *Urbian* joined the Marine Corps and later owned his own construction company in South Carolina. Lorraine and* Lucy attained "registered nurse" degrees from Central Maine General Hospital. Alfreda settled in Gilead, while Alfred built an underground house exactly on the Old Parsonage homestead site. Today, *"Deepwoods Farm"* is located at the exact spot of the 1838 parsonage / A.E. Leighton Farm and is now owned and operated by DiAnne Leighton Ward (Alfred Thomas Leighton's first wife).

The Albany School System

Up until the end of the Great Depression or the beginning of WWII, there were still one-room schools in Albany. The Dresser School had eight pupils and one teacher for grades 1-8. There was also The Town House School, Songo School, Round Mountain School (formerly Picnic Hill School and moved to Round Mountain) and the Clark School. By ten years later, we all had to go to school in Bethel.

Taking the bus was an important bonding experience when all of the students had to take one to get to school. Friends were made with people who lived a long walking distance away. Students who lived the furthest away from school got to know the most people and what their families were like as there was almost always a parent waiting for the bus to arrive.

Albany's last one room schools.

Top = Songo School;
second = Clark School;
third = Town House School;
fourth = Dresser School;
fifth - Crocker Pond School
bottom Round Mountain School.
Top four photos courtesy Barbara Inman.
Bottom Picture courtesy Sid Gordon.

Chapter Twelve: Gold, Silver, Mica and the Four Davis Brothers of Woodstock by John R. Davis, South Paris, Maine

Mining in rural areas is frequently an activity shared in families. One enthusiastic person convinces his relatives that they should join in. Much of the interest eventually gets lost in family folklore, especially beyond one's grandparent's era, but County Records maintain the legal details. The following is the result of researching my family's mining ties. A large part of Woodstock lies in the Bryant Pond 15' Quadrangle which was geologically mapped by Charles V. Guidotti during 1959-1961 (Guidotti, 1965). Guidotti believed that the presence of a fault and associated fractures, recognized by him, the Moll Ockett Fault, named for the nearby Moll Ockett Mountain, could have been conduits through which mineralizing solutions penetrated the local rocks and deposited sulfide minerals (personal communication C. Guidotti to V. King, 1989). There were supposed silver ores among these minerals in the nineteenth century.

The Maine Silver Mining Boom is generally thought to have started in the coastal towns with the very first mine in Sullivan, Hancock County, discovered by May, 1877 (King, 2000), but that source also indicates that there were a few metal mines in Maine pre-dating the Sullivan lode. Mining fever had spread across Maine from Sullivan and, in 1878, Acton in York County had a *bona fide* wildcat mining district with many places in between. The same fever that struck coastal Maine also progressed inland: to Penobscot, Piscataquis, and Somerset Counties. Oxford County was involved in its mining boom just a wee bit earlier than in Sullivan, according to the record, and as such was much more influential in the boom that followed than has previously been realized. Much of the evidence of Oxford County's silver boom lies in the paper trail. Only a few of the proposed mines were ever opened and none yielded any refined metal. The fact that no ore was ever shipped, sold, or smelted, of course, means that all investors lost every cent they invested.

Prospectors would bring in highly purified samples, going to great effort to remove any worthless rock, to find out how much of value was in the metallic portion of the vein. The local investors were unaware of many aspects of

Downtown Woodstock from an 1880 stereoview.

the principles of mining and, as was the case all over Maine, the meaning of ore assays was lifted to unsupportable heights. The reader should know that in the nineteenth century, many assayers wrote about chemical constituents of minerals as though they were separate minerals. For example, galena, the common lead sulfide, always has measurable amounts of silver "impurities". However, the silver content is usually so low that the silver was formerly disregarded when lead was being refined from the ore. Historically, silver was obtained from minerals which had essential silver, not from minerals which were only slightly silver-bearing. As chemistry and technology improved, it became possible to extract small amounts of silver which were formerly ignored.

In the 1870s-1880s, however, when an assay had a separate line for "sulphuret of silver", investors usually only saw $'s and, in fact, the assayer would report dollars of a metal per ton, rather than weight of a metal per ton of rock. An ore containing a separate silver mineral could be "free-milling". That is, a free-milling ore could be crushed and the silver mineral could be cheaply separated out mechanically. A silver-bearing mineral, though, had to go through an expensive and time-consuming chemical process to recover any silver and the process could cost more than the value of any precious metals obtained. Investors also misunderstood the meaning of "per ton", as in "$98.34 of silver per ton". Virtually all of the assayers wanted business and they probably did little to discourage their customers – or to educate them. Certainly a few assayers would point out the economics to an inquisitive customer, but the evidence suggests that few actually heeded advice once they got out the door. A promoter would get pure samples of metallic minerals and let investors think that each and every ton of rock would be worth the full assay value. The economics didn't work out

Steam train coming from Bethel approaching Bryant Pond Station. From c1905 postcard.

and the investors who were otherwise good businessmen in their own fields were under the impression that when a mine got going, all of the mineral in the vein was going to be metallic. In Maine, most ores were sparsely present in the veins. If it took 200 tons to yield a ton of concentrated metallic minerals, the assay value would have to be divided by 200. A rich assay such as $100 per ton would be reduced to an actual value of $0.50 per ton. It might cost $3 or $4 per ton to mine rock under the most favorable conditions in the time period. The shipping and refining costs would have been added on top of the simple mining costs. Those miners who realized these economics were still driven by the belief that they would find a vein so full of metallic minerals that their dreams would come true. The philosopher Ralph Waldo Emerson offered great advice to the world only several decades before Oxford County's mining boom: "The greatest hopes and worst fears of men are seldom realized."

Woodstock, as is true of many other Maine towns, is blessed with a variety of named settlements. The most important, of course, is Bryant Pond village, but there is Billings Hill neighborhood, Si Gotch, Perkins Valley, North Woodstock (the ever popular Pin Hook, pronounced pinnook), South Woodstock, and the historical Stephens Mills. The last hand crank, operated assisted telephone call in the USA was made from Woodstock in 1983. Woodstock is adjacent to West Paris with about 3 km distance from South Woodstock to Trap Corner. Lapham (1982) wrote the first book length history of Woodstock and Ruby Emery labored dedicatedly to revise and up-date the written account.

From the instruments recorded in the Oxford County Registry of Deeds at South Paris it appears the initial mining effort closest to Woodstock was the search of lead in Milton Plantation by the Oxford Gold & Silver Mining Company. (Note that the name, Oxford Mining Company, had been previously used to organize a company to mine graphite (Plumbago) on Mount Plumbago, Newry in 1859.) Although the Company's Certificate of Organization is dated Novem-

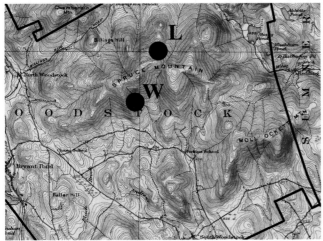

L = Lone Star Mine, W = Woodstock Gold & Silver Mine, approximate. Spruce Mountain, Woodstock. Redrawn from Thompson et al. (2000) on USGS Bryant Pond Quadrangle, 1914.

ber 11, 1876, some months before, on July 26, 1876, Charles Smith, Valentine Glines and Ancil Merrill quit-claimed all mineral and metal rights to the company on a parcel of land in Milton that Eliza Penley had then recently acquired from

Woodstock Opera House later a local store. From an 1907 postcard.

James G. Davis of Woodstock on May 5, 1875. Also on July 26, 1876, John G. Tibbetts, Samuel E. Heath and James P. Chamberlain, also jointly executed a quitclaim to Oxford Gold & Silver Mining Company for the same parcel.

On February 22, 1881, Oxford Gold & Silver Mining Company, quit-claimed its rights to all minerals and metals on the land deeded it May 5, 1875, to the Champion Mining Company for $100,000, this being the past due indebtedness. (Remember that stock ventures "capitalized to $100,000" didn't mean that $100,000 had yet to be purchased in stock or that the Company had actually raised that amount. The amount stated was the limit to the amount that could be collected from investors. Even the number of shares could have been bartered or negotiated as investments "in kind". It is never possible to assign a true accounting of money based on informal statements.) Only one entry appears in the indexes at the Registry of Deeds for Champion Mining Company, the aforementioned quitclaim from Oxford Gold & Silver Mining Company, and even after one backtracks through numerous transfers of ownership, the exact location of the mine remains vague at best.

In Woodstock proper, the Lone Star Gold & Silver Mining Company was first to set up shop, with a meeting on February 21, 1877 of Charles and Thomas Bradbury of Woodstock, William D. Chase of Lewiston, and William Small, Jr. of Lisbon, in which it was agreed to form a company for mining minerals and metals in Woodstock. The Certificate of Organization shows a capital stock of $7900, at $50 a share. Charles was elected President and as a director, Thomas and Small as the other directors, and no one was named as a Treasurer at that time. Charles had 101 shares, Small 53, and the other two held 2 apiece.

This mine was located on the part of Lot 74 in the East Division of Woodstock that Charles and his wife Melissa quitclaimed to Lone Star Gold & Silver Mining Company on March 1, 1877, and on July 26, 1877, William

Stephen Chase Davis [May 23, 1829 - May 27, 1893], Davis Family Archives.

Hemingway of Milton Plantation quitclaimed a one-part of his Lot 53 in the East Division for $300, with a provision that it be crosswise through the middle of it. On September 6, 1877 Charles sold Lone Star Gold & Silver Mining Company a right to cross his homestead farm from Sam Doughty's land to that of Melona Bradbury for $1, and Samuel Doughty also quitclaimed a right of way across his farm the same day for $1, as did Cyrus Tucker.

Possibly the assay reports from the Lone Star mine on Spruce Mountain prompted three prominent Lewiston men, Milton Wedgewood, Franklin Drew and the aforementioned William Small, Jr., and William Whitehouse of Augusta to agree to the formation of a Woodstock Gold & Silver Mining Company? According to the March 24, 1880 Certificate of Organization, the company would mine minerals and metals from "15 mines in Woodstock". Capital stock was stated as $500,000 at $5 a share with $600 already raised. Each of the four men held 20,000 shares. (One can see that if the officers gave themselves stock as a kind of "in kind" arrangement that a successful mine could reward them handsomely for a small investment.) Wedgewood was President and a director, Drew the Treasurer and a director, with Whitehouse and Small were directors. Apparently no official transac-

tions occurred with the "15" property owners, at least under the company's name. The prediction that there would be "15 mines" was certainly a sales ploy to entice investors. The suggestion that there would be so many mines was certain to play on the avarice of any potential investor.

Although there are no Certificates of Organization appearing in the records at the Registry for either the Harris Woodstock Gold & Silver Mining Company or the Tucker Woodstock Gold & Silver Mining Company, on Mar 22, 1880, Henry Harris sold William Small, Jr. the east half of Lot 80 in the East Division, which Small then sold to Harris Woodstock Gold & Silver Mining on October 4th, 1880; and on April 17, 1880 Cyrus Tucker sold Small the east half of Lot 75 and the west half of Lot 80 in the East Division, which Small then sold to Tucker Woodstock Gold & Silver Mining Company the same day. No other instruments were executed, or appear to have been recorded at the South Paris Registry, at least bearing the name of either the Harris or Tucker Woodstock Gold & Silver Mining companies. It is also of interest to note (*Maine Mining Journal*, May 14, 1880) that William Small, Jr. is shown as Treasurer and a director of both these companies in addition to being a director of the Lone Star and the Woodstock Gold & Silver Mining companies. It is further interesting to note that nearly all those serving as officers and directors in the four previously mentioned companies engaged in mining or attempting to mine Woodstock property. They were speculative businessmen from "away" who were clearly caught up in the talk of gold and silver in "them thar hills" sweeping across Maine in the late 1870s, especially Oxford County's hills.

At this time, we need to mention the four surviving sons of Si Gotch's early settler Benjamin Davis, Sr. [b. ~1797-December 21, 1876], that led the way in organizing a company with its officers all being current or former residents of Woodstock. Herrick [November 5, 1825 - 1910], Benjamin Jr. [August 11, 1827 –March 6, 1909], Stephen C. [May 23, 1829 – May 27, 1893], and Henry [b. May 11, 1838] were the Davis brothers. Their organization was the

Stage coach at Bryant Pond. c1907 postcard.

Union Mining Company. According to its Certificate of Organization signed April 21, 1880, at a meeting held in the Alpine House in Bryant's Pond village on the 15th, participants agreed to form a company for mining minerals and metals on parcels in Woodstock, with capital stock of $500,000, at $5.00 a share.

Of the brothers, Herrick was an Attorney in Bryant's Pond, later in Paris Hill, and then serving as Oxford County Register of Probate, having been appointed in 1872, as well as being a former Representative in the Marine Legislature, and he was a member of the Oxford County Mining Board. Herrick initially was a teacher at the Chase School on the Cushman Road, and eventually became Oxford County Judge of Probate. Benjamin Jr. farmed his late father's homestead near Concord Pond and had just been elected as a Representative in the Maine Legislature in 1879. Benjamin Jr. was also a member of the Oxford County Mining Board, and had a vein of galena and "sulphuret of silver" in the ledge of his abutting lots in Milton Plantation and Woodstock. (There is no existing record concerning the Oxford County Mining Board and its duties are similarly unrecorded.)

Stephen blacksmithed and farmed a sizable tract in Perkins Valley, Woodstock, purchased from his father-in-law, Jotham Perham, one of Woodstock's earliest settlers. The woodlots and pastures were on Perham Mountain, and across one pasture ran a vein containing galena, sulphuret of silver that also traversed several of his neighbor's lots and a deep gully on Andrews Brook. (Perham Mountain is an eastern summit of Spruce Mountain.) Henry was a clerk and also farmed a small tract in the south part of Woodstock, but his mining involvement appears to have been only as a stockholder in Union Mining Company.

Of Union Mining Company, the officers and directors were all native sons of Woodstock. Herrick was elected President; Fred Bartlett, Treasurer; Benjamin Davis, Jr., while Peter C. Fickett, Isaac Curtis, and Simeon Curtis were Directors. (Peter Fickett may have been a brother-in-law of Benjamin Davis Sr's, nephew Aaron Davis III who married a Lucy B. Fickett of West Paris in 1860.) Fickett was also an Attorney at Law and had removed to Paris to set up practice there. All held 30 shares of stock: Stephen Davis, Charles A. [b. 1849] and Chester D. Fickett [b. ~1826], Alonzo Felt [b. ~1834] and Granville Felt [b. ~1834], and Rufus K. Dunham [b. ~1824], the Bryant's Pond Stationmaster (1851-1896) and postmaster (probably 1854-1855). Others holding shares in the mining company were Aurestus Perham of Lewiston, Erastus Abbott of Franklin Plantation, Joseph Penley of Paris and Abner Chase of Norway.

The mineral vein that bisected Stephen's pasture also coursed through the Felt, Curtis, and. Fickett lots, and by May 1st, Betsy Curtis, Granville Felt, Chester and Charles Fickett, and my great-grandfather, Stephen, had all sold mining rights on their property to Union Mining Company. Benjamin Jr., sold Union the mining rights on his property May 6th, and he underlined the last word in his conveyance, "meaning to convey all minerals in said lot and right to enter and operate for minerals in said lot forever"

Although the *Maine Mining Journal* of May 20, 1881 cites a Portland *Argus* article that the Simeon Curtis/Fickett mine was at a depth of 12 to 15 feet, I cannot say whether it continued to be worked for any length of time nor if there was an attempt to remove any minerals from the Felt property. As a youth I recall my father speaking of there being a gold and silver mine in the pasture up on Perham Mountain, but that the cost of getting it out was more than what it sold for. Had I known back then that someday I would get myself involved in such matters, I'd have asked him to take me up where it was located.

The same article in the May 20th, 1881 *Maine Mining Journal* also mentions that at 12 feet Benjamin's "Little Ben" mine had uncovered sulphuret of silver, and galena in a surface outcropping on his Si Gotch tract. I don't recall my father ever talking of mining around Concord Pond on the family lots and Davis Mountain, and it has been forty years or more since the heirs of Benjamin Jr. divested themselves of the property.

The *Maine Mining Journal* of May 6, 1881 cited an *Oxford Democrat* article that Register of Probate, Herrick C. Davis, "has a full line of Oxford County ores in his office." It was customary to have a display a "collection" of ores from your mines to show the potential investors. More than likely it is has long been scattered into the collections of hobbyists earlier in the last century, their origin lost to the ages. More than likely the ore samples were discarded in the time period.

Mica – A New Woodstock Boom

Pond and Dickvale Road. Davis Mountain in background.
Davis Family Archives.

While the proposed gold and silver mining in Woodstock may have faded in subsequent years and more money lost than made, a renewed industrial interest in mica during the 1890s and the presence of it in the same mines and mineral veins, saw Winfield Scott Robinson of Hartford approaching parties across Oxford County known to have such deposits on their properties (See Chapter Three: Early Mining in and Near the Bumpus District, this volume) .

In Woodstock, Robinson leased mining rights in his name alone in 1899 from Charles Bisbee, Alonzo Felt, Abby

Curtis, and my grandfather, Ronello Davis [November 26, 1866 – May 30, 1936], the son of Stephen, on their Perham Mountain properties, being the same tracts as earlier mining rights that had been sold to Union Mining Company. Union Mining being a defunct company rather than one formally dissolved didn't have an advocate to contest these sales. In leasing to Robinson, the terms were 20 years at $6 per year, plus 25 cents per ton of waste or scrap mica shipped or sold, and 5 cents per pound for each finished mica square 3 1/2"x4" or larger, with exception of Bisbee. His lease, in the book, reads 25 cents per pound for each finished mica square, but that could have been an error by the Registry Clerk in copying the instrument. (I know, from being employed there for ten years following a Naval career that these kinds of errors are part of the public record.)

On September 28, 1901 Consolidated Mica Company's Boston office executed a quitclaim of its mineral rights for all mica, mined and unmined, to the Consolidated Mica Company's Saco, Maine office on the property of John Elliott, on Mount Dimmock near Rumford Point, which "Consolidated Mica Company having a place of business in Boston had acquired by a quitclaim from John E., John W., and George Elliott of Rumford on February 2, 1901." The transfer of mineral rights did not last long and was subsequently taken by an Oxford County Sheriff's Deed dated October 24, 1902 for the "seizure of all rights, titles, interests which Consolidated Mica Company of Maine has May 2, 1902 at 1:50 P.M. in the Elliott quarry in Rumford" and was executed on December 10, 1902. It's not sure why there was such a long delay between May 2 and December 10.

Harold Hastings entered into four mica leases. One being with my grandfather Ronello Davis on May 10, 1904 for the mining rights to all mica, feldspar, and other minerals and metals on his 160 acre parcel on Perham Mountain. The terms were for 20 years at $6 per year, 25 cents per ton of scrap mica and feldspar and other minerals sold, and 2 cents per pound for every finished square of mica.

The other three Hastings leases were with Gilbert Mills of Bethel, Freeman H. Bennett of Albany, and Melville Hamlin of Waterford, although they had previous leases with

Ronello Chase Davis & Oneida York Davis.
Davis Family Archives.

Truesdell and Robinson. Hasting's lease of mining rights on Hamlin's Black Hill parcel was for 17 years, Mills and Bennett 25 years, with the monetary amounts being the same as in his lease with Ronello Davis. Apparently the leases were allowed to lapse as nothing further regarding the transactions appeared in the Registry indexes. More than likely, they simply never paid the succeeding year's fees.

There were also two land transfers involving Benjamin Davis, Sr., and Stephen Davis, with somewhat ironic twists. On October 6, 1859, Benjamin sold. the west half of Lot 80 in the East Division to Cyrus Tucker, who would later sell the mineral rights to William Small, Jr. of the Tucker Woodstock Gold & Mining Company. Stephen purchased Lot 45 in the East Division on April 14, 1877, from his father-in-law, Jotham Perham [March 22, 1784 – September 24, 1864], and soon thereafter found the lot had a mineral vein across the pasture on Perham Mountain. Jotham, for those who may not be aware of such matters, was none other than Harold C. Perham's great-grandfather!

The author wishes to acknowledge discussions about historical details of Oxford County and the use of assay reports with Van King. Note: Davis family dates listed in Lapham (1882) are sometimes at variance with Davis family records.

Unissued Union Mining Company
of Woodstock, Maine Stock certificate

Chapter Thirteen: Lester E. Wiley, Mining and Construction

By Laura Wiley Ashton

In the fifth grade I lost the grammar school spelling bee to my sixth grade cousin Peter Wiley. That wasn't humiliating by itself because he was bright, but the word was "construction". Classmates made it a point to mention my father's dump trucks all over town with "Lester E. Wiley, Mining & Construction" on their doors, and if anyone had visited his shop on the Waterford Road, or noticed his equipment on a job, they could report I'd had quite a few other chances to get it right. That was in 1959 and since then I've learned to spell construction and several other things!

In spite of blowing the spelling contest at the Guy E. Rowe Elementary School, in Norway, Maine, I do have some great memories of those years, especially of the many people from all over the United States, indeed the world, who came to Norway to work for my father or who bought the mica, beryl, quartz, and feldspar that he mined, or who visited the Bumpus Quarry in Albany, Maine as tourists and rock-hounds. (They were sometimes called "dump-pickers" because that's where they set out to broaden their rock and mineral collections—the quarry dump.).

Families from North Carolina and West Virginia moved to Maine lock, stock and barrel to work in these New England hard-rock quarries, not only for my father, but for the other entrepreneurs in the business as well. As to how my father got from "Lester E. Wiley, Construction" to "Lester E. Wiley, Mining & Construction"— he met this guy...

Hitchhiker – 1957

John Robinson was an experienced and knowledgeable miner from North Carolina. Dad picked him up on the Lake Road in Norway one afternoon and brought him to the Waterford Road shop; Robinson was hitchhiking from downtown Norway. I don't know where he was going or where he'd been, but I assume now that he was in our area looking for work at one of the several quarries in operation at the time.

Robinson was a type of man that I hadn't seen before: kind of a black-haired Jerry Lee Lewis! He wore pleated-front pants and loose, long sleeved and solid colored shirts. His southern accent added to the impression that he was a real "rare bird", one of my father's favorite expressions. It seems odd to remember irrelevant details like that, but he was different. He could sing pretty good and played a guitar, accompanying my father who could play a banjo. They were both pretty good whiskey drinkers and both liked to spin a yarn and talk! And talk they did. Before many days had passed, my father made a decision to branch out from his road building and excavation business and embrace hard rock mining as a lucrative addition to his company.

Lester Wiley was never one to wait for the rest of the world once he'd made up his mind to do something. Our small world soon became a whirlwind of air compressors, southern miners and their families, as well as dynamite, blasting caps and detonators, not to mention drills, lawyers and mining leases. Excavation equipment was put to work in a way it never expected to be. New additions to our world included Patsy Cline music, stockpiles of mica, beryl and quartz, and feldspar hauled by the dump-truckload to the West Paris grinding plant. The enterprise even made the local paper about the time the mica business expanded (*Lewiston Journal,* undated 1959):

"Two years ago the Pechniks ended their many years of mining and Lester E. Wiley, a Norway contractor, leased the mine

Lester Wiley with Laura, around 1952
Road construction on Orchard Street, Norway.

from Wardwell.

He continued to haul the feldspar to the West Paris plant, but he also started processing the raw mica into saleable blocks for the electronic industry.

Until four months ago Wiley kept only two men working at the trimming of the mica, shipping small amounts to various points along the eastern seaboard.

In August he contacted two expert mica men in North Carolina and engaged them to come to his Norway shop to devote their time to preparing the raw mica for shipping. When the men arrived Wiley discovered he had engaged the third and fourth generation of a mica trimming family. …

Lloyd Stewart had practiced his profession for 45 years at Spruce Pine, N.C., learning the trade from his father, who in turn had been taught by his father. …

Since the North Carolinians arrived, Wiley has kept a crew of four trimmers busy eight hours a day. The mica is mined at Albany, hauled in raw state to the shop on the Waterford road at Norway. There the raw mica is cobbed or sorted, to eliminate the scrap mica, which is shipped in bags to Pennacook [New Hampshire] for wet grinding processing [the scrap mica only].

After cobbing, the mica is then rifted, or sheeted, ready for the trimmers to cut the mica into thickness from 0.007 inches to an eighth of an inch.

The finished pieces are shipped weekly to Franklin, N.H. to go onto the U.S. Government stockpile, where it is immediately classified as strategic defense material.

The mica streams out of the Albany mine at the rate of 2,500 lbs. daily …"

At the time, the federal government was paying big money for mica used in the space program. This was the hook that drew in newcomers to the mining field—and it was the real thing. Over the next several years tons of this complex silicate, valued for its thermal stability, incombustible and non-flammable properties, and other positive technical attributes, were blasted out of Oxford County mountains and hand cut (trimmed), boxed up and shipped or delivered, or was hauled to processing plants, primarily in North Carolina. The checks rolled in and big profits were made by those who had been willing to open themselves to the initial risk.

My father said …"real gamblers risk their money in business, not cards or horses," which didn't, no doubt, set too well with my mother's family who were great horse racing people! He lived that statement, as evidenced by his history to this point. He had owned three successful businesses by the time he met up with John Robinson. At age 18 he was

Paul Hoppes on rail car with bags of mica at Wardwell Quarry. Unidentified newspaper clipping.

a road boss, and later on worked on the "Quoddy" Tidal Power Project. When I was born, he was a cattle buyer and he had two meat trucks on the road making household deliveries. He traveled from Canada to Vermont to Massachusetts buying cattle, and worked long hours in his slaughterhouse across the road from our farmhouse on the Waterford Road in Norway.

He also had his own successful logging business, running crews, equipment and trucks, and then he started the construction company. Apparently he was ready for the next big gamble! As with all he did, he put his heart, mind, and every cent he had into this business, making sure he had the right equipment, the right quarry, and the right people to succeed.

He learned from people like John, and the buyers and other contacts he soon met. His construction crew went along with this change, no doubt looking forward to something new themselves. Winfield Rolfe and Oliver Frechette were long-time employees who were joined by John Robinson and a succession of southern miners with years of valuable experience.

My father employed skilled mica trimmers—notably Lloyd Stewart and his son Jack—who sat at a drafting table bench, wearing heavy leather aprons, and using vicious looking hooked knives, trimmed the crusty edges off big plates of mica that had been split from large blocks blasted out of the quarries. Small boxes of this high-grade mica sold for thousands of dollars.

High-Grade and Low-Grade and the Pickins'

Good miners maximized the output from their quarry. While high-grade, plate mica was the big profit driver, burlap bags full of scrap mica were also marketed by the truckload, as well as several grades in between. Beryllium was another profitable item. In addition to deliveries to North Carolina with mica, there were trips to northern Ohio with beryl. Quartz and feldspar were mined and sold, again by the dump-truckload. There wasn't a whole lot left once a blast had been sorted and stockpiled and loaded for sale. But there was some, and that was the lure that attracted the rock-hounds to the local quarries, including the Bumpus Quarry!

From 1959 until 1962 my father mined Roy Wardwell's property in Albany. That was a tunnel operation, where the product had to be hauled out in skip-pans, using cables. The Bumpus Quarry was also in Albany. Once that quarry was drained of water, which was a constant problem, the drilling began. Rock was drilled and blasted and mined

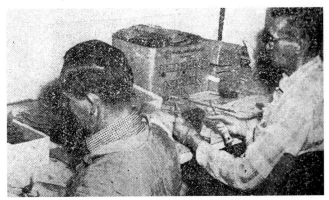

Lloyd Stewart (R) and his son, Jack Stewart.
Lewiston Journal, 1959.

from the bottom and sides of the quarry, lifted up to ground level in skip-pans with a crane. There it was readied for market, loaded into trucks right away or stockpiled on site. Anything not loaded or stockpiled was moved to what is referred to as the "dump".

Prospectors

By the mid-1960s the mica program that was so profitable in the prior decade was history. There were still profits to be made in mining, just not as great, so the incentive was there to maximize output even further. Areas that may have been considered more nuisance than value, a few years earlier, were suddenly small profit centers to be cultivated. Rock collectors and wanna-be prospectors were just such a market!

The Bumpus Quarry was more accessible to casual visitors than the Wardwell Quarry, indeed most of the quarries. The entrance was right on Route 35 and easy for an ordinary family car to pull into. As word got out that the quarry was in operation, people who

were hobbyists and collectors began to stop and inquire about possibly "collecting the dump". As interesting a concept as this was, and as interesting as some of the people were, it was a diversion from the main function of the operation, which was commercial mining. But, was this a possible avenue to maximize a dwindling profit picture?

It would seem so, because fairly soon after setting up operation at the Bumpus Quarry, my father built a small structure at the right of the entrance off Route 35, and in addition to using this as an on-site office, started selling nice specimens of beryl, rose quartz, and crystals in matrix that these collectors were anxious to acquire. He also began charging those who wanted access to the "dump" a small fee (as I remember, it was $2.00 a person) for the day. There was plenty of parking, not miles of walking, a crew to watch actually quarrying the stuff, and a really fruitful "mine dump". For those who had been bitten by the prospecting bug it was a great experience with not a lot of cost.

It sounds as though my father had never heard of rock-hounds before, but that wasn't the case. He spent a fair amount of time with Stan Perham, and knew from his success that this was a popular and profitable area. I do think the persistence and dedication of the collectors as a group did surprise him. People invested in the hobby wanted to find their own prizes and were willing to work quite hard to do it. And they were from all over the place, all walks of life, all income levels, all professions. For someone who really enjoyed meeting new people, this sideline operation provided my father many enjoyable conversations! It did my mother and I, as well.

Once visitors paid their fees, signed the guest book, and chatted for a few minutes, they continued up the hill to the mine and the dump, excited to be there and anticipated that perfect aquamarine crystals, pure chunk of gorgeous smoky quartz begging to become jewelry, or specimens resting in their native matrix and ready to be displayed in a

Anne Gregory and Laura Wiley (L-> R) in front of rose quartz
stockpile. Bumpus Quarry , Gregory (1968)

prized collection were waiting to be found.

The entrance road curved right, and headed for the mine dump, a few hundred yards further along. The quarry itself was on the left side of that road, but far enough back that the casual visitor wasn't encouraged to get too close to the operation, endangering themselves or disrupting the work in progress. There was a small lunch shack and the dynamite storage area between the quarry and the road, several

Log cabin built (on Route 118. East Waterford) from "scratch" by Lester Wiley in the late 1960s. Lucy had a shop where she sold mineral specimens and gifts, for several years. L-> R: Laura, Lucy, and Lester. Photo by Dolores Stanley, Lester's niece.

pieces of heavy equipment here and there, dump trucks, and the vehicles of the miners who were working that day.

The dump itself was a big sprawling area that gleamed in the sun with the reflective white of feldspar and quartz, dotted green here and there by clumps of bushes. Walking wasn't too bad, but since the rock was spilled from dump trucks, it created piles and obstacles that one had to pay attention to! The guests really loved it when there was a fresh mound to start working through.

During the years my father operated the Bumpus Quarry and allowed visitors to prospect, my mother Lucy and I minded the store in the summer. That meant I got to meet all of these people and gain a little first-hand knowledge about what collecting meant to them, why it was such an important part of who they were. Reggie Gordon was a most interesting man from the Brunswick area who had been a banker and later a dealer in antiques and oddities and who was a maestro at the piano. He had contacts in the mining business in the Down East areas of Maine and became an acquaintance, then a close friend to my father. He had a strong point of view on several things, but as far as rock collectors were concerned, his opinion was they were "... trying to fill a void—something is missing in their lives..." I suppose that's true for all collectors in a way, but I think there's a little more to it than filling a hole! (Reggie also referred to my mother's painting and sculpting as "housewife art", which did not faze her any nor stop her from selling it fairly rapidly at her own shop in later years.)

A day outdoors, during a Maine summer—bending,

kneeling, digging, walking, and finding truly beautiful works of art from nature—can't be all bad, and if it fills a void in your life, so much the better! A couple from Haifa, Israel who kept in touch with my mother and father for several years, as did school and camp kids on field trips, serious geologists from the Colorado School of Mines, city dwellers out of New York and Philadelphia, day-trippers from the area, who all explored, found and enjoyed, and went home happier for the experience. Did I ever collect anything? Of course! I delighted in finding tiny aquamarine crystals, a smoky quartz crystal maze, rose quartz suitable for cutting, dendrites in feldspar, just beauty in the rough. It was easy to get inspired when you spent a lot of time with the prospectors.

As far as profitability is concerned, it was a fair sideline and nothing more. It controlled access to the quarry area and provided some interesting conversations with some interesting people. Somewhere I've read that a magazine reporter interviewed me during those days when I was at the "rock shop" and that I told him that the shop would make enough money to pay my college tuition! Well...possibly I was putting him on a bit...just not the case. But it more than paid my way to a unique life experience.

The End of an Era...

The Bumpus Quarry years ended sadly with a tragedy. July 21, 1966 Oliver Frechette was buried in a wall collapse while he was working on a water pump at the bottom of the quarry. He had come to work early in the morning, was alone at the quarry, doing his job at 110% like he always did, when this awful thing happened. When we got there, and the other miners arrived, it wasn't immediately apparent what had happened. The wall collapse was obvious, however, and Oliver's vehicle was there and he was nowhere in sight. So they started digging.

I remember my father in the cab of the crane on the shallow edge of the quarry, bringing up rock, moving it aside, bringing up more, for what must have been the most hellish hours of his life. He knew what he would eventually find at the bottom, and hoped against hope for the impossible. Oliver's wife Thelma and their children were at the shop with my mother and me. As people who knew them or us or were just curious found out what had happened, they stopped by. There is no describing the anguish felt by that family and my father on that day. Watching the 9-11 aftermath on television would provide a magnified analogy. Oliver was 55 years old when he died. My father was 50. I was 18 and just out of high school.

My memories of the mining business, as it involved my family, are mixed. The late 1950s and early 1960s were good years, fun years with a new direction, new and interesting people, a lot of productive activity. The later 60s were slower, harder to make a living but harder still to give up the dream. The death of a good friend and a good person was a bitter pill to swallow and a very sad time.

Conclusion

I can say without equivocation that the best part of my association with the mining industry was the people I met. A Tennessee Williams novel couldn't begin to do justice to the John Robinsons of the world, no matter how richly developed the character - or a Reggie Gordon with his droll, dry humor and spicy way with words. The Stewarts, Lloyd and Jack, were tall, thin stereotypical southern men perpetuating their family tradition and having a little nip now and then.

One family I felt especially close to was the Buchanans who were already well-known for their mica mining in their home state. Frank Buchanan came from Spruce Pine, North Carolina to work for my father. His wife Allie and their children moved with him. They rented a house on Hobbs Pond in Norway, and the kids went to school with me. Brenda and Gaylene were about my age and we spent many great hours together. They are a Southern Baptist family, and were bound by restrictions that seemed so odd to me. But they were lovely people and I learned from their friendship that sometimes differences are a good thing.

Bascom Henline was another really good man from Spruce Pine. He and his family moved to Norway, too, and kept in touch with my parents for years after the area mining (for all practical purposes) went belly-up and everyone moved home!

Lin Carpenter was a mica buyer from Spruce Pine. He and his wife, Hess, rented an apartment in Norway at the head of Main St, across from the Advertiser-Democrat. Lin was a white haired and portly Southern gentleman who smoked cigars and sipped bourbon. Hess was peppery, but sweet, and a great cook. Their granddaughters were often with them, and Patsy became a friend much like Brenda and Gaylene.

Paul Hoppes, who most always had a big cigar in the corner of his mouth, was another prominent transplant from the Carolina group. He, his wife Midge and their family, lived in the area for several years. Some of the families ended up staying in Maine, like the Hardy family from West Virginia. Clifford and Eula Hardy lived in Norway and pretty much raised their family there. Some of their children, now parents and grandparents themselves, still live in the area. There were many others, with memorable personalities and great stories.

In 1995 I drove a gas-guzzling SUV full of high school kids to Disney World in Orlando: my son James, his best friend from Norway, Greg Sessions, Istvan Lakos from Hungary and Tatsuhiko Umino from Japan. We were showing our foreign guests as much of the US as we could in ten days. On the way back to Maine I drove through Atlanta, on to Asheville and up the backside of a mountain with many switchbacks, to Spruce Pine. We stopped at a little diner in town and I got on the phone. Within 10-15 minutes we were visiting with old friends! The boys were great and enjoyed every minute of our short visit, and I was delighted to be with people I hadn't seen since I was a kid.

My father was not well enough to travel that far, by then; he was in his late seventies and had several chronic physical problems. My mother died in 1984, and he lived the last twelve years of his life with me and my son, whom he loved dearly. We lived at Norway Lake, on what was then called Tucker Road. That neighborhood was very familiar to my father; he and my mother lived in an upstairs apartment at another house on Tucker Road when they were first married.

He maintained a vegetable garden, and entertained friends that came to visit him on the covered breezeway in the summer and in the kitchen beside his little wood stove in the winter. He had a very un-politically correct sense of humor and usually had an audience! He loved to cook and enjoyed having a hot soup or stew ready for my son and his friends on weekends after they came in from snowmobiling or skiing. They learned to like cayenne pepper!

Despite grumbling about the dogs and cats that seemed to accumulate at our house, he could usually be found scratching one behind the ears and coaxing it to sit, or eat, or whatever he thought it ought to be doing. He enjoyed watching my son and his buddies grow up. We hosted several foreign students on weekends and vacation weeks

Lester Edward Wiley

while James was a day student at Hebron Academy, and my father spent hours talking with them. He was interested in young people and what they thought and how they reasoned. He liked my friends, and got a kick out of hearing about their families, their relationship issues, their work calamities—the good gossip of everyday life!

He died on February 16, 1996, less than 20 days from his 80th birthday on March 5th. He lived long enough to learn that my son would be going to Montana to college the next fall, which was some last good news to take with him on his last 'gamble'. We miss him.

Chapter Fourteen: Mica Mining at the Wheeler Brothers Quarries, Gilead, Maine. Photoessay by Addison Saunders

Wheeler Brothers Quarry area, Gilead, near West Bethel Village. Redrawn from Google Earth.

Wheeler Brothers #1 Quarry, Gilead, near West Bethel.

Wheeler Brothers #2 Quarry, Gilead, near West Bethel.

Wheeler Brothers #1 (right) and #2 (left) Quarry, Gilead, near West Bethel.

Wheeler Brothers #1 Quarry, Gilead, near West Bethel.

Wheeler Brothers #1 Quarry, Gilead, near West Bethel.

Upper three quarry views = Wheeler Brothers #1 Quarry
Lower two buildings view = Wheeler Brothers #2 Quarry.

Rear: Cleveland Lovejoy, Arden Andrews.
Front: Albert Wheeler, Roger Wheeler,
George Laxton at Wheeler Brothers #1 Quarry.
Courtesy Millie Jackson.

Mica Splitters at
Wheeler Brothers #1 Quarry fieldtrip.
L-> R: unknown, Sidney Murphy,
Millie Jackson, Kitty Fox.
Courtesy Millie Jackson.

Mica splitters on fieldtrip: Ivy Bartlett,
Daffy Brook, Millie Jackson.
Courtesy Millie Jackson.

Drilling at Wheeler Brothers Quarry.
Albert Wheeler and Millie Jackson.
Courtesy Millie Jackson

Main Street, Bethel. Left view = 1920s postcard showing Colby Block on right, just behind Elm tree and next to tall store building; Right = view in 2008 of former site of Colby Block next to tall store.

Mica books of varying grades. The top specimen shows excellent light transmission and amber color. The top of the specimen shows a perfectly flat cleavage surface and is ideal as a source of sheet mica.. Lower right shows mica book that is nearly perfect, while lower left shows many imperfections. Specimens from Raymond Woodman collection. Photos by Van King.

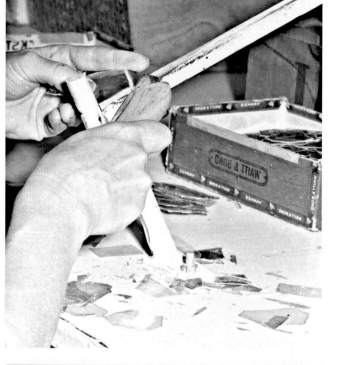

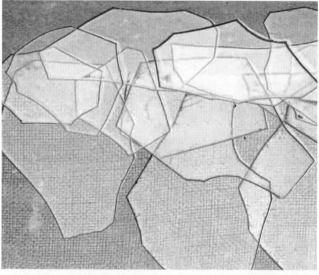

Top Left: Bill Anderson with Wheeler mica.
Top Right: Jig to trim mica of imperfect edges.
Center left: Thomas Snyder demonstrating using skiving knife to rift (split) mica at the Wheeler Mica Shop in the Colby Block, c1959.
Center Right: Trimmed and split mica. Lower Left: Splitter at work bench. Lower Right: Processed Wheeler Brothers mica cut or punched into specific shapes for final approval by end-users. Center three photos by Ben Shaub. Lower Right: Specimens by Van King. Top and Lower Left Photos courtesy Addison Saunders.

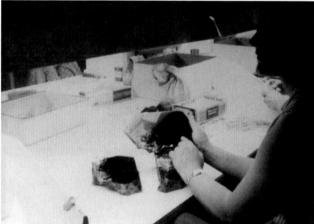

Chapter Fifteen: Localities and Minerals of the District by Van King

Albany

Freeman H. Bennett Mica Prospect (Prospected by Winfield Scott Robinson, probably about 1900. Dana Douglass, Jr. leased the mineral rights of this locality, located on the northern side of Pattee Mills Road, from Ralph H. Kimball on July 3, 1943.) [53, 337] – albite, microcline, muscovite, quartz, and schorl.

Bumpus Quarry - albite, almandine, annite (biotite), apatite-(CaF) (and manganapatite), apatite-(CaOH), autunite/meta-autunite, bertrandite, beryl (also aquamarine), chalcopyrite, chamosite, columbite-(Fe), cryptomelane, dumortierite, foitite, goethite, hematite, hydroxylherderite, kaolinite, malachite, microcline, muscovite (and sericite), nontronite, pyrite, pyrrhotite, quartz (also rose, smoky), rutile, schorl, scorzalite, siderite, titanite, todorokite, torbernite/metatorbernite, uraninite, vivianite, zircon.

Coe Ledge Quarry [171] – albite, microcline, muscovite, quartz (also rose)

Walter Conwell Prospect (Miller and Wing (1945) visited this small prospect "three miles" north of North Waterford village" on the land of Walter Conwell: "Neither the size nor quality of the mica warrants mining operation.") – albite, microcline, muscovite, quartz

E. O. Donahue Mica Prospect, (This locality, on land formerly owned by E. O. Donohue, was opened in July, 1942 by Arthur Kimball of Bethel, who later leased it to Douglass Mining Company in 1943) - albite, almandine, beryl, microcline, muscovite, pyrite, opal (hyalite), quartz, schorl. Miller and Wing (1945) were cautiously optimistic about the locality for its value for mica and feldspar production.

Douglass Road Metal Quarry (Opened about 1998) – granodiorite, pegmatite (albite, annite (biotite), microcline, muscovite, quartz)

Farwell Mountain localities (Nineteenth century collectors' site.) - albite, beryl, microcline, muscovite, quartz

Fleck Quarry - albite, beryl, microcline, muscovite, quartz, zircon

Preston Flint Quarry (Miller and Wing (1945) noted that this prospect was 1 mile south of Hunts Corner and had been prospected in 1941, but suggested that "The limited amount and quality of the mica and feldspar do not make mining operations advisable.") - albite, annite (biotite), microcline, muscovite, quartz

French Hill (Mountain) Prospect ("The locality is in the woods in a sag between two knobs of the hill crest…" [53]) - albite, beryl (and aquamarine), microcline, muscovite, quartz (and rose)

Isabelle Foster Quarry [171] - albite, beryl, microcline, muscovite, quartz

Aquamarine 0.65 carats, Songo Pond Quarry. Courtesy Nathaniel Edwards.

Chester Holt Prospect (Miller and Wing (1945) noted that the prospect was a mere test pit in biotite gneiss with small pegmatite "stringers".) - albite, biotite, microcline, muscovite, quartz

Guy Johnson Mica Quarry - This locality was first opened by Guy Johnson for feldspar about 1938, but it is uncertain for how many continuous seasons. It was leased by the Douglass Mining Corporation July 10, 1943 [338] and was worked for several months. Lawrence A. Anderson operated the quarry briefly in 1944. Although a rum-mica producer, it also had commercially valuable feldspar and some beryl. - albite, almandine, apatite-(CaF) (exceptional transparent complex crystals to over 1 cm), apatite-(OH), autunite, beryl (to 1 foot by 0.75 foot), microcline, muscovite, quartz, schorl, torbernite, unknown mineral #5 (probably uranophane), zircon.

General Electric or Lovejoy Mountain Glass Quartz Quarry - Opened by General Electric Company perhaps as early as 1927 for quartz. Thompson et al. (2000) placed the locality generally on the north central part of the low shoulder of Lovejoy Mountain, but it is on a one acre lot near the summit on the north face. [413] The General Electric Company had a presence in Albany, according to the *Maine Register*, from 1934-1943. The December 8, 1927 *Lewiston Journal* revealed: "The mining of quartz is becoming more and more extensive as new uses are found for the mineral. General Electric Company is working a quartz vein at Paris Hill, and other electrical deposits at Waterford, Norway and other western Maine towns. Fused quartz and quartz glass are still being subjected to much experiment, and quartz cores heated to 2000 degrees are used in experimental radio beacons in aid of aviation." - albite, almandine, beryl (and aquamarine), microcline, muscovite, quartz

Charles P. Pingree Ledge Quarry - Worked 1878-1879 by Maine Mica Company, also prospected in 1900 by Winfield Scott Robinson - albite, apatite-(CaF), bertrandite, beryl, columbite-(Fe), microcline, muscovite, quartz, schorl, unknown mineral #5, uraninite

Round Mountain Prospect - Morrill et al. (1958) placed this locality about 3.2 km due east of the Bumpus Quarry, generally on the southeastern slope.) - albite, beryl (and aquamarine), microcline, muscovite, quartz (and rose)

Fred Scribner Ledge Quarries - On September 18, 1935, OMMC leased the Fred E. Scribner property in order to explore for minerals. The four quarries area was later leased March 24, 1943 [339] while quarrying started around June 12, 1943 by the Douglass Mining Company and worked into the autumn. The localities were ruby-mica producers, but were of insufficient grade to support mining even during a period of wartime price subsidies. - albite, almandine, annite (biotite), apatite-(CaF), beryl (and aquamarine, golden beryl, microcline, muscovite, quartz (also rose and radiation burns), schorl

Songo Pond (Kimball Ledge) Quarry - On September 25, 1930, Loris M. Rollins leased the mineral rights to the Kimball Ledge from Abner B. Kimball. [340] The Kimball Ledge revealed excellent bright blue beryl crystals, but relatively little work was done, as the locality was essentially a gem quarry. Addison Saunders worked a locality to the extreme southern end of the Kimball Ledge Pegmatite in 1958. Leased by Charlie Bragg in the 1980s. Bought and reopened in 1993, by Double Diamond Corporation (Jan Brownstein and Alan Obler) for gems and minerals and continuously from that time, except for the year 2000. Initially, Jim Mann of Bethel was the driller for Double Diamond Corporation, using a portable gasoline-powered drill, and, later, equipment such as an air compressor or excavator time, were rented. The Corporation later reorganized and owned its own equipment and corporation members became certified blasters. When gem specimens and crystals were located, feather and wedge hand techniques were used to extract specimens from the ledge. - albite, almandine, annite (biotite), apatite-(CaF) (and blue, green, colorless, Hebron-style, manganapatite), autunite, bertrandite, beryl (also aquamarine), chalcopyrite, chamosite (chlorite), cookeite, elbaite (green), foitite, goethite, hydroxylherderite, magnetite, malachite, microcline, molybdenite, muscovite (and sericite), opal (white "hyalite"), powellite, pyrite, quartz (also rose, smoky, and amethyst), rutile (and ilmenorutile), schoepite, schorl, siderite, uraninite.

Square Dock Mountain Area - outcrops visited by prospectors only, no quarrying - albite, beryl, microcline, muscovite, quartz

Roy G. Wardwell
[April 23, 1882 Albany-
November 4, 1969 Bridgton].
Courtesy Barbara Stearns Inman.

Hugh Stearns Quarries - Hugh Wallace Stearns married Edith E. Cummings, Laura's sister, on October 1, 1927. He opened his two quarries beginning in 1938 and these were worked for feldspar, scrap mica, and beryl. Sampter (1947) wrote: "... Joe [Pechnik] and his two sons have recently opened a new pegmatite vein on the Stearns farm, where they have good spar, for the United [Feldspar and Minerals] Company, but so far no crystals for collectors. Went to this new opening several times ... and it looks like a good locality for next season, when they will have opened more ground and perhaps struck some good pockets.") - albite, annite (biotite), apatite-(CaF), beryl, microcline, muscovite, quartz (also rose), schorl, uraninite, zircon.

Larry Stifler-Mary McFadden Prospect - Opened in the summer of 1986 by Tony Waldier, Jim Mann, and others: expanded September 11, 2004 by Jim Mann, Dennis Powers, John Cluckey, and Tom Ryan. – albite, microcline, muscovite, quartz (also rose)

Roy Wardwell Quarries - These quarries were begun before 1941 and were owned by Roy G Wardwell. From 1941-1945, 1947, 1948 or 1949, also, at least, in 1952-1955, 1957, the Wardwell #1 Quarry was worked mostly by Joe Pechnik, and later by Lester

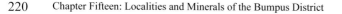

Wiley in 1959-1962. The workings consist of five open cuts and several prospect pits lying within 700 feet of one another. By 1962, the pits were coalesced into three pits, each with an underground portion. Minerals found include: albite, almandine, annite (biotite), apatite-(CaF), autunite, beryl, chrysoberyl(?), "gummite", microcline, muscovite, quartz (also rose), schorl, torbernite, unknown mineral #5 (probably uranophane), uraninite, uranophane, uranophane-beta.

Some interesting comments of the Wardwell Quarry's recent work were written by Sampter (1946):

"... heavy rains in August. This kept the pit full of water, which came in faster than the

Hugh Stearns
[October 21, 1906 Albany-
November 5, 1999 Norway].
Courtesy Barbara Stearns Inman.

pumps could get it out, and also made the road up the hill too slippery for our car to maneuver. ... It is a one man activity, being worked by a Mr. Joe Pechnic [sic], whose sons are working Mt. Newry. This quarry produces the finest grade of feldspars, beryl crystals, rose quartz and many large black tourmalines, with fine shiny terminations ... Fine beryl crystals from one-half to about three inches in diameter are very plentiful and fun to dig out of the dumps, and break out of the freshly blasted rocks. They are not clear, but of good color and fine crystal shapes, with smooth faces. The rose quartz is deeper in color than that from Bumpus.

Mr. Pechnik lives in hermit style at the quarry all year round, in a neat, cabin and does everything himself. He blasts, piles up the fine feldspars, beryl, mica and rose quartz and dumps out the worthless rock. Runs a derrick

by using an automobile engine still housed in the old car's body. About once a week a truck comes up and carts away the feldspar and brings in to Mr. Pechnic [sic], whatever he has ordered from the town the trip before. This one man operation goes on day after day, winter and summer. Except for collectors and dealers, who know the place, Mr. Pechnic's money crop is feldspar, for which I estimate he gets only about $4.00 a ton net, now, but he is very happy and seems to thrive well on his solitary miner's life."

Sampter (1947) wrote of this quarry for the Summer of the next year: "My friend, Joe Pechnic [sic] – the lone miner of the quarry on the Roy Wardwell farm in Albany, Maine – blasted every few days and produced plenty of nice beryl, but as usual, only broken crystals; also beautiful pale rose quartz, black tourmalines with shiny faces and terminations, and others of the usual pegmatite minerals. I found one piece of an 18-sided beryl there too. After about six years['] work, feldspar finally ran out and Joe and his two sons have recently opened a new pegmatite vein on the Stearns farm, where they have good spar, for the United Company, but so far no crystals for collectors."

The Wardwell Quarry was not always worked by local miners or companies. An article in the *Portland Sunday Telegram,* September 16, 1956, revealed some of the most interesting details of the quarry's history:

"Crew Bores Into Ground At Albany To Reach High Grade Mica Deposits

The farm of Roy Wardwell has taken on an aura of the old West as a crew of miners takes more than two tons of the high grade mineral [mica] each week...

Behind the project are Paul Hoppes, Lin Carpenter and John Phillips, all of Spruce Pine, N.C. ...

The Albany mine now has three headings off the main tunnel, and workmen are starting a second shaft. A vein 787 feet long has been surfaced out and the three tunnel 'headings' are 45, 75 and 85 feet long, and up to 35 feet high.

Operating with modern mining equipment, a crew of eight men mine [mine] an average of two tons [of mica] a week. Drills are of Swedish tungsten, and Hoppes believes his crew is the first in the state to use a chisel bit.

A dolly truck on a truck is used to haul the mined mica to the surface and to lower supplies [into the quarry]. Power comes from a gasoline powered winch, operating steel cables. A special automatic dump was devised by Hoppes by which the cart, reaching the end of the track, automatically trips its load into a waiting truck, requiring no attention from the winch operator. ...

Underground, to prepare for blasting holes are drilled vertically into the face of the heading, two feet apart. Ten of these are drilled, each six feet deep and large enough to handle sticks of dynamite, one and one quarter inches in diameter. The pneumatic hammer, with chisel bit drill, eats into the ground at a foot a minute.

When it is time to set off the blast, everybody goes to the surface. 'We could go into one of the other headings [tunnels] and be safe,' a workman explains, 'but it plugs your ears for a little while.'

Mica usually emerges from the blast in 'blocks' with six inches the average. They have had some ten by 12 inches, and the largest was one huge slab 16 by 24 inches.

It is then bagged, hauled to the top, unbagged, cleaned of rocks, and is ready for shipment to the depot.

Winter conditions have closed down operations only one day since work started. Water seepage, 'Seventy-five per cent of it comes from surface water.' Hoppes says. It is overcome by two pumps that can handle 20,000 gallons an hour.

An attraction for strangers, Hoppes says that 'no visitors' rule has to be enforced for safety's sake.

Hoppes' assistant in the bookkeeping department is his wife, Eula, at their Norway Lake residence. Eula, a true southerner, doesn't care for the state's winters, although her husband didn't think last year's snow was severe."

Barton and Goldsmith (1968) reported that annual production from the Wardwell quarries, 1952 – 1961, was about 1,000 pounds of beryl and "several tons of book mica" each year. The report seems to have been skewed to indicate only the highest quality of mica, as the reasonable reports of the quarry's production is much higher. Lester E. Wiley, of Waterford, opened a mica trimming shop. An undated, *Lewiston Journal,* clipping [1950] indicated the activity at the quarry:

"The Wardwell mine at Albany, owned by Roy Wardwell, was first opened in 1935 by Joseph Pechnick [sic]. He and his sons hauled feldspar by the tons to the processing plant at West Paris.

Two years ago the Pechniks ended their many years of mining and Lester E. Wiley, a Norway contractor, leased the mine from Wardwell. He continued to haul the feldspar to the West Paris plant, but he also started processing the raw mica into saleable blocks for the electronics industry.

Until four months ago Wiley kept only two men working at the trimming of the mica, shipping small amounts to various points along the eastern seaboard.

In August he contacted two expert mica men in North Carolina and engaged them to come to his Norway shop to devote their time to preparing the raw mica for shipping. When the men arrived Wiley discovered he

Transparent crystal of muscovite.
2.4 x 2.2 cm.
Wheeler Brother Quarry, Gilead, Maine

had engaged the third and fourth generation of a mica trimming family. It was then he learned that this art was handed down from generation to generation just as the clock makers in Switzerland handed down their trade to each generation.

Lloyd Stewart had practiced his profession for 45 years at Spruce Pine, N. C., learning the trade from his father, who in turn had been taught the art by his father.

When Lloyd's son, Jack, enlisted in army service in the World Yar [sic] II, he had been working at the mica trimming trade since 1941. ...

Since the North Carolinians arrived, Wiley has kept a crew of four trimmers busy eight hours a day. The mica is mined at Albany, hauled in raw state to the shop on the Waterford road at Norway. There the raw mica is cobbed, or sorted, to eliminate the scrap mica, which [the scrap mica] is shipped in bags to Penacook, N.H. for wet grinding processing.

After cobbing, the [sheet grade] mica is then rifted, or sheeted ready for the trimmers to cut the mica into thickness from 0.007 inches to an eighth of an inch.

THE FINISHED PIECES ranging from

one inch to three square inches in size are packed and shipped weekly to Franklin, N.H. to go onto the U.S. Government stockpile, where it is immediately classified as strategic defense material.

The mica streams out of the Albany mine at the rate of 2,500 lbs. daily and moved to the Norway shop where the trimming processed is conducted by the expert mica trimmers."

The experienced mica trimmers were probably recommended by Paul Hoppes. "The Hoppes brothers knew no fear when it came to mining." (Neil Wintringham, personal communication, 2007).

In 1962, the USBM, in attempting to prove Oxford County beryl reserves, collected drill core samples from the Wardwell Pegmatite.

Lloyd Stewart (sight) and son, Jack (left) trimming mica. From 1959, Lewiston Journal.

Wentworth Quarry - Opened before 1950, some beryl produced in 1953. - albite, annite (biotite), beryl, columbite-(Fe), apatite-(CaF) (and manganapatite), microcline, muscovite, quartz.

Batchelders Grant

The **Peabody Mountain Quarry** is located on the north spur of Peabody Mountain, not far from the Wheeler Brook Trail and near the head of the southwestern tributary of Wheeler Brook. Morrill et al. (1958) called this location the Nason Quarry. Cameron et al. (1954) related: "[Floyd] Mason discovered the pegmatite in 1928, and obtained the mineral rights in 1937. He opened the quarry in 1938, operating it intermittently until the fall of 1940. It has been idle since. The quarry face, extending northeast along the base of the sparsely timbered rock knoll, is 90 feet long and 20 to 30 feet high, and has been cut into the mountain a maximum of 20 feet. … The pegmatite has a distinct, sharply defined zonal structure." There is known, so far, a small suite of minerals at this location: albite-oligoclase, almandine, annite, dumortierite-like mineral, apatite-(CaF), muscovite, pyrite, quartz (milky, rose, smoky, crystal), and schorl. Morrill et al. (1958) noted a Binford Quarry in Batchelders Grant that was explored for mica in 1955, but that location may be a mistake for a Binford Quarry in Rumford.

Gilead

The **Wheeler Brother #1 and #2 Quarries** in Gilead, sometimes mistakenly given as in West Bethel, were productive localities, particularly for mica from 1954-1962. Lasmanis (1958), later State geologist for Washington, wrote about mineral finds made there. Minerals found at Wheeler quarries include: albite (exceptional crystals and cleavelandite), almandine, annite (biotite), apatite-(CaF) (and manganapatite crystals to 3 cm), autunite, beryl, chalcopyrite, chrysoberyl, columbite-(Fe), cryptomelane, goethite (limonite), microcline, muscovite, quartz (rose, smoky), pyrrhotite, schorl, and torbernite.

The three small quarries on **Peaked Hill** were prospected in "a few years prior to 1942", but a quarry was actually operated in 1880 by the Maine Mica Company. [64] There is a small list of species known: albite-oligoclase, almandine ("Red garnets, one-half inch across, occur in the coarse pegmatite near the hanging wall."), microcline, muscovite (flake and plumose), and quartz.

Mason

The first mention of the **Brown Quarry**, located below a side summit to the north by northeast of the main summit of Pine Mountain in Mason township was in 1956, when it was worked for mica by Lawrence Anderson. Caldwell and Austin (1957) called this the Anderson Quarry. Robert C. Tibbetts worked at the Brown Quarry in 1957. The partners, "Tyler and Rich" searched for mica there in 1958. The location is so off the beaten track that there is no reasonable species list known for this location except for muscovite. It certainly had albite, annite, microcline, and quartz and may have had a rich species list as did the nearby Wheeler Brothers Quarries to the northeast. The quarry was worked during a time when there was a subsidy being paid for exploration work and production of muscovite for the Federal strategic minerals stockpile.

Waterford

A mica quarry was established near the southern summit of **Beech Hill** in 1900 and was worked by the Beech Hill Mining Company in 1902, which company was sold to "New York investors": "About a ton of thumb-trimmed mica was marketed at prices ranging from 8 cents to $1 a pound, and about 10 tons of scrap mica was sold. The remainder of the material quarried was still in the mine buildings." (Bastin, 1911, nearly *verbatim* Sterrett, 1923). Clarence Leslie Potter was a miner and Sandy Morse may have been the mine foreman (Joseph Nile, taped interview with Ben Shaub, 1961), while Dr. Hiram Francis Abbott of Rumford Point seems to have been the Company president. The pegmatites, located on the George L. Kimball farm, were relatively simple and only several minerals are reported: albite, almandine, annite (biotite), microcline, muscovite, and quartz. When it was visited by Sterrett in September 1906, the quarry had been idle, but his impression, later stated in his report, was that the Beech Hill Quarry "... has been operated for mica on a larger scale probably than any other deposit in the State." Bastin (1911) also mentioned that microcline was present in masses to 5 feet, while some mica books were a "foot across. ... Much of the mica is worthless for anything but scrap because of the prevalence of ruling, wedge structure, and twinning. ... The equipment includes a steam drill and boiler and a shed where the trimming was done." Wentworth (1977) indicated that:

> "... George, Prentiss and Bill Kimball prospected the first one on the south side of Mr. Kimball's farm. ... A corporation was formed and mining shares were sold in anticipation of greater success. Around 1912 or 1914 they leased the mine to a Mr. Hettersheimer whose luck likewise soon ran out.
>
> In 1918 {actually 1913-1915, [414]} Mr. [Fremont] Westcott [of Portland] leased the mine and ran it for six or seven years. He had as partners Mr. Hill and Mr. Libby. ...
>
> Doctor Logan, a mineralogist from St. Lawrence University, Canton, New York, came and appraised the minerals from the different mines on Beech Hill."

However, Pink and Delmage (1957) do not acknowledge "Dr. Logan" and it may have been that the "mineralogist" came from another school, or may have actually been a student. Ober Kimball (taped interview with Ben Shaub, 1958) remembered the New York City "investor" as Dr. Hettersheimer, but the name may have been misremembered as this name is so rare. Kimball indicated that the Beech Hill Quarry lay idle until WWII and it was prospected by Stan Perham. The Beech Hill Quarry was re-opened by the Pechnik Brothers in the last of November of 1957, but "had rather hard luck" with snowstorms, ice, etc. Several reasonably large cuts were made but the weather prohibited further development: "My sister and I got two checks, at 10% [royalty], it came to $28."

Burnell Hill Feldspar Quarries on the west side of Blackguard Road, north of the former Blackguard School, were opened, perhaps in the 1950s in search of feldspar. This is a simple pegmatite with albite, annite (biotite), microcline, muscovite, and quartz, supposedly including quartz crystals some of which contained amethyst. [171] The same reference lists an occurrence, on the east side of Blackguard Road, of amethyst crystals with phenakite crystals on their faces, and, if true, would be virtually the only such specimens of this type known in the world. Although very nice Waterford amethyst is known, the phenakite is unbelievable and is certainly an erroneous report.

C. P. Saunders Quarries are located just below the northern spur of Beech Hill and were opened about 1940, worked by United Feldspar & Minerals Company in 1943. Minerals reported include: albite, annite (biotite), beryl, microcline, muscovite, quartz (smoky), and schorl (crystals). The eastern pit has small amounts of chalcopyrite, mag-

South end of Streaked Mountain showing streaks on ledge, Paris, 2007.

netite, and pyrrhotite. [171]

Additional small feldspar prospects are located on **Thunder Mountain**, opened in 1955, possibly near the southern crest of the hill, as well as the **Willis True Feldspar Quarry** just on the north side of the Goshen Road and west of the Blackguard Road. Several other small occurrences are in Waterford. [171]

A mica prospect was made on **Stearns Hill** in South Waterford in the mid to late-1910s: "They had a boiler in there [for steam powered drilling]. When I was a youngster you could look across on the other side of the road and see that boiler there." (Ober Kimball, taped interview with Ben Shaub, 1958).

Bear Mountain Quarry (Prospect), also known as the South Waterford Mica Quarry, is mapped both by Caldwell and Austin (1957) and Morrill et al. (1958), the latter saying that is was operated in 1900. Bastin (1911) visited the quarry in September, 1906 and noted it was 40 feet long, 15 feet wide, and 15 feet in depth, located on an eastern hillside." Despite Morrill et al.'s (1958) date of operations, Bastin (1911) merely called it "An old mica mine". It is reasonable that it was contemporaneous with the activity at Beech Hill and could have been worked by the same company. Bastin (1911) listed: annite (biotite), microcline, muscovite, oligoclase, and quartz.

Winfield "Winn" Knight opened the **Knight #1 and #2 Quarries** on his property along the Lovell Road just southwest of North Waterford in the mid-1950s, and the quarries were mentioned as beryl producers in 1956. The locality was mapped by Caldwell and Austin (1957) and Barton and Goldsmith (1968). The later reference reported that the #1 quarry produced 3 tons of beryl, 2 tons of sheet-quality mica, and 150 pounds of columbite-(Fe), the latter an amazing quantity for the Bumpus District. One beryl crystal was 175 pounds and over 4 feet long; additional minerals found include: albite, almandine, annite (biotite), microcline, and quartz. The #1 quarry was about 55 feel long and 20 feet wide. The #2 quarry was almost the same size as the #1 quarry, but was much shallower and worked briefly in the mid-1950s, but only 2 tons of sheet mica and only 25 pounds of beryl were found. Four additional small prospect pits are near the two large ones.

Appendix A. Early Oxford Mining Companies
by Van king

Mining Companies in Oxford County

Since the Maine Silver Rush of 1877-1884, perhaps hundreds of mining companies have been incorporated in Maine. However, a great many of the mining companies incorporated in Maine, were never intended to operate in the State. Maine's laws for incorporation were formerly very favorable as well as there being favorable laws in New Jersey, Delaware, and a few other states. For the most part, the only evidence for incorporation, of "Maine" mining companies, was the proliferation of stock certificates, now a curiosity and collectible hobby. Historians should not confuse the incorporation in Maine with any intention of a company doing any mining in the State, and companies, such as the Magma Mining Company, with mines in Arizona, or even the Gardner Mountain Copper Mining Company of Winterport, Maine with its only mine in New Hampshire, were nothing more than easily formed Maine corporations. Occasionally, companies would open an office in Portland or Bangor and there would be a brief mention of the fact in the *Maine Register.* However, "Boston interests" would frequently organize with "offices" in conveniently located Kittery. Occasionally, Maine mining companies were not incorporated in Maine.

More often than not, real Maine mining companies did not have a legal existence and were DBA's (Doing Business As). The earliest Oxford County mining "company" was the Holmes/Hamlin discovery, in the autumn of 1821 (Sturtevant, 1948; Perham, 1987; [123, 124] (Hamlin (1873, 1895) wrote that the discovery was made in 1820, but the date is inconsistent with contemporary statements and was much more than likely an approximate date that became "set in stone". Actual blasting and mining was done by brothers Cyrus Hamlin, age 19 or 20, and Hannibal Hamlin (age 12 or 13), in 1822.

The next mining company in Oxford County, *National Tin and Mining Company,* was formed by Samuel Carter in 1862. Carter married Elijah Hamlin's daughter on February 27, 1857 and was a graduate of Union College. Carter was probably influenced by noted miner, mineral collector, and College president, Eliphalet Nott. Carter's first registered mining lease, in 1862, was at the Mount Rubellite locality in Hebron, although he is better known as a casual miner at Mount Mica. Carter's brother-in-law, Augustus Hamlin "leased" Mount Mica in the late 1860s and had partnerships with several famous Pennsylvanians, Dr. Joseph Leidy ("The Last Man Who Knew Everything") of the Philadelphia Academy of Science, as well as noted mineral collector and wealthy industrialist, Clarence S. Bement. [123]

About 1870, an unnamed company mined at Mount Mica for mica. They were more famous for having discovered that Mount Mica was not "exhausted" and that gem tourmaline was still to be found there, than for their success in recovering mica. Much has been made of tourmalines having been thrown away on the dumps, but we must imagine that the farm owner, Odessa Bowker, a mineral dealer of sorts, recovered a significant amount of the discarded tourmaline, himself.

The Maine Silver Boom
an Outside Influence Reaching into Oxford County

Beginning about May, 1877, a silver mining boom began in Maine. [123] In that year, silver ore was discovered in Sullivan, Hancock County, Maine. Soon, there were reports of "silver ore" from all over Maine, from Acton, York County, in the West to Lubec, Washington County, in the East. (Actually, silver had already been known in Lubec in the early 1830s and there was an attempt at mining silver in Guilford, Piscataquis County, in central Maine in the early 1850s.) There were gold and silver mines in Oxford County before the hoopla of the Sullivan District really fired the public's imagination.

It can not be emphasized enough that almost all of the Maine metal mining attempts during this silver boom were organized by amateurs, with little advice from knowledgeable professionals, and almost all of the 1880s metal mines were total failures. The so-called ores were little more than quartz veins with tiny amounts of sulfides. The boom hysteria was enough to convince people that the few active mines with glowing reports would somehow have their luck rub off onto them. Even the Douglass Copper Mine, with its fully operational refinery in Blue Hill, was a financial loss. However, a great many so-called mines were never dug, but there are still stock certificates in old collections to testify to the sale of stock to fund the adventure.

Although most of the mining boom occurred relatively near the coast, if not within sight of tidewater, there

Harvey-Elliott Mining Company stock certificate. This company was a child of the Down East Silver Boom of 1877-1882.

were "inland" mines attempted in Oxford, Piscataquis, Somerset, and Penobscot Counties. The *Maine Mining Journal* cautioned businessmen, professionals, town leaders, et al., from being "used" by outside promoters. The presence of noted local citizens on mining boards gave an air of legitimacy to a company's existence and led local investors to gamble their money on a mine's future based primarily on the promoters' putting prominent figures into positions of conspicuous, but ineffectual, power. While the Lewiston-based mining companies were mostly governed by local investors, many of the coastal companies were manipulated by "outsiders". Almost always, the promoters reserved the positions of treasurer and secretary for themselves.

The town of Woodstock, Oxford County, had at least two metal mining attempts. One was the Woodstock Gold and Silver mine, while the other was the Lone Star Gold and Silver Mine, both on Spruce Mountain. The two mines were located near the NE-SW trending Mollocket Fault. [403] It was noted (*Maine Mining Journal*, April 9, 1880) that the officers of the Woodstock Gold and Silver Mine were "all from Lewiston": Dr. [Milton Curtis] Wedgewood (president and director), [Franklin Melon] Drew, city justice and later lawyer; and Company treasurer, secretary, and director), William Small, Jr. (city justice and director), and W. P. Whitehouse (director). F. M. Drew was listed in the 1880 *Maine Register* as the contact person for the Woodstock Gold and Silver Company, with office in the Central Block, while William Small, Jr. was listed under "Mining Exchanges" in the Odd Fellows Block.

The earlier formed Lone Star Gold and Silver Mining Company had Jonathan Levett Haskell Cobb, of Lewiston, as treasurer. (Cobb, an "industrialist", president of Lewiston & Auburn Horse Railroad Co., was known for having one of the finest mansions in Lewiston and in 1870 was elected to Lewiston's Fat Men Club.) The *Maine Mining Journal* (February 6, 1880) cited the *Lewiston Journal*: "A special meeting of the Lone Star Gold and Silver Mining Co., was held Friday evening last, at the office of Judge Morrill, in Auburn. Hon. B. D. Metcalf [owner of a steam-powered sawmill] of Damariscotta, presided, and the directors of the company were instructed to proceed at once to sink a shaft and open the mine belonging to the company, at Woodstock. There have been twelve assays of ore from this mine, made by different mineral assayers, which gave an average yield of $464.38 per ton, and the stockholders are jubilant." These first two Oxford County metal mines were incorporated, as was the Oxford Silver Mining Company. As they are mentioned fairly early in 1880, it is certain that sampling for assaying work was made in late 1879 and the assayers would have had to have time to submit their results. The simple organization of a company probably required several months. (Readers should be alerted to the fact that assays were usually made on highly purified small samples and are useless in evaluating a property; if a one ounce samples are compared with the tons of rock that need to be moved to recover the miniscule amount of valuable mineral in a huge vein.)

The provocatively named, Oxford Silver Mining

Company, with mine in Milton Plantation, had John Tibbetts as president and F. B. Wheelock as treasurer and was headquartered at 7 Lisbon Street, Lewiston. Although the company was listed in the *Maine Register,* it did not incorporate until late (*Maine Mining Journal,* December 24, 1880): "The Oxford County Mining Board was recently organized at Bryant's Pond, Oxford county, with the following list of members: Col. F. M. Drew, Lewiston; Hon. [Judge] Nahum Morrill, Auburn; Dr. M. C. Wedgewood, Lewiston; Dr. [Josiah Carr] Donham, Lewiston [Maine Legislator 1899-1900]; Herrick C. Davis [lawyer], Paris Hill; Hiram L. Libby, Norway; P. C. Fickett [lawyer], West Paris; R. K. Dunham [justice and dedimus], Bryant's Pond, and Benj. Davis, Woodstock. Col. F. M. Drew has been elected president, Fred M. Bartlett [general store owner], Bryant's Pond,

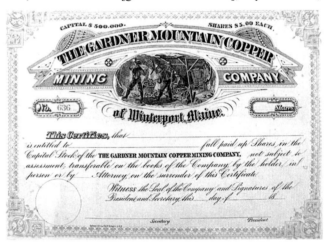

Stock certificate of the Gardner Mountain Copper Mining Company. Although incorporated under favorable Maine laws, the mining property was located in New Hampshire near Littleton.

Secretary, P. C. Fickett, Assistant Secretary, and S. E. May, Esq., Lewiston, Treasurer. The Board has for its purpose the exploring and developing the mineral lands of Oxford County." With winter abated, the *Maine Mining Journal* (May 13, 1881) cited Joshua Seitz's newspaper in Norway, *the New Religion:* "Dr. J. C. Donham of the Champion mine says his company will put on a large force of men during the month of May and that they will immediately erect suitable smelting works on the south side of Mt. Zircon, convenient for working ore in the Zircon mining district. Mr. J. H. Miller has bought the Quillen Mine and is organizing a company to work it. – Mr. T. H. Thornton is organizing a company to prospect and work a mine in Franklin Plantation [now = Peru]. It is reported that work will be commenced on the Wyman mine in Franklin Plantation this season. – We were shown, this week, by Mr. H. C. Davis, some very fine specimens of galena and sulphuret of silver ores, the latter from the Little Ben Davis mine. We are having more and more faith in Oxford County mines." The 1880 *Maine Register* listed C. T. Wyman and Erastus Abbott as people with interest in "Gold and silver mines" in Franklin Plantation. The exact location of many of these mines is not generally known

(Thompson et al., 2000). In 1904, when George Otis Smith (1904) visited William McCrillis' Mount Glines Mine, Milton Plantation, there was no mention that his mine, in the "Zircon Mining District", was resurrected from the ruins of a previous mine. Oftentimes, newspaper notices are not reliable recorders of events. Donham traveled with "Mr. Marble" to examine mines in Quebec (*Maine Mining Journal,* June 24, 1881).

Despite the present lack of knowing exactly where the Champion Mine was located, there are indications that a substantial amount of work was completed. The *Maine Mining Journal* (May 6, 1881) recounted:

> "The Oxford County Democrat, after noticing the production of silver bullion at Sullivan, thus speaks of the minerals of Oxford County: 'While other sections of the State have been carefully worked, Oxford County has not been left entirely in the cold. There are a number of mines in this section which have been opened with flattering prospects. Last week we saw a large box of hand-dressed ore, which had just been taken from the Champion mine in Milton Plantation. Old miners say that the ore is rich in silver. The mineral comes in, as they claim it should, in paying mines. First there are the coarse lumps of lead and silver then a finer mixture, and then still finer grades appear. At the depth of 25 feet, rock is taken out which assays $45 of silver to the ton. This mine has been worked for a year, and many tons of good hand-dressed ore are now on the surface. Those interested in the mine claim that they have demonstrated, by work already done, that mineral exists there in paying quantities. It is expected that foreign [out-of-state] capitalists will open the mine still further the coming year, and, if the yield holds good, will be soon shipping bullion from Oxford County. We shall welcome foreign capitalists to this work. Our people have little spare capital to divert from their general business, and therefore cannot develop the mines; but they will reap a rich harvest for their labor and produce, as soon as work is inaugurated with plenty of money behind it. The Register of Probate for Oxford County, H. C. Davis, Esq., is interested in Woodstock, Milton and Paris, and in his office may always be seen a full line of Oxford ores. Many of these specimens astonish old miners and mineralogists, who say 'if Oxford County can show such rock as this, there is no need of going to Colorado to do our mining.' We believe that the coming summer will demonstrate the value of Oxford mines, and settle the question finally as to whether or not they can be profitably worked."

In the springtime 1880, it was reported (*Maine Min-*

ing Journal, May 14, 1880): "The mining fever continues unabated in Lewiston. Messrs Ricker & Spaulding, mining brokers, report large sales of stocks at good prices. ... Two new mining companies have been organized in Lewiston to operate the lode in Woodstock. The Tucker Woodstock and Harris Woodstock [mines]. The following are the officers: Tucker- President, S. W. Cook; Secretary, L. H. Hutchinson; Treasurer, Wm. Small, Jr.; Directors, S. W. Cook, L. H. Hutchinson [Lewiston alderman and justice], W. Small, Jr., O. R. Small. Harris- President, E. F. Packard; Secretary, John B. Cotton; Treasurer, Wm. Small, Jr.; Directors, E. F. Packard, John B. Cotton, Wm. Small, Jr., E. Gile. A company has been organized at Norway to work a vein in the same town. –[*Boston Post.*" Apparently, cooler heads prevailed and these three announced companies do not seem to have actually started digging or blasting. Florian B. Maxim wrote to the *Maine Mining Journal* (July 9, 1880) indicating the mining excitement was high in Oxford County and related about individual hopes of finding gold or graphite. The thought of a gold mine in Norway was still alive, but seems to have been no more judicious than most of the mining companies: "There is quite a mining fever here just now and several parties have been blasting open ledges where they have found gold-bearing quartz, graphite, etc., and a company has lately been formed in the adjoining town of Norway, to work a ledge thought to contain gold in paying quantities."

There was also a "Union Mining Company" listed in Woodstock in the *Maine Register* in 1880. As the *Maine Register* frequently appeared in June of the year, it was still a full year before there was newspaper coverage of the company (*Maine Mining Journal,* May 20, 1881)):

> "The Portland Argus says the ores of Oxford county, Me. and Coos county, N. H. are attracting much attention among mining people. The veins there are very massive and the ores largely free smelting, and as such are worth a high price as fluxes for other ores, and are well nigh indispensable for working the more refractory ores from other sections. Only a beginning has as yet been made toward developing their resources.
>
> A West Paris correspondent of the Norway New Religion, writes concerning the Union Mining company (whose organization was noticed last week in these columns) and other mineral properties in the vicinity, as follows: The members of the company were in fine spirits and sanguine of ultimate success and came forward and offered their money to be expended in mining operations instead of selling or putting stock up on the market for sale. One member present offered to buy out the interest of any member of the company with a good big bonus over and above the cost but found none that were inclined to sell. Immediate operations are to be commenced on

Douglass Copper Mine, Blue Hill.
~1882. From an old postcard.

what is known as the Ben Davis mine, the Sim Curtis or Fickett mine near West Paris, and a mine in Si Gotch [Woodstock]. Since the organization of the company, one year ago, many encouraging developments have been made. At that time the Champion was the only mine worked, and that a mere experiment. Now the Champion is a grand success a wealth to its owners. The Ben Davis mine, at the depth of 12 feet, has come upon a sulphuret of silver, assaying $500 to the ton. At the mine in Si Gotch, galena is found cropping out at the surface among the crystallized quartz. The Sim Curtis or Fickett mine, and the depth of 12 to 15 feet assayed $32 to the ton. So much for this company. Other companies are finding like encouragement. Galena has been found on what is known as the Waterhouse mountain, [43] at a depth of only six or eight feet, and work is soon to be commenced there. The mining companies of Oxford county are owned and operated by some of our best man who are determined that no wild cat business shall be done here, but, like good farmers, they will warrant a good return for the amount of labor performed, and for this they put no stock in the market and ask no help outside of themselves."

(It is interesting to note that many of the officers of the companies mining in Oxford County had also graduated from Bowdoin College.)

The metal mining boom in Maine was at its height in 1880-1881, however, it became increasingly apparent that the veins that were thought to be so promising had been vastly over-rated and the enormously rich assays frequently had been made on carefully selected and very purified samples. The identification of the minerals, present in the ores,

was frequently fantastic as the inexperienced miners would claim that they had particularly rich and rare silver minerals among their ordinary minerals.

Despite the bravado and notoriety of Oxford County metal mines in the early 1880s, no gold or silver ores were shipped for refining and the mined rock lay beside the mine unprocessed and eventually forgotten. There was seldom mention of a failed mine in print, but the word of mouth must have been full of the trivia associated with Maine mining's demise. Many attributed the mine failures to a lack of processing facilities, although a mill was built in Portland to receive ores for refining. Some ores were shipped to New Jersey. (The Deer Isle Silver Mine shipped ore to England for processing, but the load was shipwrecked off the coast of Iceland, where it remains to this day.) Other miners said that their ore was valuable, but it was too intermixed with other valuable minerals to be economically processed. While the statement was partly true, and the separation process called "floatation" was only a dream at the time; the real cause of failure was the assay process.

The proper method of assaying included taking a large sample, homogenizing it, and analyzing a <u>representative</u> sample of the rock that had to be removed. Assayers were never ones to turn away business and when they were provided a sample for analysis, they rarely gave advice to their customers as to the meaning of their analyses. An ore mineral that assayed $400/ton usually meant that if you had a pure sample, that's what you would get. Unfortunately, Maine "ores" were rarely as rich as 1% in ore minerals. That is, for every ton of rock mined, there would be twenty pounds of pure mineral. The assay should have been translated to mean that the ore was worth $8/ton. The distinction between "ore mineral" and "ore" was lost on the naive mine owners, but they learned their lesson the hard way. Mines simply stopped running as the money ran out or the mine owners realized that the definition of a mine was "a hole into which you threw money". Most of Maine's mines were trivial in size, but it would not have helped if they had been bigger or better funded. Only two of the 1880 Maine metal mines ever became genuine economic successes. The Mammoth Mine in Blue Hill was operated in the 1960s to early 1970s as the Black Hawk Mine, while the Cape Rosier Mine of Brooksville became the Callahan Mine, in the 1960s-1970s.

The general absence of mining news in 1882 was telling. Only the mines in Sullivan and Blue Hill, both in Hancock County, were the few districts that constantly received the public's attention, but by 1884, their voices were silent. The *Maine Mining Journal* recognized the shrinking mining news market. At first, it reported on mines further a field, both in New Hampshire and in Eastern Canada. There were articles about the Western mines, but increasingly, the newspaper included information about agricultural products, railroads, even the Paris Manufacturing Company and, by mid 1882, altered its name to become more inclusive: the *Maine Mining and Industrial Journal*. Eventually, in 1884, it became the *Maine Industrial Journal* and survived until 1918.

Oxford County's Mica Quarries

Although Mount Mica had once been a bona fide mica quarry, there was no impulse to explore for mica elsewhere in the County for another ten years. It was only when the entire State was gripped in a Silver Boom did it occur to people to begin exploring for other minerals in the rest of Oxford County. The mica industry did not need an expensive mill. Pegmatites could be drilled and blasted and any mica could be separated from its enclosing rock by using small hammers, etc. The resulting mica, set free from its matrix could be shipped directly to market in relatively small packages or crates, due to its high value. For these reasons, Oxford County pegmatites were first explored as mica-producers.

The first mining in Albany was at the Charles P. Pingree Mica Quarry in 1878-1879 and may have inspired a subsequent search resulting in the discovery of the Peaked Mountain Mica locality in nearby Gilead. By the turn of the nineteenth to the twentieth centuries, the rise in electrical use of mica, including the rapid expansion of markets associated with electrical apparatus, including radios, telephones, switchboards, etc. resulted in an enormous demand for mica insulation. Previously, the demand for mica was relatively low, mostly for use in fireproof windows in stoves, lamps, etc. Unfortunately, the information on the Pingree Quarry is minimal. The Peaked Mountain locality in Gilead was more fortunate. The *Maine Mining Journal* had begun publication January 2, 1880 to report on all Maine mining activities.

On April 30, 1880, the *Maine Mining Journal* reported:

"The discovery of clear white mica at Gilead, Maine, and the success with which it is being quarried, add another interest to the mineral wealth of that State, For sixty-five years the major part of the mica used has been secured in one mine in [Ruggles Quarry, Grafton] New Hampshire and one other in North Carolina, so that prices have ranged from 80¢ a pound for the smallest stove sizes upwards. The new mica quarry now brought to light is on Peaked Hill, on the An-

droscoggin river, three miles north of the West Bethel station on the Grand Trunk [rail]road. The property has been visited by some of the best geologists in the country, who have been employed by a number of the leading business men of Portland, and the vein is reported to be from five to eight thick for three hundred feet, with surface indications that it continues equally rich about 1-8 of a mile further, The mica is a clear dead white, free from flecks and cracks, and is taken out in sheets as thin as writing paper up to fifteen and twenty inches square, The property has been worked for two months, and the best of merchantable mica in sizes 3x1 1-2 feet secured. A company to be known as the Maine Mica Company is being organized by Portland people. – [*Boston Advertiser.*"

The June 11, 1880 *Maine Mining Journal* announced the incorporation of the mica company. It is particularly interesting to note that Hugh J. Chisholm was president of the company. The youngish Chisholm also was president of the unsuccessful Lebanon-Acton Silver Mining Company (incorporated February, 1880) in Lebanon, York County, Maine (*Maine Mining Journal,* July 9, 1880; [123]).

"The Maine Mica Mining Company of Gilead, Me., was organized in Portland, June 5, with the following officers: President, H. J. Chisholm; Treasurer, A. K. P. Leighton [owner of a shipwright company]; Secretary, Robinson Williams; Directors, Wm. Reed and Wm.

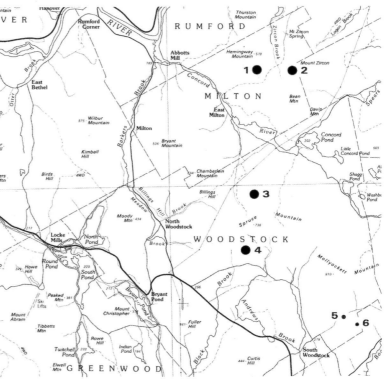

Map of "Silver" Mines in Woodstock and Milton. Thompson et al., 2000.

P. Gould of Boston, H. J. Chisholm and R. Williams of Portland, and C. W. Pierce of Bethel, Me."

The short article extolled the virtues of the mica "in large sizes, of good quality[,] is present in large quantities, also feldspar of superior quality, together with many curious and precious stones, but that of commercial value is mica and feldspar ..." The Gilead mica and locality had been evaluated by Bowdoin chemistry and physics professor, Henry Carmichael. Carmichael also was an "Assayer for the State of Maine". [Hugh Chisholm was close friends with Adam Philips Leighton Sr. who eventually became known as the father of the American postcard industry. Leighton was a clerk in Chisholm Brothers' book business beginning November 9, 1867. Hugh Chisholm Leighton was the son of Adam P. Leighton, Sr. Leighton became vice-president of a company that had the concession of selling newspapers on the Grand Trunk Railroad. Coincidentally, Chisholm and Thomas Alva Edison both sold newspapers on a Grand Trunk Railroad run from Detroit to Toronto, about 1860 and were lifelong friends and one may see a connection of the Chisholm paper companies' roots. Similarly, the Leighton postcard company certainly used card stock manufactured by International Paper, as that was an important early product at the Rumford mill. Leighton was president of the Chapman National Bank, for four years, and director of Casco-Mercantile Bank, and Fidelity Trust of Portland, as well as Portland mayor in 1910. Leighton was also a chewing gum manufacturer and became president of the Sen-Sen Chiclet Company and, later, director of the American Chicle Company]

A short note in the Maine Mining Journal (September 3, 1880) indicated: "A letter from Wm. D. Gould of Gould & Watson, Boston, to Hugh J. Chisholm, Esq., of this city, speaks of two boxes of Mica sent there as of splendid quality and will cut good sizes. He advises Mr. Chisholm to put on more men and push work rapidly. – [Portland Argus." The next report came several months later (Maine Mining Journal, December 24, 1880): "The Maine Mica Mining Company, now operating its mines at Gilead and Albany, Me., exhibit at their Portland office a large quantity of merchantable mica, cut to order in sizes for use. They also have orders for next season's work aggregating some $12,000. The ease and facility with which this product can be prepared for market gives its position in advance of other classes of mining. Two of our prominent business men are directors in the company, - W. P. Gould and William Reed. None of the stock has yet been offered to the public, only 20,000 shares ever having been sold. It is an established enterprise. – [Economist."

The February 25, 1881 Maine Mining Journal had an informational article about mica:

"The market value of mica is great, and at the present time the demand is such that it cannot be supplied. A piece of mica 4 in. x 4 in., ⁵⁄₈ of an inch thick will weigh one pound, and its market value is $4.10. A piece 3x6 in., ½ inch thick, will weigh one pound, and has a market value of $4.60; and a like relative proportion of value exists except in the smallest pieces, say 1 ½ x 2, which even has a value of 50 cents per pound; and all the refuse and trimmings of the mica are readily sold to parties who pulverize the same for use in connection with the manufacture of nitro-glycerine and other dangerous explosives."

(It should be clarified that mica was one of the constituents as the filler used in making dynamite. In 1881, Chisholm was an investor in the Umbagog Pulp Company in Livermore Falls and a decade later organized the development of Rumford Falls and the town's supporting infrastructure, including railroads, to create a paper mill town, leading to the formation of the International Paper Company in 1898.)

Oxford County wasn't the only place to have a search for mica. There was also a Beacon Mica Prospect in Waterboro, York County, developed by "Boston interests", but the effort did not last long (Maine Mining Journal, October 1, 1880). A mica occurrence, which may have been later developed as the Poole or Williams prospect, both worked sixty to seventy years later, was announced (Maine Mining Journal, April 8, 1881): "A very valuable deposit of mica has been discovered in Edgecomb, Lincoln County. The vein is about half a mile in length, and a company is soon to be organized to work the property." Another company, the Farmington Mica Mining Company, organized in Portland, December 1, 1880, may have intended to work in Oxford County, as well, but there is no indication that they operated a true quarry. The news from the mica quarries disappeared as more and more mica began to be imported from India. By 1884, Maine mica quarrying had nearly become extinct.

William N. McCrillis was listed as the president of the U. S. Nickel Mine in Rumford by the Maine Register 1905-1914 and president of the Mount Glines Gold and Silver Mining Company of Milton in 1905-1908. By the turn of the nineteenth century, there were many Oxford County mining companies and the reader is directed to Bastin (1911), Sterrett (1923), Cameron et al. (1954), Perham Stevens (1972), Perham (1976), King and Foord (1994), and King (2000) for additional information.

Mica books. 6 x 8 x 1.5 cm.
Bumpus Quarry.

Appendix B Summary of Maine Pegmatite Mining by Mineral Feldspar, Mica, and Beryl

The business trivia of mining feldspar, mica, and beryl explains much about what was going on, at least in terms of the mill management as well as the various independent mining contractors. Shrinkages in one market might be offset by expansion in another mineral. By following the economic details, the rationale for many miners' efforts may be understood. The following sections also contain news and developments of the various mineral industries.

The most important section relating to the Bumpus Quarry in Albany is, of course, the beryl section. The rise and fall of beryl's fortunes were tied to the Cold War and the production of nuclear weapons. The annual reports of the U. S. Geological Survey did not hide the fact that beryllium was used in weapons, but tried to spread the uses of beryllium around so that it wasn't immediately possible to calculate the maximum number of weapons that could be built. There is no doubt some incompleteness or disinformation in those reports.

Feldspar in Oxford County

In the 1920s, crude feldspar was worth a little under $10 per long ton. Feldspar had a wide variety of uses ranging from the starting material in the manufacture of china enamelware: plates, cups, and food serving dishes, as well as tiles, bricks, porcelain glazed electrical insulators, and bathroom fixtures. Impure grades of feldspar, naturally mixed with quartz, were used in making abrasive wheels, roofing compound fillers, facing compounds on concrete walls, bottle glass, opalescent decorative glass, poultry grit (filler in feeds). The opalescent glass was becoming popular as an Art Deco medium and was increasingly used for glass partitions in public businesses such as hotels, restaurants, etc. where light penetration was valued along with a visual barrier, including applications as varied as lamp light diffusers to "frosted windows". Quartz-free feldspar had a market with the Bon Ami Company for non-scratching scouring cleaners. Feldspar has a Mohs' hardness of 6 while quartz has a hardness of 7. Porcelain, being slightly harder than feldspar, could be scoured by feldspar powder without destroying the bright finish. Bon Ami's slogan was: "It hasn't scratched yet!" Pure feldspar was a requirement to support that claim, because porcelain would quickly become dull by using a powder harder than the glaze. It was for the last mentioned application that the mill's, then important customer for ground feldspar, Bon Ami company, wanted #1 grade feldspar. At this writing, Bon Ami still manufactures cleaning compounds, particularly for home use. (At one time, ground feldspar was sold as a mineral fertilizer, but the effectiveness of the fertilizer, as a source of potash, was very low and the few available nutrients were soon leached in the first season of application.) The principal manufacturing center for porcelain, china, tile, etc. was in the East Liverpool, Ohio and nearby West Virginian potteries. Trenton, NJ and New York City were also important manufacturing centers consuming ground feldspar. (Most production records and values of products will be calculated from tabulated data from USBM reports.)

Maine had long been among the USA's leading feldspar producers. In 1913, Maine, with production of 38,114 tons, ranked first in the nation and accounted for nearly one-half of the value of all feldspar produced in the USA (Katz, 1914). The previous year, feldspar production of Maine was only 19,091 tons and the State ranked second (Katz, 1913). Although Maine didn't produce the greatest amount of mined feldspar in 1912, the value of her feldspar was still about one-third of the value of all of the nation's ground feldspar. The secret for the success was in quality. Maine produced almost all high-grade, low impurity, pottery grade feldspar, while New York, the runner up in value, but first in crude feldspar output, produced both pottery and lower grades. Manufacturers were also willing to pay a premium for some Maine ground feldspar, particularly from Auburn, as it had better kiln firing characteristics related to the frequent presence of rare elements making up the feldspar. (The up-swing in production was due to the new feldspar mill of the Maine Feldspar Company at Topsham, first operated in 1912, that was in high-production and several new quarries were opened. That year, Oxford County feldspar of Hedgehog Hill and Lobikis Quarries, both Peru, were tested for feldspar grinding quality.)

In 1924, there was a bumper crop of feldspar in the USA as there had not been enough feldspar stockpiled over the previous winter to meet demands. The average USA price for crude feldspar was $7.37 per long ton (2240 pounds), an increase of 1% from 1922, although the range in price for crude feldspar in New England in 1924 was $3.73 to $16.74 (Middleton, 1927). Maine's crude feldspar averaged $9.19 suggesting that it was virtually all of "pottery grade". Twelve states produced feldspar in 1924, South Dakota being a new producer. North Carolina was the new leading producer of feldspar with 57,622 tons, while Maine was third, after New Hampshire which had a 70% production growth more than offsetting Maine's 20% growth, with 25,029 tons. The average USA value of ground feldspar, the output of the West Paris mill was generally of this type, was $16.84 per short ton (2,000 pounds), an increase of 46¢ from the previous year. New England ground feldspar varied from $14.15 to $20 per short ton with the Maine average on the high side at $19.05 per short ton. There was, therefore, a 10% difference in the way crude feldspar was bought and ground feldspar was sold. There were three feldspar grinding mills in Maine in 1924 servicing 15 producing quarries: Oxford Mining & Milling Company, West Paris, Maine

Feldspar Company (the largest producer in the State for that year), Danville Junction in Auburn, and Trenton Flint & Spar Company, Cathance River, Topsham. It wasn't possible to deduce the prices obtained at the various Maine feldspar mills.

In 1925, the first year the West Paris Feldspar Grinding Mill was in operation, Maine feldspar production was up by nearly 5,000 tons (29,912 tons), but the value of crude feldspar remained about the same at $9.07 per ton, while Maine ground feldspar dropped a bit to $18.76 per ton. Both North Carolina and New Hampshire dramatically increased their production, yet Maine remained in third place (Middleton, 1928). By 1926, Maine production continued upward to 33,827 tons crude feldspar with a value of $9.07 per ton, while ground feldspar averaged $18.31 per ton. Maine crept into second place with just over more than 600 tons more than New Hampshire, but a new grinding mill in North Carolina predicted changes (Middleton, 1929). There were 4 grinding mills in operation in Maine, suggesting commencement of the operation of the J. W. Cummings feldspar grinding mill in Bath. In 1937, the *Maine Register* indicated that the Cummings company had been sold to the Ceramic Feldspar Company, but there was not a subsequent listing of that company. Thomas J. Cummings, president of the J. W. Cummings Company, surrendered his mining lease on the Lester P. and Cora M. Twitchell mining property in Paris on June 4, 1937. [350]

Overall crude feldspar production in the USA decreased in 1927 except by the leading states North Carolina and Maine (34,328 @ $8.72/ton). This was Maine's highest output, although it was probably not Oxford County's highest output. Most of Maine's feldspar still came from the Auburn and Topsham Districts. Ground feldspar remained at about $18.28 (Middleton, 1930). As 1927 saw the Bumpus Quarry coming online for the West Paris Mill, one wonders if the average cost per ton of crude feldspar dropped partly because of additions from Albany?

Because of a trend in the formation of larger feldspar companies in 1928, two mergers affected a large part of Maine's feldspar industry (Middleton, 1931). The Oxford Mining and Milling Company was acquired by Charles H. Pedrick, Jr. of New York City who had also founded the United Feldspar Corporation, also acquiring the Tennessee Mineral Products Company of Spruce Pine, NC, and the United States Feldspar Corporation of Cranberry Creek, NY. H. P. Margerum of the Goldings Sons Company of Trenton, NJ purchased a number of companies to form the Consolidated Feldspar Company by uniting: Maine Feldspar Company, Brunswick, Maine (including feldspar mils at Auburn and Topsham), Bedford Mining Company, Bedford, NY, Dominion Feldspar Company and New York Feldspar Company, both of Rochester, NY, Isco-Bautz Company, Murphrysboro, IL, and the Erwin Feldspar Company of Erwin, TN, as well as properties near Keystone, SD and Kingman, AZ. The Oxford Mining and Milling Company continued to operate under its own name until January 9, 1940, when the OMMC officially became property of the United Feldspar Corporation. [351]

In 1940, the United Feldspar Company was "reorganized", changed its name to United Feldspar & Minerals Corporation, and had a new president, Francis E. Haag, of Pelham, NY. The OMMC continued to be listed by the *Maine Register* until 1944. Actually, the "reorganization" was a sale on February 21, 1938 (OCRD v416 p618-620) of the Oxford Mining and Milling Company, by a five-year $40,000 mortgage, to Francis Edgar Haag and Winifred L. Haag, copartners in F. E. Haag & Company. The Haag Company also controlled the Feldspathic Research Company, which also engaged in the feldspar business. The sale included: mineral rights to the A. C. Perham Quarry and nearby land, the land and feldspar mill and nearby lands on Maple Street, West Paris, and included all of its mineral leases and other agreements.

The year 1928 also saw a drop in feldspar production and a drop in prices. Maine dropped to third place, 25,063 tons ($8.07/ton), behind New Hampshire and the price of ground feldspar averaged about the same as the previous year, $18.23/ton. In effect, the small miners were getting squeezed by lower crude feldspar prices paid by the newly consolidated companies.

Although there was a reduction in national production in 1929, Maine remained in third place and rebounded with 33,897 tons of crude feldspar with a value of $9.02/ton, but the ground feldspar plummeted to $17.14, a trend seen across the country (Bowles and Middleton, 1932). Merrill and Perkins (1930) wrote about the price of ground feldspar: "The present price runs about $19 to $20 a ton. A crushed spar war is on but it is expected that this will soon be settled and the price will go up."

In 1930, the effects of the Great Depression were beginning to be felt. The entire USA industry of 34 mills had the capacity of three times the actual annual consumption of ground feldspar (Hughes and Middleton, 1933). Probably because of the sensed decrease in feldspar consumption, a tariff of $1/ton was enacted to protect domestic producers from imported feldspar, much of which came from Ontario and Quebec, Canada. It was also reported that the feldspar consumers were beginning to specify the fine grinding size of the feldspar they bought and that many mills were being upgraded to take advantage of the strengthening requirements to have a competitive edge. Mesh-sizes of ground feldspar were frequently narrowly specified and there must have been a need for washing or otherwise cleaning ground feldspar to remove fine, unwanted particles. The West Paris mill was still relatively new and may have been in a better technological position than the other mills in the State. Experiments were also noted that were designed to remove iron minerals, including black tourmaline, as well as impurity muscovite from the ground feldspar. Not only was there a demand for exacting grinding sizes, there was also a demand for a purer product. Despite all of the adjustments being made in the year, Maine was the nation's second largest feldspar producer with 22,738 tons with a value of $7.11/ton. Oddly, the Hughes and Middleton (1933) report downsized

Maine's previous production records and indicated that Maine was one of the only major states to increase production, with a 14% increase in value. The changes in figures cannot be reconciled. Maine's ground feldspar was worth only $15.90/ton and the amount of ground feldspar produced was down 25%. The real loss in income for the mill is apparent, as well as for the miners and land owners who were no longer getting paid for feldspar production.

In 1931, feldspar mineral had taken a disastrous downturn in Maine with production off by over half and falling to fourth in production after North Carolina, New Hampshire, and South Dakota with only 10,220 tons of crude feldspar mined with a value of only $5.42/ton (Hughes and Middleton, 1933a), although the particular producers were not identified, some probably faring worse than others. The mill didn't fare as badly as the miners, percentagewise, as the ground feldspar from Maine averaged $14.61/ton, however, Sagadahoc County accounted for two thirds of the State's feldspar production in 1931, although it isn't known if the Consolidated Feldspar Company combined the figures for the Topsham and Auburn grinding mills. The value of Maine's ground feldspar was third in the nation despite the lower price of its crude feldspar. One problem was a price war that was begun by the larger feldspar producers in an effort to stay in business, but everyone suffered from lower prices. (The downturn in mineral production across the country allowed the USBM to catch up on its reporting and there were two issues of minerals reports published in 1933.) Because of the downturn in the housing construction industry, sales of bathroom fixtures, a large consumer in the ground feldspar market was down as well. Demand diminished for porcelain products for a number of years.

In 1932 and 1933, demand for ground feldspar continued to diminish in the products used in the housing industry, although the manufacture of china tableware was not nearly as seriously affected (Rogers and Galiher, 1934). Feldspar mining in New England was severely affected with a 75% decrease in production, but Maine was third in the nation during the year, but with an output of only 8345 tons of crude feldspar worth $5.02/ton. Ground feldspar continued to decline to $13.18. Part of the decline in value was because of competition, increased efficiency, and lower prices, from new grinding mills constructed in North Carolina and Virginia. (Rogers and Metcalf (1934) stated that ground feldspar from Maine averaged $16.79/ton in 1932, but the calculations used here have been based on total reported tons of ground feldspar produced and total dollar value reported.)

The passing of the Cullen Act of April 7, 1933. permitting the sale of beer, and, eventually, the repeal of the 18th Amendment to the Constitution on December 5, 1933 were very important to the feldspar industry and brought some minor relief to Oxford County. Rogers and Metcalf (1934) reported on the situation:

> "Overshadowing all other factors influencing the feldspar industry in 1933 was the legalization of beer and the consequent rise in demand for bottles, glass-lined tanks, enameled vessels, and allied specialties. The total output of crude feldspar, including both potash and lime-soda spars, was 150,633 tons in 1933, an increase of 43.9 percent over 1932…"

Maine's increase in 1933 was only 35%, but it must have been welcome. Maine crude feldspar production was fourth in the nation, with 11,273 tons with a low value of $4.29/ton, as Virginia catapulted into second position between leading North Carolina and third place New Hampshire. Despite the desirability of Maine's feldspar, it was still more expensive than the "Southern" producers. Ground feldspar value was not reported by state.

The next year, 1934, Maine feldspar production rebounded to 14,685 tons worth $5.64/ton a price that hadn't been seen for five years. Due to lowered production in Virginia and New Hampshire, Maine reoccupied second place behind megaproducer North Carolina. Maine's ground feldspar brought $15.22/ton (Metcalf, 1935).

In 1935, there were big increases in the amount of housing construction and remodeling and there was a consequent large rise in demand for porcelain plumbing fixtures, etc. Maine feldspar mining rose to 17,103 tons worth $5.83/ton, but the increased production in Colorado and South Dakota dropped Maine to fourth. Maine ground feldspar was $15.01/ton (Metcalf, 1936).

In 1936, glass production, particularly for bottles, consumed half of the nation's feldspar production, whereas it was about a third several years before. The repeal of Prohibition brought prosperity back to feldspar mining, but Maine production slipped slightly to 16,392 tons, just good for fifth place, but given the reported weakness in the Sagadahoc County feldspar production, the West Paris mill, and therefore Oxford County, may have been "up" for 1936 (Metcalf, 1936): "The Trenton Flint & Spar Co., Trenton, N.J., with grinding plant at Cathance, Maine, is reported to be out of business." and "The Ceramic Feldspar Co., Bath, Maine took over the plant formerly known as the Cummings Feldspar Co., and started grinding operations early in 1936." From the notice, we might gather that the Topsham/Bath District suffered a collapse in feldspar grinding facilities and that there must have been a consequent collapse in Sagadahoc County mining as crude-feldspar purchasers disappeared. It must be also imagined that the West Paris mill represented most of the State's ground feldspar production as the formerly major millers in Sagadahoc County were no longer among the strong players. Feldspar mills were built in 1936, but not in Maine. The Consolidated Feldspar Company had opened a new mill in Erwin, TN and there was a new mill, one each, in Colorado, New Hampshire and South Dakota and a mill was revitalized in New York. Sagadahoc County's fate may have rested with a change of philosophy with the Consolidated Feldspar Company. Metcalf (1937) wrote: "Normally, the tonnage of ground feldspar produced from domestic crude is about 87 percent of the crude-spar output, the remaining 13 percent representing spar sold for purposes that do not require fine grinding and that lost or discarded during the grinding process." Maine ground feldspar sold for "$14.65/ton."

The year 1937 was a perplexing year for Maine feldspar mining. The USA production for the year was at an all time record, yet Maine dropped in position down to sixth place with 20,191 tons, only slightly fewer tons than the year before, suggesting the importance of the new mill capacity in other states (Metcalf, 1938). Nonetheless, the competing feldspar-producing states were generally larger and much more populous than Maine, excepting New Hampshire, which was in third position. Crude feldspar in Maine brought $5.49/ton, while ground feldspar was $13.74.

All feldspar markets were down in 1938 (Metcalf, 1939). Maine feldspar production plummeted to 13,764 tons, but without the former large production of Sagadahoc County quarries, Maine was fifth in the nation as there were larger drops in mining output in other states. Innovations included the use of feldspar in making window glass. Refrigerators and stoves, both with porcelain coatings, increased in production. Spodumene, primarily mixed with feldspar, saw greater use in the ceramics industry as did imported nepheline from Canada. Maine's ground feldspar also loss pricing level and dropped to $12.55/ton. The Oxford Mining and Milling Company also began to import nepheline from Canada for grinding. Four feldspar grinding mills were listed for Maine: Ceramic Feldspar Company, Consolidated Feldspar Company, Oxford Mining and Milling Company, and Topsham Feldspar Company. Ground feldspar consumption was: glass manufacturers 55%, pottery manufacturers 35%, and enamel manufacturers 9%. Methods of purifying feldspar had been discussed in detail for many years, employing a method called floatation. Magnetic separators had already been in use at some grinding mills, but Keene, New Hampshire's feldspar grinding mill was reporting success by using a new method utilizing selectively oiling particles on a shaking table and obtained greater than 98% pure feldspar with saleable quartz as a by-product. Oxford County depended on its high grade feldspar. Electrostatic separation was receiving attention and there was discussion of obtaining high grade feldspar concentrates by processing dump minerals.

Metcalf (1940) announced that the USA feldspar industry had rebounded in 1939 and experienced its second best year on record. Although producing 18,109 tons of crude feldspar with an average value of $4.10/ton and ground feldspar at $12.68, Maine dropped to sixth place. The new owners of the West Paris Feldspar Mill were specified: "The Oxford Mining & Milling Co., West Paris, Maine, and the Tennessee Mineral Products Corporation, Spruce Pine, N.C., former subsidiaries of United Feldspar Corporation, were operated after January 1, 1940, as the Oxford Division and Minpro Division, respectively, of the United Feldspar & Minerals Corporation." Nepheline was imported from Blue Mountain in Canada and was "blended with granular glass spar."

In 1940, USA crude feldspar mining reached yet another output record (Metcalf, 1943). Maine managed to keep pace with the previous year producing 18,390 tons and dropped to seventh place, while South Dakota, New Hampshire, and Colorado established new state records for themselves. The value of Maine crude feldspar rose to $4.61/ton and ground feldspar $13.06.

Maine's feldspar production increased slightly to 22,566 tons in 1941, but she was in fifth place (Metcalf, 1943). Crude feldspar also rose, to $5.17/ton, a welcome boost to miners. Ground feldspar was $13.72/ton, a welcome boost for the mill. (Up until this time, most crude feldspar processed in West Paris was quarried by mill employees. However, there was a growing contingent of independent miners and the price subsidies of WWII, greatly increased the ranks of the independent and contract miners. In particular, Oxford County people quarried small crops of feldspar from their lands to raise extra money during otherwise idle time. The WWII miners were usually seeking beryl and mica, but feldspar was a by-product, even if it might exceed the value of the mica and beryl found.)

Mote (1954) listed: "In 1953, 17 mines were active [in Maine] – 2 in Androscoggin County [LaFlamme and Sturtevant, both Bell Minerals], 3 in Oxford County [Foster Quarry (Mount Marie Quarry Group), South Paris, leading producer in county, others included Perham and Tamminen quarries, all by Bell Minerals], and 12 in Sagadahoc County [with the largest quarries being those of the Consolidated Feldspar, as well as the Purington, Aldred, and Diamond Match quarries, all Topsham]. All feldspar mined in 1953 were [sic] ground locally at the three mills active during the year. Topsham Feldspar Co. operated its mill at Topsham entirely on purchased crude during the year. The Topsham mill of Consolidated Feldspar Dept., International Minerals & Chemical Corp., treated both company-owned and purchased ore, whereas the Bell Minerals Co., West Paris, Oxford County, mill treated company ore only." Oxford County production was down 56 percent from 1952.

Kaufman and Cleary (1957) noted of 1954 that feldspar was produced from the LaFlamme and Sturtevant, Mount Marie, Foster, Perham, and Tamminen quarries by Bell Minerals Company. (Sagadahoc County data were more explicit: "Eight feldspar mines were active. The major producers were the Consolidated Feldspar Dept., International Minerals & Chemical Corp., Topsham mine, David Ponziani, Purington mine; Ray C. Leavitt, Diamond Match mine, Cesare Trusiani, Powers mine; Joe Palozzi, Willis mine; Topsham Feldspar Co., Trenton mine, and Charles Chapman, Chapman mine. … Cesare Trusiani reported a small production of columbium-tantalum concentrate as a by-product of feldspar mining. This was sold to the GSA Depot at Franklin, N.H.)

Metcalf and Otte (1958) briefly noted for 1955 that Maine feldspar "rose substantially" and was 26,282 long tons with average value of $7.19/ton down 2% from the year before. There were 13 Maine feldspar producers in 1955 including: Dunn, Foster, LaFlamme, Perham, Nevel, and Tamminen Quarries. In addition to feldspar, spodumene and columbite were recovered from the Nevel Quarry at Newry by the Whitehall Company. "R. C. Benson, West Paris, subleased the [Tamminen] mine and sold the output to Bell Minerals. Pechnik Bros. Dunn mine at Norway, and Whitehall, Inc., Newry mine at Newry reported output of feldspar. In addition, feldspar mined by a number of small producers, was sold to Bell Minerals Co. ... [In Sagadahoc County] The principal producers were Consolidated Feldspar... Topsham and Mount Ararat mines; Frank DiBassio and David Ponziani, each of whom worked the Purington mine during the year; Ray C. Leavitt, Diamond Match mine; and Joseph Paulin, McIver mine. ... Topsham Feldspar Co., Topsham, produced and ground feldspar for sale as poultry and pet grits and for soaps and abrasives." The high graphic granite content of the Topsham pegmatites meant that they produced a lower grade ground feldspar than did Oxford County.

In 1956, Maine feldspar production declined 15%, but it was revealed that there were only two feldspar mills in operation (Metcalf, 1958a). Average crude feldspar prices fell from $7.19/ton to $6.46/ton. Maine feldspar quarries producing during the year included: Dunn (Pechnik Brothers), LaFlamme, Gildings, Glover-Wilbur (South Paris), Maine Mica Company, Nevel (Whitehall), Perham, Tamminen, Tiger Bill (Buck & Baker), Alex Cunningham[,] Brown (Sagadahoc) quarries. ("The Sturtevant Quarry was not worked.") The Whitehall Company also recovered and sold spodumene and columbite at the Nevel Quarry, Newry. Sagadahoc County had relatively few producers: Augustin Carter, Henry Haskell, and Frank DiBiasso (Aldred Quarry), near Topsham; Alex Cunningham (Brown Quarry), Georgetown; Ray C. Leavitt (Diamond Match Quarry), John H. Palozzi (Fisher Quarry), Victor Ponziani (Tedford Quarry), James Russo (Russo Quarry), and White's Service (Addition Quarry), all Topsham; Cesare Trusiani (Georgetown Quarry). All of the Sagadahoc County feldspar was ground at the Consolidated Feldspar mill. The Topsham Feldspar Company mill did not operate.)

The year 1957 saw some changes in Maine pegmatite mining. Feldspar saw a large decline (36%), but three grinding mills were active: West Paris and the two mills in Topsham. In Oxford County, there were four producers who worked at six different quarries, while Sagadahoc had 12 producers at 13 quarries. In Oxford County, Bell Minerals Co. and R. C. Benson were the principal companies involved in feldspar. Alex Cunningham worked in Georgetown; White's Service and Cesare Trusiani worked in Topsham, while James Russo worked in Brunswick: "Feldspar was mined from open pits by R. C. Benson (Conant and Tamminen mines), Buck and Baker (Perham Mine, subleased from Bell Minerals in early part of 1957), Pechnik Bros. (Pelletier Mine, formerly the Dunn mine), Bell Minerals Co. (Perham mine), and unidentified producers who sold to Bell Minerals Co. Operations at the Whitehall Co., Inc., Newry mine at Newry (leased from International Paper Co.) were discontinued in 1956 and the lease was cancelled later." Interestingly, it was mentioned that: "A small quantity of gem stones was recovered by a New York mineral specimens collector [Ken Carr of the town of Round Lake] from

two mines [Phillips and Tamminen Quarries] in [Oxford County] (Metcalf and Otte, 1958)." Charlie Bragg was reported, also, to have sought minerals and gems ("triphylite, beryl, amblygonite, apatite, aquamarine, topaz, tourmaline, and zircon") commercially in Stoneham, Waterford, Newry, Hebron, Greenwood, and Lovell. (In later years, the annual Maine mineral reports did not specify who the professional collectors were and, in many instances, target minerals were listed that parallel reports of mineral discoveries made by amateur naturalists in *Rocks and Minerals*.) In Sagadahoc County, Cunningham, Russo, White's Service, and Cesare Trusiani were the major miners of the "11 principal producers" who supplied the two Topsham feldspar mills.

The Sagadahoc County miners and mills may have suffered more by a considerable decline in orders for ground feldspar in 1958: "The drop was due primarily to a slackened demand from soap and abrasives manufacturers and to decreased ceramic demand. ... All crude feldspar was mined from open pits. The chief producers were R. C. Benson (Conant, Forest, and Tamminen mines) and Bell Minerals Co. (Perham mine) (Metcalf and Otte, 1959)." (The Topsham Quarries, in particular, produced a high proportion of graphic granite yielding #2 grade feldspar, the material particularly favored by the soap and abrasives industries. (there were still three feldspar grinding mills active in the State. One Topsham mill ground feldspar only for poultry grit.) State feldspar production was the lowest since 1945 and the average price per ton declined from $6.41 to $6.33. The largest Sagadahoc County producers of feldspar remained Alex Cunningham (near Georgetown), White's Service (near Topsham), and James Russo (near Topsham).

In 1959, feldspar mining in Maine "continued to decline" and feldspar prices dropped slightly to $6.31 with Oxford County accounting for 75% of the output despite there being 4 quarries producing in Oxford County compared with 12 in Sagadahoc County: Russell Garland, James Russo, Adolfo Ponziani, and White's Service were the leading miners (Metcalf and Otte, 1960). One of the Topsham grinding mills was inactive.

The decline in Maine feldspar in 1960 was 16% from the previous year with prices dropping to $6.09/ton, and it was the lowest production year since an earlier milestone, 1944 (Metcalf and Otte, 1961). Ninety percent of the State's feldspar was going through the West Paris mill, all from Oxford County: six producers from West Paris, West Sumner, Hebron, and Norway. Only one grinding mill was operating that year in Topsham. There were six producers from nine quarries in Oxford County, while there were five still active in Sagadahoc. The Consolidate Feldspar mill did not operate in 1960, but the Topsham Feldspar mill did.

The Sagadahoc County feldspar industry collapsed entirely in 1961. Both grinding mills were closed and no quarries produced feldspar (Metcalf and Otte, 1962). As a result, State production was at its lowest since 1906, but the State's market-share of the national output was nowhere near the early glory days.

Androscoggin County was reported to have feldspar

production again in 1961 at the LaFlamme Quarry by Pechnik Brothers and the Phillips Quarry by Bell Minerals Co. Four miners operated five quarries in Oxford County at West Paris, Hebron, and Buckfield (Metcalf and Otte, 1962).

Feldspar was not quarried in Androscoggin or Sagadahoc Counties in 1962, but was produced by seven quarries in Oxford County, principally by Bell Minerals Co. at West Paris and Hebron (Krickich and Otte, 1963).

The ending of the subsidy purchases of mica and beryl may have influenced the feldspar industry. As there was no longer a reason for a miner to aggressively pursue the formerly subsidized commodities, miners who wished to stay in their field apparently switched to feldspar mining: Reversing an 8-year downward trend, production and value of feldspar increased slightly. Production was limited to three principal mines [Perham, Waisanen, and Conant] in Oxford County. Average unit value for crude feldspar remained at $6.00." There were also feldspar shipments to West Paris from the Ruggles Quarry, Grafton, New Hampshire (Krickich, 1964). Mineral Materials Co. produced crushed quartz for three months, but no source was identified.

Kusler (1965) wrote that feldspar continued an "upward trend" in 1964. Bell Minerals worked the Conant, and Bessey Mines, while it subleased the Bessey, Conant, and Perham quarries and bought feldspar from the Biron Quarry in Sumner. (The Silva Biron Quarry probably produced much low-priced #3 grade feldspar, but had an advantage of being located very near a tar road and not far from the feldspar mill, otherwise it could not have afforded to operate. [401] Feldspar production in Oxford County was up 18% in 1965. [83] Feldspar was bought at $6/ton and there were shipments from Ruggles Quarry to West Paris. In 1965, the operating feldspar quarries included Perham, Conant, Bessey, Lord Hill, Bumpus, and Newry Quarries. Bell minerals Company operated the Bessey, Lord Hill, and Newry Quarries and subleased the Perham and Conant Quarries.

Ela (1967a) reported that feldspar production decreased by 7% from the previous year. There were shipments from the Ruggles Quarry and most shipments went to usual customers in Pennsylvania, New Jersey, and Ohio. There were eight operating feldspar quarries in Oxford County in 1966 including the Bumpus Quarry, Albany, A. C. Perham Quarry, West Paris, the Silva Byron [Biron] Quarry, Sumner, a quarry in "Bethel" (probably the Wheeler Quarries), and a quarry in Hebron.

Ela (1969) noted that feldspar production was down in Oxford County due to the retirement of experienced miners in 1967. Crude feldspar was $6/long ton and shipments were made to Pennsylvania, New Jersey, Ohio, "other states", as well as exports to Canada and Africa. Feldspar was shipped to West Paris for grinding from the Ruggles Quarry in Grafton, New Hampshire.

In 1968, production of crude feldspar increased in Oxford County by 60 percent and ground feldspar increased by 40%, but there were important shipments of feldspar from the Ruggles Quarry, Grafton, New Hampshire. [86] See also

remaining Oxford County West Paris Feldspar Mill production in "Lester Wiley – 1965-1966 Bumpus Miner".

Mica in Oxford County

Despite the Mica Rush of the late 1870s and early 1880s, as well as the one in the late 1890s to early 1900s, Maine was not particularly a mica producing state and, while there may have been a small production at times, the quantity was usually so small that no production was listed for the State, virtually all USA mica coming from North Carolina and New Hampshire. The USA was never self-sufficient in mica and usually up to 80% of its needs were supplied by India, Brazil, Madagascar, Republic of South Africa, Canada, etc. The uses of sheet mica for windows in stoves, lanterns, lamps, etc. has been widely noticed but were becoming obsolete usages by the 1930s as electrical demands increased. Ground mica from "scrap" was usually used as a filler in electrical insulation, refractory ceramics, wallpaper, plastics, rubber, house paints, axel greases, asphalt shingles, or roofing materials (the end use of about 65% of the early 1930s production).

Of the 10 listed mica producing states in 1924, Maine was not among of them, but there was a tiny, unspecified production in 1925 (Myers and Stoddard, 1928). In 1926, there was no Maine mica reported.

On August 12, 1925, Harold Perham signed a five-year, renewable, mineral lease for the J. Alton and Nellie J. Sawyer Hibbs [29] property with the provisions: "Operations hereunder are to begin in the early part of summer of **1926**." [352] A one year lapse of mining would allow cancellation of the agreement. The lease specified payment of a $0.50/ton royalty on the feldspar and quartz removed, and a 50% of the selling price royalty on "mica, gems, and other minerals". There was a $100 advance on royalties required with monthly accounting and payments. Because of the generous royalty for the "other minerals" than feldspar, it is reasonable that the original intent was to have a feldspar quarry than a mica quarry, as Harold was then foreman of the nearby Bennett Quarry in adjacent Buckfield. The lease specified: "After beginning operations said Perham is to keep three men or more employed on said operations at all suitable seasons; winter work not being obligatory." The delayed date for the start of mica mining may have been related to Harold's first wife's, Phyllis, illness. She died of pneumonia within three weeks of the signing of the Hibbs' lease. No Maine mica production was reported in 1926, however, and it might be assumed that the mica was sold to a Boston firm and the weight was assigned to New Hampshire or remained unassigned. It is unlikely that Harold stockpiled his mica for sale in the next year, although it is possible.

By the year **1927** Maine's mica production was noticed for the first time in many years (Stoddard, 1930):

"Harold C. Perham worked the Hibbs mine, 1 ½ miles north of Hebron, Oxford County, under lease from J. A. Hibbs, of Hebron. This mine was first opened in 1906 and at that time a promising return was indicated for both feldspar and mica. ... Mr. Perham reports that the mica is amber-colored, very clear, and is successfully used for chimney canopies and for other products requiring exceptionally clear, flawless mica. He also states that there is a large amount of 'book' mica, from which 1 to 7 inch squares can be cut. The product is sold f.o.b. cars Hebron, in the following forms: Mine run, 'cobbed', and scrap."

In his "first" year, Harold sold mica in its least expensive, essentially unprocessed, forms. There was no other mica producer in Maine in 1927, although there were six producers in New Hampshire. As Harold adjusted to his life as a professional miner, he had to look at his accounts and determine if he had made a profit or a loss in his business. Apparently in view of this accounting, he had to establish a new five-year renewable mineral lease [353] between himself and Alton Hibbs. He reduced the royalty on mica to 25% of the sales price and "gem material" would have a royalty of 33%. Crew size, feldspar royalty, and other conditions remained essentially unchanged. However, the new agreement provided a bonus royalty at the end of "each and every five year period" to increase the royalty equal to one third of the net profits received from mining at the Hibbs Quarry. The lease also provided that a guaranteed $100 would be paid to Hibbs, even if the quarry was not operated and for succeeding inoperative years, he would be paid $250 for each succeeding unproductive year, although the lease could be "surrendered" without liability. These last two provisions were probably made as a compromise for requiring a new agreement. On August 27, 1927, the Hibbs-Perham lease was amended to require accounting for and payment of royalties essentially on a biannual basis. [354] Occasionally, Harold also had to lease a right of way for easier access to a quarry. For example, Harold agreed to pay Alice S. Perkins, of Hebron, $25.00 per year to cross her land in order to better work the mica quarry. [355]

The Hibbs Quarry was the only listed primary mica producer in Maine in 1928 (Stoddard, 1931), and mineral production figures were not published as they would reveal the income of the producer, Harold Perham. (Coincidentally, the Maine Feldspar Company, in Auburn, processed scrap mica imported from the Cardigan Quarry, Grafton, NH. Scrap mica brought $10 to $30 per ton from the Cardigan quarry, while punch mica sold for about $112.47 a ton. Sheet mica could bring up to $4000/ton if it would all yield sheets 3 x 4 inches. These prices would probably have also applied to mica from the Hibbs Quarry.) The national average for the year for uncut mica that would yield larger than punch sizes was worth about $1323/ton.

Stoddard (1932) briefly described the quality of the Hebron mica produced in 1929:

"... Hibbs mine ... yielded several tons of good quality mica, which sold in mine-run form to manufacturers for preparation and use in the stove industry and for use as mica chimneys and disks."

If the mica could have yielded 2 x 2 inch sheets, the price per pound for "rough-trimmed uncut sheet mica" of that size varied from $0.25 to $0.50 per pound, while 2 x 3 inch could bring $0.60 to $0.95 per pound. In 1929, the value of "uncut mica larger than punch" was about $1.51 per pound. The difference in the prices of uncut and cut mica is obviously the amount of labor and skill required to process the mica into usable form. The measurements were unforgiving and mica that could yield sheets even a fraction below a stated size category had to be included with the smaller size. The mica would be sorted according to potential sheet size and prices were paid accordingly. The likely purchasers would have been either of the two principal New England mica buyers: the Macallen Company of Boston or the New England Mica Company of Waltham, MA. Even if the average size was small, the "several tons", perhaps 5000 pounds would have meant the Hibbs Quarry may have yielded $3000 to $7500 or more in gross sales of mica. The probable profit was probably good in a time period when an annual wage for a Maine miner was probably about $1,000 annually. If the grade of mica was lower, the quarry would have to cease operation.

Mica was reported to have been produced in Maine in 1930 and 1931 without any specifics (Stoddard, 1933, 1933a), except in one of the great under-statements: "In 1931 the mica industry in the United States experienced unfavorable trade conditions which caused declines in both quantity and value of marketed production of unmanufactured mica." (Scrap mica from the Bumpus Quarry and from the Black Mountain Quarry (collected by his brother Stan) may have been lumped together in sales made by Harold Perham.) Horton and Stoddard (1934) did not detail mica production from the USA for 1932 or 1933, but Maine was listed as a mica producer in 1934 and 1935, but not in 1936. [106, 107, 108] Harold Perham had ceased mining at the Hibbs Quarry in 1932, by his own account, and one may be led to suspect that Maine's continued mica production was through Stan Perham's actions.) Mica values for most sheet and ground products were down by about half from 1931 and that about 90% of all sheet mica was used in "electrical applications", although toasters and other appliances were consumers of mica. Roofing and siding materials continued to use ground mica, to prevent asphalt shingles and siding from adhering to each other, but mica fillers in house paint which had been somewhat experimental, previously, was on the upswing as a consuming use for ground scrap mica.

Maine mica production was again noticed in 1937-1938 with a slight increase reported for 1939, but, as usual, there were no details as Maine was such a minor producer (Johnson and Cornthwaite, 1937; [239, 240]

Mica production got a huge boost in 1953 when the General Services Administration began paying a premium for "block, film, and hand-cobbed mica resulting in an eight-fold increase in USA production (Mote, 1954): "mica produced in Maine was obtained largely as a by-product of feldspar or beryl mining and was predominantly scrap mica, although some output was suitable for use as sheet mica. The largest mica producer was The Beryllium Development Co. Scotty mine on Plumbago Mountain from which scrap mica was recovered as a by-product of beryl mining. Other leading producers during the year included Stanley E. Perham's BB-1 and 7 properties at Norway and Wm. Pechnik's Wardwell mine at Albany.

Quartz (Silica). – A small tonnage of silica for use in manufacturing sandpaper was produced in 1953 at a quarry [of Jeffrey LaChance] near Brunswick, Sagadahoc County."

Kaufman and Cleary (1957) wrote about mica production of 1954: "Value of production of sheet mica increased substantially in 1954, primarily as a result of the Federal mica purchasing program. Sales to the GSA Materials Purchase Depot, Franklin, N.H., increased from 2,189 pounds of sheet mica in 1953 to 3,923 pounds in 1954. ... Over 30 operators produced from 17 different mines in Oxford county, some of whom mined from the same mine. The largest producer was the Roger W. Wheeler mine near Gilead, Oxford County. Most of the output of this mine was scrap mica. ... Twelve mica producers were active in Oxford County in 1954. These included Lawrence Anderson, Howard M. Irish, John L. Maderic, Sabon T. Milligan, Pechnik Bros., Stanley I. Perham, Donald N. Rich, H. L. Robinson, Strafford Mines, Inc., T.C. Mining Co., Roger W. Wheeler, and the Whitehall Co., Inc. The various mica mines were active in Norway, Newry, Rumford, Albany, Paris, and Greenwood areas."

Metcalf (1958) announced that: "The quantity of sheet mica produced in Maine in 1955 more than doubled compared with 1954, and the value more than tripled. ... Output from 18 companies at 19 mines, chiefly in Oxford and also in Sagadahoc Counties was sold as sheet and scrap mica. In addition, other producers not reported by county, sold punch, sheet, and scrap mica to GSA, including some mica mined in Maine and sold in Spruce Pine, N.C., Government Purchase Depot. Several miners obtained mica from more than one mine, and several also from the same mine. The chief producers were Maine Mining Co. at Norway, Pechnik Bros., also near Norway; and Wheeler Bros. at Gilead. The first two produced both sheet and scrap mica. ... Thirteen operators in Oxford County producing mica were: Lawrence Anderson, Beryllium Development, Inc., Elmer Daggett, W. Philips Cole, Gerald Harrington & Elgen Tibbetts, Harry E. Leach, Maine Mining Co., Sabon P. Milligan, Donald M. Rich, Pechnik Brothers, Sparks and Buchanan, Roger W. Wheeler, and Wheeler Bros. These mines were mostly in the following places: Albany, Bethel [sic], Gilead, Newry, Norway, and Stoneham. Beryllium Development, Inc. Elmer Daggett, Maine Mining Co., Pechnik Bros., and Sparks & Buchanan produced scrap mica."

In 1956, the value of mined mica increased indicating that a higher average grade was found (Metcalf, 1958a). The value increased from $27.07 received in 1955 to $28.18 per ton paid in 1956. There were 12 operators working in Oxford County at 13 quarries, but not all locations were precisely identified: Lewie Aldridge, Pechnik mine at Norway,

Lawrence Anderson, B.B. No. 1 and Dunn mines near Norway, Big Deer mine at Stow, Brown mine at Mason, Waisanen Quarry at Greenwood, and Wheeler Quarry at Gilead; Paul Carpenter, Pechnik Quarry at Norway; Marjorie Milonich, B.B. No.1 Quarry at Norway; Pechnik Brothers, Pechnik Quarry at Norway; Wheeler Brothers, Wheeler Quarry at Gilead; and Whitehall Co., Inc., Newry (Nevel) Quarry at Newry. "Two of these producers – Pechnik Bros., and Whitehall Co., Inc. – reported output of scrap mica." Most of the mica was sold at Franklin, NH while some was shipped to Spruce Pine, NC.

Mica fared better than feldspar in 1957 and rose in value 38 percent, again rising in average quality: "The average value per pound in 1957 rose to $7.94 from $7.35 in 1956 and $6.09 in 1955, reflecting increasing costs of mining, including higher wages. Scrap production almost ceased and was reported by only one producer. About 20 miners produced mica at 15 mines in Oxford County, and only 2 mined in Sagadahoc County." (Metcalf and Otte, 1958). The production was sold at GSA Depots in Franklin, NH, Spruce Pine, NC, and Custer, SD: "The leading suppliers were Maine Mica Co. (Hibbs mine at Hebron, Pelletier mine at North Norway, and Wardwell mine at Albany); John Maderic (Wheeler mine at Gilead); Paul Carpenter (Pechnik mine at Norway); Robert C. Tibbetts (Pine Mountain at Mason); and Lawrence Anderson (Wheeler mine at Gilead). In all, 20 producers mined at 15 locations; more than one miner often worked the same mine at various times." (Metcalf and Otte, 1958).

Mica mining fared better in Maine than either feldspar or beryl: "Sales of full-trim mica decreased 17 percent in quantity, although the value increased 14 percent because of better quality material and higher prices. A small quantity of scrap was also sold by six producers, principally to private industry. The chief producers of sheet mica in the county were Bernice and John Maderic (Wheeler mine near Gilead), B & L Mining Corp. (Wheeler mine near Gilead and Pechnik mine at Norway), and P. E. L. Mining Corp. (Pechnik mine at Norway and Wardell [sic] mine at Albany). Indicative of the greater interest in mining was the fact that 33 miners worked 18 mines in 1958 compared with 20 producers at 15 mines in 1957. Several miners worked the same mine at different times during the year." (Metcalf and Otte, 1959). Even Sagadahoc County experienced some increase in mica production: "Earl Williams and Willard Titcomb sold full-trim mica from Trott Cove mine [Edgecomb] near Woolwich to the GSA Franklin (N.H.) purchase depot."

While feldspar mining was shrinking in Maine in 1959, mica mining continued to increase in value by 10%, but all production was from Oxford County. There were 24 miners (companies) that worked at 18 locations: "The five mines yielding the largest tonnage of mica during the year were: Wardwell (Albany), Pechnik and Cliff (both at Norway), Hibbs (Hebron), and Wheeler (Gilead). Among the mica producers were Wheeler Brothers, who have been removing mica from a 400- by 75- by 50-foot hole in Wheeler Mountain, near Bethel. A tunnel driven into the mountain has exposed other mica-bearing areas. A labor force, skilled in trimming and rifting the mica recovered from the mine, has been developed in Bethel [sic]. … Discovery of a large deposit of high-quality cesium in Oxford County was claimed by T. C. Mining Co., West Paris. This firm also produces mica." (Metcalf and Otte, 1960). Mica was also produced at the Trott Cove Quarry, Edgecomb, by Earl Williams and Willard Titcomb.

Sheet mica in Maine increased by 20% in 1960 due to increased orders for mica for the Federal stockpile, to "nearly 27,000 pounds… Leading mica-producing operated included the Wardwell (Albany), Wheeler (Gilead), Pelletier and Cliff (both at Norway), Tyler and Rich (Mason), and the George Elliott (Rumford)." (Metcalf and Otte, 1961), as well as the Trott Cove Quarry operated by Earl Williams and Willard Titcomb. Thirty-three individuals or companies working at 13 quarries sold mica to the GSA depot in NH.

The Maine mica industry in 1961 also suffered a catastrophic year with production down 75%. Metcalf and Otte (1962) revealed: "Foreshadowing the cessation of the Government purchase of sheet mica, for the national stockpile at the end of June 1962, sales to GSA depots in 1961 were about one-fourth of 1960 sales and totaled less than 7,400 pounds in terms of full-trim mica. This decline was attributed to the lack of incentive for exploration and development of new prospects and mines.… Twenty-two individuals or firms reported sales of sheet mica to GSA from eight mines. … Full-trim mica recovered from the Trott Cove mine near Woolwich was sold to the GSA (Franklin, N.H.) purchase depot."

Despite the "foreshadowing", mica purchases continued by the GSA. Krichich and Otte (1963) revealed: "Mica mining activity decreased sharply during 1962. Sheet mica was recovered by producers from the cliff, Wardwell, Wheeler, and Rich mines. Lester E. Wiley recovered hand-cobbed beryl as well as mica from the Wardwell mine near Albany. Earl F. Williams sold full-trim mica from Trott Cove mine near Woolwich to GSA (Franklin, N.H.) purchase depot." The reason, of course, was that the General Services Administration ended its purchase program for mica, as well as beryl, on June 7, 1962.

Beryl Production in Oxford County

World War II Beryl Production

Much has been made about the USA's search for beryl and consequent production. Because of the increased demand for beryl, the U. S. Government began a program of price subsidies to stimulate mining production of minerals it defined as "strategic" for military uses. In anticipation of a need for "7,000 short tons by the War Production Board and at 5,000 tons by the Army and Navy Munitions Board." (Matthews, 1943a). The U. S. Geological Survey visited Maine and many other states, beginning in 1942, to search for strategic quantities of beryl and mica for the war effort.

[64] The Maine geologists visited innumerable locations including prospects to discover indicators that untrained eyes may have missed. The geologists had records of mineral production as a guide and several of the geologists were "local". For example, Joseph Muzzy Trefethen [May 27, 1906-July 3, 1990 Rockport] lived his life in Maine and he was the State Geologist, also professor at the University of Maine. John B. Hanley wrote his Ph. D. thesis on the Poland 15' Quadrangle, and Vincent E. Shainin, born in Shanghai, China and a resident of New York City, was professor of geology at the University of Maine in 1945 to his death in 1950. The Page brothers, James and Lincoln were from nearby Melvin Village, Tuftonboro, New Hampshire and "Linc" became New Hampshire's State Geologist.

Although some of the geologists were locally connected, the various quarry locations may have initially required guidance to locate. The managers of the feldspar mill certainly helped to gain access to their properties as the Survey's work was free and benefited them. The same co-operation would have been extended by independent miners such as Stan Perham, Dana Douglass and others. However, many locations probably required knocking on doors to locate land owners in order to get permission to examine a site not currently under lease. Significant properties were described in detail and property owners and lessors were advised about the advisability of mining and what the indicators predicted. Teams of geologists swarmed through the various pegmatite districts. Most states in New England were investigated, although they had a large area to cover. Fourteen different geologists examined pegmatites in Maine and 31 detailed maps were released for public view. [64] One of the very first Maine localities visited and mapped by the Geological Survey was the Bumpus Quarry in May, 1942. The quarry was idle at the time and filled with water.

Through all the efforts designed to increase beryl production, the result was a 71% increase in domestic production, to 269 tons with Maine appearing in the list of producers and presumably South Dakota remaining a principal producer. The predicted drop in foreign imports was 27%. High BeO beryl increased in price to $75-$100/ton. It was announced that Be oxide was being used in fluorescent lights. Much of the imported beryl was from the western hemisphere: Argentine and Brazil.

Matthews (1945) indicated that domestic beryl production was up another 40% and that imports also rose substantially. It was summarized that Maine beryl production was 25 tons in 1938 and was 45 tons in 1942 and only 2 tons in 1943. In 1937, 1939-1941, production was grouped with "other" in order to prevent release of proprietary information, suggesting that there was only one producing locality in Maine those years, but it could also mean that all beryl was purchased from one Maine supplier.

The USA continued to increase 9% over the previous year's beryl output, but with cancelled import contracts, the 1944 overall acquisition of beryl was down (Nighman, 1946). However, Maine's 1944 production was also the same as the previous year's: two tons. Nighman (1946) provided an eclectic list of record size beryl crystals, mostly from South Dakota:

> "In what is called the [Black Hills Keystone Corporation] No. 2 Dike, one crystal mass that contained 24 tons of beryl was mined in 1933. Subsequently three others that provided 16, 11, and 6 tons, respectively, were excavated. In July 1942 a crystal 19 feet long was uncovered. At the small it was 18 inches in cross section and at the other slightly over 5 feet. This crystal weighed 34 tons. Not until November 1943 were any large crystals or masses found in the No. 1 Dike. Then, one mass without definite form produced 26.5 tons. Late in 1944 a gigantic crystal was mined. The cross section was not determined; but the length was 28 feet, and it provided 61 ¼ tons of beryl."

The huge masses were undoubtedly made up of a number of crystals. No photographs are known of the huge South Dakota beryls. It was also reported: "Miners disposed of their inventories in anticipation of a price drop after termination of the Metals Reserve Company purchase program on December 31, 1944." However, the Metals Reserve Company increased its purchase price from $12 per percent BeO to $14.50 in May, 1944.

Price subsidies continued to be paid until June, 1945 (Nighman, 1947). However, beryl production was down 75% everywhere and no beryl sales were made from Maine beryl. Prices reached $9-10 per percent BeO.

Demand for beryl continued to diminish and there were no sales of Maine beryl in 1946 (Matthews, 1948), however Matthews (1949) indicated that in 1947 some beryl had been produced in Maine, but that New Hampshire experienced a 12-fold increase in its production. Prices crept up to higher than the former Government price subsidies and sales were made from the beryl stockpile as mines could not keep pace with demand. There was a growing concern about health risks of BeO used in fluorescent lights. Clark (1950) indicated that demand for beryl was "keen" in 1948, but that although prices were still high, production was half of the previous year's. Nonetheless, Maine produced "important" quantities of beryl from locations of "uncertain identity", probably Newry, and due to proprietary information, that is, production by only one producer, data were grouped with "other" producers: "Because of the apparent importance of beryllium to the atomic energy program, the activities of all refiners were necessarily partly secret." Because of high demands in beryllium production, sales were made from the national stockpile and the stockpile would be depleted in the next year. Because of the finally recognized toxicity of BeO, plans were being made to phase out its use in fluorescent light bulbs. The idea of Nuclear Power Plants was also being discussed with its consequent need for beryllium. The price of beryl, based on percentage points of contained BeO, rose to an all time high of $26 per unit, potentially equal to $364/ton. These prices for beryl, double those received

under the price subsidy program of WWII, began to attract attention. Almost independently, the Federal government made plans to increase the output of beryl and miners began to plan to do the same. Imports had increased slightly, but it was revealed that India had experienced a beryl embargo, all Madagascar production went to France, and that Argentina had resolved "complications which had hindered" its exporting of beryl to the USA, although the USA had its own export restrictions on beryllium metal. The increase in African beryl production may have had unwanted consequences.

In 1949, Northern Mining Company worked at the Bumpus Quarry and at Black Mountain in Rumford. The giant Bumpus crystal was mentioned "27 feet 7 inches long with end diameters 39 inches and 11 inches." (see also Maillot et al., 1949) and New Hampshire beryl production was lumped together in "other" localities with a total of 169 tons, although it is known that the "spectacular" giant beryl from the Bumpus Quarry accounted for almost 20% of the year's total of the two states. As a curiosity, chrysoberyl became an ore of beryllium as 200 pounds were produced from Scotts Rose Quartz Quarry in South Dakota, 10.5 km southeast of Custer. University of Maine Professor Vincent Shainin's reports on Maine beryl-containing pegmatites were noticed. Shainin had been employed summers by the U.S. Bureau of Mines to search for beryl-bearing pegmatites in Maine as part of a larger investigation of New England beryl resources. [416] It was also revealed that fluorescent light bulbs ceased to contain BeO, used in the manufacture of the beryllium zinc phosphors, by June 30, 1949 by "mutual consent" of the various bulb manufacturers. Beryl prices continued to increase to $390 -$455/ton, based on 13% BeO as was the case for Bumpus beryl. Incipient studies were under way that eventually evolved into the "beryllometer" used for detecting cryptic beryl concentrations in rock. The uranium rush was blamed for the lack of new prospecting in pegmatites.

In 1950, there were two Maine producers of record for beryl, Northern Mining Company's Bumpus Quarry and the Whitehall Company's quarry at Newry. Although Maine and New Mexico were lumped in "other", it was revealed that of the 260 tons of this category, 200 tons had been produced by the Harding Pegmatite in New Mexico. There was also an unknown, undoubtedly small, contribution of beryl from Arizona. Of the 60+/- Maine tons, undoubtedly the lion's share was produced at the Bumpus Quarry. The year saw the largest USA beryl production to date. However, the year also saw a dramatic policy change:

> "Toward the end of 1950 demand for beryllium products, principally beryllium-copper, began to outrun supplies of beryl ore available to consumers. ... Quantitative data on industry or Government stocks of beryl are not available for publication. The year 1950 saw a drop in Government holdings and a correspondingly sharp drop in industry stocks. Although the aggregate of Government-industry stocks was favorable from the standpoint of numerical requirements, the fact that beryl

contained in the National Strategic Stockpile was not available, without Presidential sanction, to an industry faced with mounting military orders created a serious supply situation near the end of 1950."

From these comments, one might conclude that the USA military was converting beryl ore into beryllium for armament uses. Beryl of Bumpus Quarry quality was at least $339/ton and certainly higher. The USA was also sending geologists to Africa to help in the exploration for beryl and the Republic of South Africa was the world's largest producer of beryl ore. The world supply for the year was the largest to date and even Afghanistan showed a small beryl output for the first time.

In 1951, Beryllium was given its own chapter in the USBM Minerals Yearbook. The year was also one of crisis regarding beryl. By that time, almost exactly 90% of the USA's beryl was imported, but Brazil, the largest producer, also began export restrictions. Maine produced more beryl than either North Carolina or Arizona, but was lumped in "other' with 58 tons, or about 12% of the USA total beryl production (Needleman, 1954). The price of Bumpus-grade beryl escalated to about $481/ton.

New Beryl Subsidies – 1952-1962

Beryl output in the world was a record in 1952, especially as export restrictions in Brazil expired. Domestic production increased, but a large fraction of the available beryl was less than 10% BeO (Griffith, 1955). Maine and New Mexico accounted for about 26 tons of beryl. However, Griffith (1955) noted:

"A program for exploration, development, and mining of pegmatites in the Newry Mountain district, near Andover, Maine, was initiated by Beryllium Development, Inc., a wholly owned subsidiary of Beryllium Corp., Reading, Pa. A large production of beryl was reported by this company from the Scotty mine on Plumbago Mountain near Bethel, Maine; however, this production was not shipped and is not included in the 1952 totals. (American Metal Market, v. 59, May 27, 1952, p. 1)"

Griffith (1955) also reported that beryllium was used in missile combustion chambers, because of its highly refractory nature when made into ceramics and that 65% of all beryllium was used in "defense". On October 7, 1952, the new government subsidy program was announced, exempting beryl from price controls and the flat rate for beryl with at least 8% BeO was $400/ton, but if the producer had an analysis of the beryl being shipped, a price of $50 per percent BeO would be paid for grades higher than 12%. By the time subsidies were announced (October 7, 1952), beryl was already bringing up to $47.50/unit/ton. Maine's beryl reserves were estimated at 2,800 tons with about a potential 45% recovery rate. The entire USA reserves were estimated

at about 268,000 tons.

Mote (1954) reported on the beryl production of Oxford County in 1953 and the beryl subsidies influencing this production:

"Beryl was produced at three mines [in Maine] in 1953; the largest was the Beryllium Corp. Scotty mine on Plumbago Mountain, Oxford County. Beryllium concentrate produced at this property was trucked to the depot at Franklin, N.H., for sale under provisions of GSA beryl-purchase program established in October 1952.

In this program, designed primarily for small domestic producers, purchase depots at Franklin, N.H., Spruce Pine, N.C., and Custer, S. Dak., were authorized to accept up to 25 tons of beryl a year from an individual producer. Producers wishing to participate in the program, which would terminate June 30, 1957, or when 1,500 dry short tons of ore was received, had until June 30, 1955, to notify GSA of their intent.

Specifications for acceptable ore were as follows: It must not contain less than 8 percent beryllium oxide by weight and must be clean crystals cobbed free of waste.
2. Shipments up to 500 pounds each are accepted or rejected on the basis of visual inspection.
3. Ore accepted on visual inspection will be purchased at $0.20 a pound or $400 a ton.
The price of ore subjected to chemical analysis will be based on the number of short-ton units of beryllium oxide contained in the ore. Ore containing 8.0-8.9 percent BeO will be purchased at $40 per unit; 9.0-9.9 percent BeO, $45 per unit; 10 percent or more BeO, $50 per unit.
A comprehensive review of beryllium prepared by the Federal Bureau of Mines and the Federal Geological Survey in cooperation with the National Security Resources Board stated that, although beryl resources have not been studied as thoroughly in Maine as elsewhere in New England, large deposits occur in Oxford County, especially in the Main pegmatite on Newry Hill, the Bumpus pegmatite, and the West pegmatite [Rumford]. ... Beryllium resources in Oxford County were estimated in this report to be 2,700 short tons in deposits containing 1.0 percent or more beryl and 3,800 tons in deposits containing 0.1 percent or more beryl."
Mote (1954) also recorded: "Beryl concentrates were recovered at the Scotty mine and at the Wardwell and Wentworth mica mines."

Kaufman and Cleary (1957) reported that in 1954: "Beryl was obtained from the Scotty mine of the Beryllium Development, Inc., near Bethel; the Newry mine of Whitehall Co., Inc., Newry; the Mount Marie mine, South Paris, operated by Bell Minerals Co.; and the Guy Johnson mine of Pechnik Bros., near Albany."

In 1955, there were five companies producing beryl; principally Beryllium Development Corporation at Newry and Pechnik Brothers at North Norway. Beryl production decrease in value (production) from the previous year. Metcalf (1958) noted: "Beryl was produced by Richard I. Baker from the Tiger Bill mine, near Greenwood; the Scotty mine of Beryllium Development Inc., near Bethel; Howard M. Irish at the Mount Mica [Irish] mine near Paris; the Dunn mine of William Pechnik near North Norway, and the Newry mine of Whitehall Co., Inc., Newry."

As beryl subsidy pricing continued, Oxford County continued to be listed as a producer in 1956. Metcalf (1958a) reported: "Recovery of beryl concentrate was nearly half that reported for 1955; the entire output came from Oxford County. Of 5 producers, the 2 leading were Whitehall Co., Inc., at its Newry mine at Newry and William Pechnik at the Dunn mine at North Norway. … Producers of beryl concentrate in 1956 were: Howard I. Baker, from the Tiger Bill No. 2 mine near Greenwood; Mrs. Howard M. Irish, from the Mount Mica mine near South Paris; William Pechnik from the Dunn mine at North Norway; Whitehall Co., Inc., from the Wilson mine at Bethel. Most of the material was sold to the GSA Franklin (N.H.) Mineral Purchase Depot for the strategic stockpile. A small quantity of amethyst and rose quartz was produced near Stow and Albany, respectively, by Orman McAllister of Lovell."

Metcalf and Otte (1958) reported: "The output of [Maine] beryllium concentrate was less than one-third that in 1956 and came mostly from Oxford County; a small quantity was produced in Sagadahoc County. W. Phillips Cole (Mt. Adams mine at Stoneham) and William Pechnik (Pelletier [=Dunn] mine at North Norway) were the leading producers." Winfield Knight was also a producer from his quarry in North Waterford.

Maine's beryl production was low in 1957 and 18 tons total were reported for all of Maine, Maryland, Connecticut, Arizona, Idaho, North Carolina, Wyoming, and Georgia combined. High grade beryl sold at about $624/ton, but about 92% of the beryl consumed in the USA was imported.

Beryl continued to decrease "sharply" in produced amount in 1958: "Beryllium concentrate (beryl) was sold to the GSA purchase depot at Franklin, N.H., by four miners: William Pechnik (Norway), George Wiley (Albany), Winfield Knight (North Waterford), and Elmer Daggett (Canton)" (Metcalf and Otte, 1959).

Beryl reversed its downward trend in 1959 and four producers totaled a better year than previously. It was revealed that entire production of beryl averaged 11.8 [weight] percent BeO, a relatively strong value. There were four mining Oxford County companies producing beryl: Donald E. Cross at the Cross mine, Milton; William Pechnik at the Kendall mine, North Lovell; P. E. L. Mining Co. at the Pelletier mine, North Norway, and Lester E. Wiley at the Wardwell Quarry, Albany, and one in Sagadahoc County by Arthur Trusiani at Georgetown (Metcalf and Otte, 1960).

Two quarries in Oxford (Stanley Pechnik at the Pelletier Quarry and Lester E. Wiley at the Wardwell Quarry) and one in Sagadahoc County (Mrs. Francis MacDonald in Georgetown) produced all of Maine's beryl in 1960 (Metcalf

and Otte, 1961).

Beryl production increased in Maine in 1961 with Oxford County production recorded only from the Wardwell Quarry, Albany (Metcalf and Otte, 1962). Sagadahoc County's only mineral production was for mica and beryl with the latter showing a significant shipment from the Georgetown Quarry operated by Arthur O. Trusiani.

The GSA purchase program for beryl ended on June 7, 1962 and as a result, there was but one Maine producer: "Lester E. Wiley recovered hand-cobbed beryl as well as mica from the Wardwell mine near Albany." (Krichich and Otte, 1963). With the end of Government subsidies, Maine beryl production dwindled to insignificant amounts.

Beryl crystal, Bumpus Quarry.
20 x 8 cm.

Appendix C History of Maine Beryl Discoveries and the Pursuit of the World Record

The most provocative report of Albany, Maine beryl, from the nineteenth century was by Jackson (1838): "Very large crystals of beryl have recently been found in the town of Albany, between Bethel and Waterford. I have not yet visited the place, but have seen a specimen of large size, which was sent to Professor Cleaveland, in Bowdoin College." Dana (1844) probably referred to this discovery in his locality register for Albany, Maine: "*Beryl!, green and black tourmalines, feldspar, rose quartz.*" Note Dana's use of the punctuation regarding beryl. The problem is that very little "green" tourmaline is known from Albany today. We might imagine that this unusual tourmaline was correctly identified, although that cannot be completely certain. Beryl had first been found in Maine by 1814, in Topsham, [71, 72] but the report, under the headline "Emerald", stated: "This Mineral is by no means uncommon in the United States." From this first sentence, we may conclude that the old name for the mineral was used. The short report concluded: "… in some instances, in point of colour, it equals the finest Peruvian Emerald…" [8] The second localities in the State, Mount Mica and "Streaked Mountain", were both in Paris. [10] Maine's first record beryl for which a size was given, from Streaked Mountain, Paris, was announced by Elijah Hamlin (1824), [25] a lawyer then living in Waterford and co-discoverer, along with Paris Hill resident, Dr. Ezekiel Holmes, of Mount Mica: "Beautiful beryls are obtained from this mountain, sometimes so large as to measure *eleven* inches in circumference; they exhibit a *fine green color,* and are well characterized crystals." Hall (1824) did not report any "beryl" from Maine, but did report "emerald" from Topsham and Paris. Robinson (1825) listed beryls from "Bath", but that report was more properly for widely located occurrences from Georgetown to North Yarmouth, The next reported source was near Fryeburg village (Shepard, 1830). However, only one of the foregoing reports specify how large the crystals were.

The world record size for beryl was first established in New England at Beryl Mountain, Acworth, Sullivan County, New Hampshire but the record was held by that locality for only 19 years. [11] Dana (1837) wrote of New Hampshire and Maine beryls, increasing the number of published locations:

> "The United States have afforded some magnificent specimens of beryl; they are remarkable, however, only for their size. The largest has been found at Acworth, New Hampshire, about fifteen miles from Bellows Falls, where the beryls occur in an extensive vein of granite, traversing gneiss. It measured 4 feet in length; and 5 inches across its lateral faces, and was therefore 11 inches in diameter. Its color was a bluish-green, excepting a foot at one extremity, where it passed into a dull green and yellow. Its weight was about 240 lbs. This locality affords smaller beryls in great perfection; they are usually of a pale yellow color; rarely a deep honey or wax yellow. Small regular crystals of beryl occur also at Bowdoinham and Topham [sic], Me., in veins of graphic granite; their color is a pale green, or yellowish-white; also at Georgetown, Parker's Island, at the mouth of Kennebec river, associated with black tourmaline."

In lieu of information to the contrary, we may accept Dana's notice of a remarkable size for beryl, by 1837, to be 240 pounds.

By 1854, Dana noted that the largest beryl crystal in the world was from Grafton, New Hampshire, weighing 2900 pounds and 4.25 ft x 32 in x 22 in. Another remarkable Grafton beryl crystal actually made it to the Boston Museum of Natural History and was on exhibit there in 1939: "There is a beryl crystal from Alger Mine, Grafton, N.H., that is 40 inches in diameter and was originally 9 feet long." [170] The current length was unstated. Grafton's world record endured for just over 50 years. (See also Alger, 1856).

Verrill (1861) reported beryl in Oxford County from a locality about 3 kilometers from his childhood home:

> "… on the eastern flank of Furlong [Noyes] Mountain, Greenwood … At other localities on the same mountains are found beryls of large size, black tourmalines, ilmenite [probably = columbite], and black garnets."

Verrill (1861) specifically said of Mount Mica:

> "*Beryl.* Associated with the large crystals of black tourmaline and sometimes imbedded in them, very good specimens of beryl are sometimes met with. These are hexagonal prisms, generally somewhat irregular in form, and not often with the ends perfect, of a light green color, translucent, and vitreous in lustre. The hardness is greater than that of quartz."

Nathaniel True (1861) only said of Bethel: "Occasionally a beryl may be seen on the mountains." And later True (1869) commented on transparent white beryls from Mount Rubellite Quarry, Hebron, which were in the "S. R. Carter collection", but in none of these reports were sizes of crystals mentioned.

Maine's beryl specimens first received national attention in the 1880s. An undated, unidentified, perhaps 1884 Lewiston newspaper commented of the curator of minerals at the Smithsonian Institution and chief chemist of the newly formed U. S. Geological Survey:" Prof. F. W. Clark [sic], who has charge of the Mineral Department of the National Museum at Washington, spent Thursday and Saturday of last week with the Oxford County Mineralogist, N. H. Perry of South Paris, Me., says the Oxford Advertiser. He selected a large number of interesting minerals to represent this section of the National Museum and others for the International Exhibit [World Industrial and Cotton Exposition, December 16, 1884 – June 1, 1885] at New Orleans. Among those selected were the colored tourmalines of Auburn, Hebron, Norway and Rumford. Amethyst from Stow, Topaz and Beryl from Stoneham. Some of these were fine cut gems, others to be cut and have a place among the rare American gems at the New Orleans exhibit."

Maine's most famous discovery of aquamarine, to date, may have been among the cut stones. About 1882, two gem beryl crystals were found in fields on or near the Melrose Farm and one was estimated that it would cut the then USA record aquamarine of "at least 20 carats" (Kunz, 1884). Folklore relates that it was found in a freshly plowed field and was sold to the postmaster at East Stoneham, Sumner Evans, for $1 – much more than a day's wages. The discovery started a small gem rush, Maine's second, and Kunz (1885) reported that one fine gemmy crystal was 91 cm x 11 cm (!) and "of a very fair color" had been found. This gem crystal may have also been the state record for the species! In 1886, Kunz had news concerning this large crystal:

> "The large beryl mentioned in 'Mineral Resources' for 1883 and 1884, has afforded the finest aquamarine of American origin known. It weighs [137.16] carats and measures 35 x 35 x 20 millimeters. It is a brilliant cut gem and with the exception of a few internal hair-like striations it is absolutely perfect. The color is a deep bluish green, equal to that of gems from any known locality."

Conklin (1986) reprinted a letter of Augustus Hamlin's April 30, 1887, with a tally of Mount Mica treasures, to Tiffany's gemologist, George Kunz:

> "Yours rec. I sent you last fall a printed account of the find at Mt. Mica and have little to add except that we have had cut about between 100 & 200 tourmalines of the value of about $5000. Cut white beryls $250."

The mention of the faceted gem beryls alone certainly was equivalent to a year's wages for an average Oxford County farmer.

A confounded article did seem to give Maine a beryl gemstone record:

> "A pair of Maine's most wonderful and beautiful gems may now be seen in the mineral collection in the State of Maine building at Poland Spring. These are yellow beryls of 36 $\frac{1}{8}$ and 34 11-16 carats respectively, and were cut from a single crystal found in Topsham [Sagadahoc County]. They are very brilliant, of a light yellow color. Most yellow beryls have been comparatively small, and those

mentioned by Kunz [1892 edition?] in 'Gems and Precious Stones of North America' are nearly all from two to twenty carats in size. One notable exception is a yellow beryl from Pennsylvania weighing 35 11-16 carats, probably the largest hitherto known. Although these rare Maine gems came from the lapidary but a few days ago connoisseurs of rare gems have been eagerly soliciting a chance to buy." (*Daily Kennebec Journal,* August 14, 1903).

The next year, Chadbourne (1904) indicated that the golden beryls from "Maine" were from Stoneham, not Topsham. Nonetheless, although the newspaper report suggests that the Maine golden beryl gems were world records, the only localities in the world which may not have come to a "westerner's attention" were probably in Russia, east of the Ural Mountains.

The report of the discovery of a 20 foot long by 4 feet in diameter beryl crystal from at the Mount Apatite District, Auburn by Bastin (1911) has already been mentioned. The 1928 discovery of giant beryl crystals, 18 x 3.5 feet, from the Bumpus Quarry did not break the State record. In 1949, the 27 foot 7 inches long by 39 inch to 11 inch diameter Bumpus Quarry beryl became the new State record, but was only a record for the eastern USA, the world record being held by South Dakota. The 33 foot long by 6 feet in diameter beryl of 1950 finally eclipsed all previous records and the Bumpus Quarry was the world champion for at least twenty years.

Beryl crystal, Bumpus Quarry.
14 x 7 cm

Appendix D What Use are Beryl and Beryllium Alloys and what has Happened to the Interest in Beryllium?

The discovery of beryl at the Bumpus Quarry came just as beryllium began to be an important metal. Even in 1929, Kendall said in his college textbook: "The element [beryllium] is exceedingly rare, and its compounds are not of sufficient importance to warrant extended consideration."

Pure beryl theoretically contains 14.02% beryllium oxide, BeO, the usually reported constituent, but, without the oxide combination, pure beryl may be reported to contain 5.03% Be metal. Because beryl is common enough, these percentages were reasonably advantageous commercially. Bertrandite, the most utilized ore of beryllium, today, contains 42.00% BeO or 15.13% Be metal. Although beryl was the historical source for beryllium, the Spor Mountain Bertrandite mine, Juab Co., Utah started in 1969 and nearly immediately affected the value of beryl as an ore as the bertrandite was an easily mined uniform ore. Beryl in Maine pegmatites, not just the Bumpus pegmatite, occurs unpredictably in the rock. The beryl is found within particular zones of a pegmatite, but the next blast might reveal a bonanza of beryl or the next blast might go into the "barren" rock, without predictable reward for future effort. By 1979, Spor Mountain Bertrandite mine had made beryl mining in the USA uneconomical. Spor Mountain Bertrandite mine currently (2006) produces about 93% of the world's beryllium ore.

Because of beryl's crystal structure, which contains tubular openings, beryl may have impurities included in it. The impurities are usually chemically bonded to the mineral, but any water in these channels is not, but is merely "trapped". The chemical impurities greatly affect the BeO content and most beryls from Maine are 11-12% BeO. When the customer receives the beryl, he usually assays the particular shipment of beryl and pays according to the BeO analysis.

Prior to 1918, the chief interest in beryl was as a gem, a specimen, or, rarely, as a source of a laboratory chemical. Sterrett (1911) supported this observation: "Mr. Alfred W. Smith, of the Maine Feldspar Company, Auburn, Me., reports the sale [in 1909 or before] of large beryl crystals and fragments for commercial purposes. This material was not suitable for gems, but was used in the chemical industry." However, in 1918, Hugh S. Cooper patented a strong, light weight alloy, beryllium aluminum. In 1921, Charles Brush, Jr. and Charles Baldwin founded Brush Laboratories which specialized in beryllium alloy research and commercial applications of unusual materials particularly for acoustics and radio (Brush Wellman Company, accessed 2006). In 1926, Michael G. Corson patented a beryllium copper nickel alloy with remarkable properties and shortly afterwards, Siemans and Halske Company in Germany patented a beryllium copper cobalt alloy. The Beryllium Corporation of America, founded by Lester Hofheimer in 1927, was the first commercially producing company of beryllium-containing materials, but, by 1931, Brush Laboratories expanded to became Brush Beryllium Company.

Andre W. Gahagan, decided to get into the beryllium industry in 1929 and formed the Beryllium Development Corporation, and later, gained control of the Beryllium Corporation of America and changed its name to the Beryllium Corporation (BerylCo) in the 1933. The patents and founding of several beryllium corporations expanded between the years 1921 to 1932, when Brush Laboratories was started and during which time Gahagan was trying to start a small beryllium manufacturing monopoly.

In May, 1936, the Stanley Tool Company introduced a set of twenty-five non-sparking tools made from beryllium copper alloy (carpenters' hammers to sledge hammers, screwdrivers, picks, chisels, etc.; Jacob, 2000). The 20 ounce beryllium copper alloy hammer sold for an astounding $13.80 in April of the next year compared with $1.20 for a comparable drop-forge steel hammer.

Initially, in the 1930s, small amounts of Be-copper alloys, Be-oxide compounds and metallic beryllium were made but their growth in production was stimulated dramatically, through the increased demand of beryllium-copper products used in World War II. Through subsequent complicated name changes, mergers, acquisitions, divestitures, and sales; Beryllium Development Corporation, Temple, Pennsylvania branch, north of Reading, has subsequently appeared under a variety of company names including: BerylCo, Beryllium Corporation of America (merged 1968 to form KBI), Kawecki Chemical Co. (founded 1950, merged 1968 to form KBI), Kawecki BerylCo (KBI; sold to form Cabot Industries in 1978), and, most recently (1986 bought from Cabot Industries, no longer specializing in beryllium), NGK Metals Corporation of Japan.

Wheeler Brothers Quarry.
Gilead,Maine

Appendix E Beryl and the Bomb

The Manhattan Project search for Beryl

In 1938, German scientists, Otto Hahn, [6] Lise Meitner, and Fritz Strassman, were working on problems which led them to experiment successfully with nuclear fission and on December 19 were able to split uranium atoms into barium atoms. On August 2, 1939, Albert Einstein wrote his famous letter to President Franklin Roosevelt, at the behest of physicist, Leo Szilard, that "uranium may be turned into a new and important source of energy in the immediate future". On December 6, 1941, Roosevelt authorized the founding of the Manhattan Project to research and manufacture a uranium bomb. The timing, the day before the attack on Pearl Harbor, is extraordinary.

As a military project, Major General Leslie Groves was placed in charge. He supervised the search for scientists, technicians, and workers for various parts of the Manhattan Project. Tom Lewis was a young metallurgist who had enlisted in the Navy and was assigned to submarine duty. On his first and only submarine voyage, his Captain was contacted to verify that Lewis was aboard and the submarine was ordered back to port. Lewis was given a new assignment in the Manhattan Project and eventually helped develop the beryllium initiator necessary to detonate an atomic bomb. He remembered handing in requisitions for unusual research materials to his commanding office, a general, who inquired, "What is this for?" to which Lewis replied, "I can't tell you." His commanding officer said, "OK" and signed the authorization. [417]

The USA exploded the first nuclear bomb at the "Trinity Site" near Alamogordo, New Mexico on July 16, 1945. A few weeks later, on August 6, the USA dropped the "thin man uranium bomb", also called "Little Boy" (alluding to Roosevelt) on Hiroshima, Japan and on August 9 dropped the "fat man plutonium bomb" (alluding to Churchill) on Nagasaki. Although these were the only three atomic bombs that had been so far constructed and subsequently exploded, the illusion was that there were more bombs ready to be dropped and Japan surrendered, ending WWII, although there is an on-going debate whether the bombs were absolutely necessary to end the war. Of course, there were many bomb components still scattered in the facilities associated with the project, but there would have been a considerable delay to make a complete fourth bomb. Unfortunately, the hot war of WWII was soon followed by another tense period when beryl would remain in high demand. [19]

When Charlie Bragg, of Buckfield, returned from his tour of duty in the service in Korea, he went to work as a beryl miner at Newry, but was soon employed as a prospector as he knew minerals and pegmatites so well. During the Second World War and subsequently, government workers from the Geological Survey, Bureau of Mines, etc. as well as companies and individuals would explore for minerals in Oxford County. Land owners, principally farmers, were asked for permission to prospect, while larger land holdings by paper companies were usually explored until there was a need to blast an outcropping to see if it contained minerals needed for making bombs.

Cold War Search for Beryl

The Cold War [5] was a period from 1945-1991 when the USA and Russia, then the Union of Soviet Socialist Republics (USSR), were experiencing severe political tensions. The two countries responded to various perceived threats to their security from each other and both eventually engaged in a "weapons race". Abrahamson and Carew (2002) wrote of the near dismantling of the Manhattan Project. After the war, the military itself was being "dismantled" as soldiers, sailors, marines, and airmen were going home, as were scientists and technicians. To be sure, the military was attempting to round-up German scientists to become part of post-war USA research effort, in what was known as "Operation Paperclip", but Groves, uncertain about the USA's military capabilities during the post-war era did what he could to maintain the Manhattan Project in what had become a kind of authorization vacuum.

Because of the Soviet Union's failure to free Eastern Europe after WWII and as it was doing what it could to strengthen its dominance of that region, Winston Churchill made a speech on March 5, 1946 at Westminster College in Fulton, Missouri where he said that "From Stettin in the Baltic to Trieste in the Adriatic an *iron curtain* has descended across the Continent." The fear was that the Soviet Union would also develop nuclear weapons and that fear was fulfilled on August 29, 1949 when it exploded its first nuclear bomb. By that time, the USA had tested 5 post-war nuclear weapons. The next test by the Soviet Union did not occur until 1951 when they exploded 2 bombs. USSR's next tests were made in 1953 when 5 bombs were detonated and there was an escalation to 34 bomb tests in a year by 1958. Except for a few years when there were no tests, up to 1990, the USSR made 715 nuclear tests, depending on how you count (Mikhailov, 1996). The USA conducted 1054 nuclear tests (including joint tests with the United Kingdom) during 1945-1992 in Alaska, Nevada, Colorado, New Mexico, and Mississippi, as well as in several locations in the Pacific Ocean. However, the USA built an enormous number of nuclear warheads, each requiring uranium, plutonium, beryllium, etc.

According to the Brookings Institution (www.brook.edu/fp/projects/nucwcost/50.htm, accessed 2006), in 1966 alone, the year with the greatest number of constructed nuclear bombs, the USA had 32,193 active nuclear weapons. By 2002, that number had dwindled to about 10,600. All of this activity meant that raw materials had to be mined to manufacture them. In 1955, there were 925 mines producing uranium ore in the USA, however, the number of quarries producing beryl is uncertain, although 500 total tons were produced that year, at least according to public records. (Ordway (1951) reported on the results of exploring for radioactive minerals in five Albany pegmatites (Donohue, Guy Johnson, Fred Scribner, Roy Wardwell, and Ernest Wentworth quarries) as well as six other Maine locations, but the results were "largely negative".)

Because of the enormous need for beryllium, rare beryllium minerals, of the helvite group, that occurred in unusual quantities in Iron Mountain, near Jackson, New Hampshire and, coincidentally, in Iron Mountain, near Winston, New Mexico were investigated, as well.

The Search for Beryllium Ores for "Peaceful" Purposes

Although the military use of beryllium was a "government secret", the government needed to dramatically increase production of beryllium ores. Kennedy and O'Meara (1948) wrote:

"Beryllium is a strategic metal of ever-increasing peacetime importance because of the unusual properties of beryllium-copper alloys. These unusual characteristics are the combination of high strength and hardness, high fatigue resistance, and high wear resistance in alloys that have the corrosion resistance of copper.

The principal industrial source of beryllium is the mineral beryl, a beryllium-aluminum silicate that occurs in pegmatites associated with quartz, a feldspar, mica, and smaller amounts of various accessory minerals. The beryl is usually recovered in processing the pegmatites for the recovery of some of the other mineral constituents. Pegmatite deposits are found in various parts of the United States, but those in the Black Hills of South Dakota are the source of most of domestic production.

The beryl marketed from the South Dakota pegmatite deposits occurs as coarse crystals, which are obtained by hand sorting. These pegmatites also contain finer crystals and fragments of crystals, which are not recovered by the sorting practice. In addition, numerous low-grade deposits in the area contain finely disseminated beryl that could be recovered only by fine grinding and beneficiation.

An impending shortage of beryllium became apparent soon after the last war began, and it was evident from the character of the known reserves of beryl that hand-sorting could not increase production. To alleviate the shortage, the Bureau of Mines undertook to find new ore bodies of beryl or other beryllium minerals such as helvite [in New Mexico], a silicate of beryllium, manganese, and iron containing sulfur, from which beryllium could be produced. Many of the explored deposits were too low-grade or contained beryllium minerals in particles too fine to be recovered by the customary hand sorting, and both fine grinding and beneficiation were necessary to recover suitable concentrates."

Fortunately, the metallurgical uses of beryllium metal were increasing in industry and the publicly-released reason seemed plausible enough for the overwhelming majority of the public. The pegmatites studied by the USBM were in South Dakota, New Mexico, Colorado, Nevada, and Connecticut. Although no "commercial" beryl was produced by floatation techniques, the USBM was successful in its experiments to produce a high BeO concentrate. Beryl continued to be produced by hand sorting and soon the USBM and the USGS turned its attention toward the Bumpus Quarry in Albany, the Scotty Quarry in Newry, and the Black Mountain Quarry, Rumford.

The Federal government renewed its wartime subsidies policies designed to increase production. Publicity was necessary to inspire local miners and entrepreneurs to develop mining locations. There was no pay unless there was something to buy, however. Two government agencies, the U. S. Geological Survey and the U. S. Bureau of Mines, were charged, as they had been in the war, to continue searching for new sources of mineral supplies and to perform basic research so that the ores could be quickly utilized. The USGS had thoroughly inventoried strategic mineral resources during the war and that information was used to focus attention on localities that should yield the most beryllium ore. Potter (*Boston Sunday Post*, February 20, 1949) interviewed the flamboyant [12] Vinny Shainin of the University of Maine at Orono concerning the summer work he had been doing for the USGS since 1946:

"A vast deposit of minerals of great wartime worth to the armed forces has been located in western Maine as a result of a survey being conducted for the federal and State governments, according to Professor Vincent Shainin, who is in charge of the work.

The minerals, says the professor, are termed 'strategic' by the government, and the find in Maine is considered an important one. Some of the minerals are used in atomic piles, others in making up metal alloys used in the manufacture of airplane parts, and others in internal surgery. …

Another strategic mineral in the deposit is spodumene, which is used in making lithium, another light weight metal of great strength desired by manufacturers of military planes.

The two light weight metals of great strength are needed by military planes, which must fly under greater strain than do ordinary commercial planes, which at move at much lower speeds and lower altitudes. For that reason the minerals are of great value in time of war. Even though there is little demand for them in normal peacetime."

Although some readers of the government reports knew full well that beryllium was essential to the development of nuclear warheads, the repeated reminder of the use of beryllium in industry, as well as its use in military planes, meant that there would be an uncertainty of just how many nuclear warheads were being made. The beryllium for just one airplane could be made into many warheads instead and the constant misdirection of information was a major component in Cold War politics.

The USBM existed from 1910-1995 and had the responsibility to do research faster than industry would have on its own, to publish the research results for anyone to use, and to promote the commercial interest in those results. The USBM had hoped to increase production of beryl by using a process called floatation (Kennedy and O'Meara, 1948). The metals industry had begun using floatation in the 1890s and which process had become ever more popular. Most metal ores are mixed with several impurities as well as potentially valuable minerals and even valuable minerals may be mixed and require separation. Before floatation, there were several methods developed to concentrate ores before they were smelted. Concentration greatly reduced the cost of processing and would also increase the yield of the metal being sought. Unfortunately, impurities that were similar to the target mineral were frequently unfavorably mixed and most mines avoided processing badly mixed ores, either discarding unusable minerals on a dump or by avoiding the undesirable minerals in the deposit. The waste is obvious. By crushing ore minerals to a powder and by using chemicals that would froth into bubbles, it was possible to develop mixtures where undesirable or desirable minerals would adhere to the floating froth and could be physically skimmed away from the rest of the minerals being processed. The valuable component could be cleaned and the concentrated ore could be finally processed.

The floatation process allowed miners the freedom to remove tons of rock without a great deal of selection

and the mine mill would be responsible for the separation. Pegmatites, by comparison, were usually very small compared to metal sulfide mines. Nonetheless, the demand for beryllium was so great that any increase in production was highly valued. The USBM postulated that much beryl was not recovered because the usual crystal sizes were too small to bother with. If it were true that beryl was being ignored for this reason, floatation might increase USA beryl production significantly. Several occurrences were chosen for study. The pegmatites had to be large and had to have a promising percentage of beryllium available in the rock. Although the USBM demonstrated that beryllium ores could be concentrated by floatation, the Government found that its best way to stimulate production was to increase its buying price. It has already been noted that most of the USA's beryl was imported. If there had been an interruption in beryl imports, there was probably a plan to erect flotation mills to acquire more domestic beryl.

Wardwell Quarry, Albany.
Beryl and mica producer.

Appendix F – U. S. Geological Survey in Maine During WWII

1942

Spring Bennett Quarry, Buckfield Lincoln R. Page and John B. Hanley, later David M. Larrabee and I. S. Fisher

May Bumpus Quarry, Albany Lincoln R. Page and John B. Hanley

Coombs Quarry, Bowdoin Lincoln R. Page and John B. Hanley

27th Parker Head area, Phippsburg Lincoln R. Page and John B. Hanley

21st Little Singepole Mountain Feldspar Quarries Lincoln R. Page and John B. Hanley

29th Dunton, Crooker, and Nevel Quarries, Newry Lincoln R. Page and John B. Hanley

June

2nd Hibbs Feldspar Quarry, Hebron Lincoln R. Page and John B. Hanley

LaChance quarries, Brunswick Lincoln R. Page and John B. Hanley

Thomas Feldspar Quarry, Phippsburg Lincoln R. Page

21st Lobikis Mica Quarry, Peru Vincent E. Shainin and Karl S. Adams

Plumbago Mountain Beryl Prospect, Newry Joseph Trefethen

August

C. P. Saunders Quarry, Waterford David M. Larrabee and I. S. Fisher

September

5th Peaked Hill Mica Quarry, Gilead David M. Larrabee and Irving S. Fisher

Plumbago Mountain Beryl Prospect, Newry David M. Larrabee and Irving S. Fisher

September-October

Parker Head area, Phippsburg David M. Larrabee, Irving S. Fisher, and G. H, Brodie

October

Consolidated Feldspar Quarry, Topsham David M. Larrabee, Irving S. Fisher, and G. H, Brodie

Thomas Feldspar Quarry, Phippsburg David M. Larrabee and Irving S. Fisher

1943

June 9 Brown-Thurston Quarry, Rumford David M. Larrabee, W. M. Hoag, W. H. Ashley, H. R. Morris,

Wardwell #1 Mica Quarry, Albany David M. Larrabee, Louis Goldthwait, W. M. Hoag

June-July Black Mountain Quarry, Rumford David M. Larrabee, Hoag, W. H. Ashley, H. R. Morris, Louis Goldthwait

July

C. P. Saunders Quarry, Waterford David M. Larrabee,

Louis Goldthwait, W. M. Hoag

17th Gogan Mica Prospect, Mexico David M. Larrabee, Louis Goldthwait, W. M. Hoag

Scribner Ledge Quarry, Albany David M. Larrabee, Louis Goldthwait, W. M. Hoag

Trott Cove Mica Quarry, Woolwich R. Miller and Lawrence Wing

August

Bumpus Quarry, Albany David M. Larrabee and Irving S. Fisher

Barrett Beryl Prospect, Hebron David M. Larrabee, Louis Goldthwait, W. M. Hoag

17th Willis Warren Feldspar Quarry, Stoneham David M. Larrabee, Louis Goldthwait, W. M. Hoag

19th Stearns Beryl Quarry David M. Larrabee, Louis Goldthwait, W. M. Hoag

19th Noyes Mountain (Harvard) Quarry, Greenwood David M. Larrabee, Louis Goldthwait, W. M. Hoag

19th George Elliot Mica Quarry, Rumford David M. Larrabee, Louis Goldthwait, W. M. Hoag

26th Donahue Prospect, Albany David M. Larrabee, W. M. Hoag, Louis Goldthwait

Peabody Mountain Feldspar Quarry, Batchelders Grant David M. Larrabee and Irving S. Fisher

September 18th Black Mountain Quarries, Rumford David M. Larrabee, and James J. Page

September-October Black Mountain Quarries, Rumford drill cored David M. Larrabee and James J. Page

October 24th Black Mountain Quarries, Rumford James J. Page

November

LaChance quarries, Brunswick Joseph Trefethen and James J. Page

Trott Cove Mica Quarry, Woolwich James J. Page and Louis Goldthwait

13th-14th Coombs Quarry, Bowdoin James J. Page and Louis Goldthwait

In 1943, Robert Miller and Lawrence Wing, probably frequently in the company of the State Geologist, Joseph Trefethen, investigated a number of locations on behalf of the U, S. Geological Survey. The report also included a list of all properties investigated 1043-1944, including many non-pegmatite occurrences. Underlined localities indicate a map was also published.

Bowdoin - Coombs Quarry (7/2)
West Bath – Davis Quarry (7/3)
Topsham – Russell Brothers Quarry (7/4)
Brunswick – LaChance Quarry (7/5)
Woolwich – Trott Cove Quarry (7/10)
Phippsburg – McKay Farm Prospect (7/12)
Freeport – Taylor Quarry (7/14)

Bowdoinham – Jack Prospect (7/14), Trufant Prospect (7/14)

Paris – Mount Mica Quarry (7/22)

Albany – Johnson Quarry (7/21), Donahue Prospect (7/25), Wardwell Quarry (7/27), Scribner Quarry (7/28), Wentworth Prospect (7/30-31), Flint Prospect (8/5), Conwell Prospect (8/5), Holt Prospect (8/5)

Greenwood – Noyes Mountain Quarry (7/29), Noyes Mountain area (August)

Hebron – Alton Hibbs Feldspar Quarry (7/31), George Hibbs Prospect (10/2)

Waterford – Beech Hill Quarry (8/3)

Lovell – Maxim Prospect (8/6)

Waldoboro – Benner Mica Prospect (8/10)

Warren – Starrett Quarry (September)

Gilead - Peaked Hill Mica Quarry (October)

1944

April Hibbs Feldspar Quarry, Hebron Eugene N. Cameron

June George Elliot Mica Quarry, Rumford Eugene N. Cameron

Davis Mica Quarry, West Bath Vincent E. Shainin and Karl S. Adams

Russell Brothers Mica Quarry Topsham Vincent E. Shainin and Karl S. Adams

17th Matti Waisanen Quarry, Greenwood Vincent E. Shainin and Karl S. Adams

18th Guy Johnson Quarry, Albany Vincent E. Shainin and Karl S. Adams

Wardwell #1 Quarry, Albany Vincent E. Shainin and Karl S. Adams

October

Matti Waisanen Quarry, Greenwood Vincent E. Shainin and Andrew H. McNair

Wardwell #1 Quarry, Albany Vincent E. Shainin and Karl S. Adams

Davis Mica Quarry, West Bath Vincent E. Shainin and Andrew H. McNair

November

Russell Brothers Mica Quarry Topsham Vincent E. Shainin and Andrew H. McNair

1945

May

Bennett Quarry, Buckfield David M. Larrabee and Karl S. Adams

21st Willis Warren Feldspar Quarry, Stoneham David M. Larrabee and Karl S. Adams

19th-20th Bumpus Quarry, Albany David M. Larrabee and Karl S. Adams

21st Noyes Mountain (Harvard) Quarry, Greenwood David M. Larrabee and Karl S. Adams

22nd Peabody Mountain Feldspar Quarry, Batchelders Grant David M. Larrabee and Karl S. Adams

26th Black Mountain Quarry, Rumford David M.

Larrabee and Karl S. Adams

31st Perham Feldspar Quarry, West Paris David M. Larrabee and Karl S. Adams.

Frank Perham underground in the Mount Mica Pegmatite, 2007.

Beryl crystal, Bumpus Quarry. 18 x 6 cm

Canadian geologist Hugh Spence sitting on the giant beryl crystals, Bumpus Quarry. Engineering and Mining Journal (1929).

Appendix G: Description of Wages, Production, and Related Employment Factors in Oxford County and Maine in 1897 to 1907

Extracted from the Maine Bureau of Industrial and Labor Statistics (BILS) of 1898, 1900, 1908

The average monthly farm wage in 1907 in Oxford County, including room and board, was $26.56, up from the 1887 average wage of $19.07. The State's highest farm wages were $29.28 up from $17.57 for Aroostook County. The lowest wages were paid in Knox County at $22.62 up from $18.00. There were wage increases in the era from 23%-67%, indicating that there was very uneven wage improvement among farm laborers across the State. Franklin County paid the highest wages in 1887, while Washington County was the lowest. Farm laborers were seasonal and a worker had to have several jobs in a year in order to survive. Lumbering was a common winter occupation, as there were relatively few manufacturing jobs available in Oxford County.

By far, the most important farm crop in Maine in 1907 was hay ($26,250,000), followed by potatoes ($9,582,000) and oats ($2,560,000). In 1907, "The corn crop, which in the earlier years averaged about $1,500,000, has steadily declined so that the average for the last ten years is only slightly above $300,000." BILS (1900) in its *Brief History of the Corn Canning Industry in Maine* indicated that the previous year, Burnham and Morrill had two corn shops in Oxford County, one each at Paris and Denmark. Portland Packing Company had one in Buckfield and one in Canton. H. F. Webb had one in Norway, while Minot Packing Company had one in East Sumner. Snow Flake Canning Company had one each in North Fryeburg and Lovell. Fernald, Keene, and True Company had one each in Oxford and Bryant Pond. J. & E A. Wyman had one in Bethel. Maine's corn crop was valued at $550,000 in 1899. Earlier *Maine Registers* listed many corn, vegetable, blueberry, fish, and clam canners across the State.

In 1899, small wood working factories were located in Andover (2), Brownfield (2), Denmark (1), Dixfield (3), Hiram (1), Norway (1), Oxford (1), Paris (5), Rumford (1), Sumner (1), Waterford 2), and Woodstock (1). The survey, BILS (1900), did not include factories such as saw mills or companies which bought wood to assemble furniture, novelties, and toys, such as Paris Manufacturing Company or the Charles Forster Company of Dixfield: "The originator of the toothpick industry in Maine." The 1897 output of toothpicks by both the Dixfield and the larger Strong factory was 6.5 billion toothpicks. Interestingly, Oxford County had only one brick maker in 1897 and that was J. Weeks in Porter.

BILS (1898) indicated of factory workers that men earned $1.50 per day while women earned $1.50, doing the same work. An average worker certainly experienced extended periods of unemployment between "seasons". Workers in the "building trades" in Oxford County in 1907 earned a little more than twice what a farm worker earned, $2.26 per day, but there was no room and board benefit. Alternatively, builders usually had families to support, unlike the average farm worker. BILS (1908) noted the migration from rural areas to the "city": "With the exception of Aroostook county and a few towns and a few towns along the border of the wilderness in other northern counties where the settlements have been largely promoted since 1860, you will be found almost without exception that, wherever a town shows any considerable gain in population between 1860 and 1900, it is due to the establishment of manufacturing plants, the development of quarries, the extension of the summer home of industry, or to some special cause, as the Togus Soldiers' Home in Chelsea; while on the other hand, with the exceptions above noted, the purely agricultural towns for the most part show a marked decrease in population, a considerable number of fairly good agricultural towns having in 1900, less then ½ the population shown in 1860."

In 1907, Maine was one of 13 states that limited the number of working hours for women to 60 hours. Three states limited those weekly hours to 58, one was limited to 55, and Colorado and Wisconsin had limits of 48 hours. The remaining 30 states apparently had no such limits. It was noted that the advent of higher speed looms in the woolen industry, for example, demanded a higher work effort than when there were the notorious 5 AM to 7 PM factory shifts in Massachusetts. The formerly used slower equipment meant that "…they had plenty of time to sit and rest,' sometimes 'twenty to thirty minutes at a time,' was not so wearing or detrimental to health as the 10-hour day of the present time when, with machinery at a much higher speed than formerly, a woman tends ten spending frames or from twelve to sixteen Draper looms which require her almost constant attention."

Changes in Maine's quarrying industry were apparent. The granite cutters and tool sharpeners went on strike in 1908 from Jay, Hallowell, Hurricane Island, and Waldoboro. The Vinalhaven strike started during the off season, March 2, but was not resolved until August 24, with essentially little increase in benefits. The Bodwell granite company on Vinalhaven made a stipulation when agreeing to the wage deal: "There shall be no discrimination against tools for machinery now in use, or against the use of any tools for machinery tending to increase production for reduce costs, which may be introduced." The granite cutters' health concerns included the "machinery now in use" and so the strike took longer to conclude than in most other areas of the State. A statement of their concerns included:

"In the introduction of machinery and of tools operated by compressed air has wrought a revolution in the methods of quarrying and cutting granite. The policy of the union is not to object to the introduction of or discriminate against the use of any machine or other labor saving device but it does claim the right to regulate the conditions under which such machines shall be operated.

Granite cutters realize that the conditions under which they work at best are not conducive to longevity. The work is hard and the particles of dust and steel that must necessarily be inhaled are injurious to health. Machine tools create more dust than hand tools and when they are used in a shed they are very objectionable for this reason."

The agreed minimum hourly wage was 37 ½ cents per hour. Payday was every other Friday. The workweek consisted of six 8 hour days. Similar arrangements were previously negotiated with other granite quarrying companies. While there were other details to the strike agreement, the most important concern was the health issue to be resolved. Unfortunately, most of Oxford County's miners worked for themselves and, mostly because there was very little attention paid by individuals concerning workplace safety. Feldspar miners did not learn from the experience of the granite workers. Workers in the feldspar mills were not organized and there were few air quality regulations to protect them, although there were some actual improvements made.

Albany

In this town there are several farms in the market at from $1,000 to $2500. They are from 8 to 12 miles from a railroad station. There are also farm lands for sale but no prices are quoted. The demand for farm laborers is very small, and wages from $2600 o 3000 per month or from $12 to 2.00 per day. But the only opportunity for employment s in the lumber woods. Rents are from $2.00 to $6.00 per month but there is very little call. A preference is indicated for Swedes in case foreign immigrants should settle in town. The tax rate for 1907 was $2800 per $1,000 valuation.

Bethel

Bethel farms of from 25 to 250 acres are in the market but no range of prices are given. The farm lands without buildings are held at from $20.00 to $50.00 per acre. These farms and lands are from 1 ¼ to 7 miles from the railroad station. At the present high price of labor there is no demand for farm laborers by the year, and probably not over twenty additional hands could obtain work by this season. Wages are from $30.00 to $35.00 per month, or about $1.50 by the day. There is always an opportunity for those working as farm laborers in summer to work in the woods winters.

It is estimated that at least 20 house girls are needed with wages averaging $3.50 per week. The present demand for mechanics is small. It is intimated that a few additional masons might find work, wages being $3.50 per day of 8 hours. Rents command about $5.00 per month. The tax rate

for 1907 was 20.00 per $1,000 valuation.

Gilead

We are informed that in Gilead No farms are for sale. It is estimated that half a dozen additional farm hands could find permanent employment, and twice that number for a period of 6 months. Wages are reported to be very high, higher then farmers can afford to pay. Lumbering furnishes employment in the fall and winter.

Three or 4 good house girls would be all that are needed, wages being from $3.00 to $5.00 per week. No additional mechanics are needed. Wages are from $6.00

Paris

There are not many farms for sale in this town. A few would be willing to sell but as a rule the sales are brought about by the natural change in families, but whenever a cheap farm is in the market there is a Finn ready to buy it. There are quite a number of Finns in town, they are industrious and make quite good farmers, much better than poor Yankees. Probably 20 farm hands could get employment for the season and as many as 50 for a few days in haying or harvest. Wages are from $20.00 to $30.00 per month and board, and by the day from $1.50 to $2.50. There are several woodworking plants in town where there is occasionally an opportunity for employment where wages run from $1.25 to $2.00 without board.

It is estimated that there is need of 25 additional girls for house work at wages from $3.00 to $4.00 per week, but otherwise there is little demand for female labor. No additional mechanics are needed. The price of rents varies from $6.00 to $10.00 per month. For foreign settlers in preference is indicated for Finns. The tax rate for 1907 was $20.80 per $1,000 valuation.

Waterford

Farms with buildings in Waterford, situated from 4 to 10 miles from a railroad station and containing from 40 to 200 acres are offered for sale at from $1,000 to $4,000. There is a demand for a few farm laborers by the year at about $25.00 per month and board and it is estimated that 25 will be needed for about 4 months in the busy farming season, wages being about $30.00 per month, or for a few days at from $1.50 to $2.00.

Wages at house work are from $3.00 to $4.00 per week and the demand for girls is greater than the supply. Local mechanics are handling all work in their general lines. Rents are about $6.00 per month. Among foreign nationalities, Finns would be preferred as settlers. The tax rate for 1907 was $24.00 per $1,000 valuation.

Appendix H: History of the Town of Albany
by Mitchell, Harry Edward and Davis, B. V. (1906).

[Editor's note: The text has been scanned, but the pagination has been kept intact for those wanting to cite pertinent pages as they appeared in the original publication.]

History of Albany. Page 30

OXFORD PROPRIETARY .

The earliest date we find relating to the history of Albany is Dec. 29. 1788 which date is borne by an instrument drawn up at Andover, Mass., and signed by twenty-four men each binding himself to the purchase of a tract of land lying between Sudbury Canada and Waterford. Over two years passed before these men were given the deed to the coveted tract. It was probably during this time that the tract was surveyed by Samuel Titcomb and classified as "Township Number Five." The deed is dated Feb. 18. 1791, the sum paid for the land, as specified therein, was "826£ 5s in the consolidated notes of this Commonwealth, and 208£ 10s 6d in currency." The deed bears the signatures of the committee appointed by the General Court of Massachusetts for the sale of Eastern lands and shows the following men to have become the first individual owners of the land now constituting the town of Albany: Joseph Holt and John Russell, gentlemen, Isaac Blunt, hatter, Asa Cummings, - Stephen Cummings, millwright, James Holt, Jr., Jonathan Abbott, F.- and Nathan Abbott, Jr., Joseph Lovejoy, Jacob Jones.. Nathaniel Fay, Wm. Chandler Jr., Nehemiah Holt two shares, Uriah Russell, Benj. Goldsmith, and Samuel Cogswell, yeo-

HISTORICAL Page 31

men, all of Andover; Johnson Proctor and John Lambent, yeomen, of Danvers; John Jaquith, Jr., Samuel Boutell and Judo Jones, yeomen, of Wilmington; and Daniel Lovejoy, Jr., of Wilton, N. H. The township contained, (and its boundaries are substantially the same today), about 23,000 acres. Each proprietor had a right to one-twenty-third of the grant, except that four lots of 320 acres each were reserved for public purposes, viz: one lot for the first settled minister, one for the use of the ministry, one for the use of schools, and one for the future appropriation of the General Court.

Little time was lost by delinquency of the proprietors, for they saw in the new township either the possibility of founding a new home in a new country, or the realization of a considerable sum from the sale of lots to settlers. We find that many of these men settled their own lots and became leading men in the community; whether the non-resident proprietors actually gained much wealth from the deal it would be gratifying to know.

Most of the proprietors' meetings were held at Isaac

Blunt's tavern in Andover, where the first one was assembled on July 25, following the signing of the deed. Mr. Chandler became clerk, Joseph Holt was selected to handle the funds, and Asa Cummings, Nathan Abbott, Jr. and Lieut. John Russell were chosen assessors. We learn that five men, including Lieut. Joseph Holt, had visited the place the previous autumn, and had agreed to take up lots in the grant. At a mee[t]ing held Feb. 3, 1792, a committee was chosen to lot the town into twenty-four parts, and "to qualify the

ALBANY Page 32

town." The same year the name Oxford was adopted for the township.

In December a move was taken toward the erection of a saw mill, which would be of great aid to the settlers. It was agreed to give Benj. Proctor a gore of land lying in Range 11, if he would erect "a good saw mill in the town of Oxford in one year, and maintain it ten years": also if he should build a corn mill and operate it the same length of time, he should have a gore in Range 10. In 1799 he was given a piece of land for keeping the mill in good repair, showing at least one of the mills to have been erected in 1793: this was built at Lynches Mills.

In Jan. 1794, Stephen Cummings was chosen proprietors' clerk. The proprietors showed much diligence in improving the roads and offering greater inducements to settlers. Some came in from the earlier settlements where they began to feel crowded. In 1797 ten dollars was voted, and Nathan and Jonathan Abbott, Jr., were chosen a committee to purchase books for a library.

After the organization of the plantation in 1802, part of the government of the township was shared by the settlers. The proprietors had builded well and had founded a flourishing settlement which was fast increasing in number and prosperity. Gradually they disposed of their lands until the organization of a proprietary became no longer necessary. The last recorded meeting was held in March, 1815. The proprietors did much for the town, seeking rather to make of Albany a thriving community than to gratify any desire for wealth.

HISTORICAL Page 33

EARLY SETTLEMENT.

Who was the first white man to visit the lands now contained in the town of Albany, it is impossible to say. A settlement was begun in Waterford in 1775, and in Bethel two years earlier. This locality was doubtless explored soon after. Warren's History of Waterford gives the date of the settlement of Albany, then ungranted, as 1784, but this settlement may not have been permanent as a much later date if usually given. The proprietors received their grant in 1791, having waited more than two years during which time it is probable that some settlement was begun, certain it is that it was settled very soon after. Most of the settlers were natives of Andover, the home of many of the proprietors. Many of the proprietors settled their own lots, and it is supposed that a number of these men came here to make clearings as early as 1792. To Abner Holt is accorded the felling of the first tree for a settlement. John Foster, Abner Holt, a Mr. Chamberlain and Jacob Chandler came in the early spring, (year uncertain), from their homes in Andover. Coming to McWain's opening, in Waterford, they made their way to the vicinity of No. Waterford, thence up Crooked river some three miles, where they turned to the east and climbed the hill. They made camp on the hillside which was afterwards the Stephen Cummings farm, now owned by F. E. Bean. Here it was that Mr. Holt felled a spruce to get the boughs for a bed. Mr. Foster was a land surveyor and during the summer devoted much of his time to surveying the township. In the fall these men returned

ALBANY Page 34

to Massachusetts and the following spring returned here with their families, together with other settlers. Abner Holt's lot line ran but a few rods from where he had cut the first tree, the farm on which he settled being that now occupied by W. W. Bird. Mr. Chamberlain erected his cabin on lot 11, range 5, now the late John Cummings farm. Mr. Chandler settled a few rods east of Chamberlain on the late John Cummings place. Dea. Asa Cummings came from No. Andover with a party of settlers in 1798, settling on the farm now occupied by his grandson, Geo. C. He reared a family of fourteen sons and daughters and has been followed by numerous and honorable descendants, many remaining in Albany, and others taking prominent positions in other places, many have become prominent member of the Congregational, and other clergy. Abner Abbott settled where his great grandson, Frank Abbott now lives. The earliest burying ground was a family yard on the Stephen Cummings farm, near the first clearing. A stone found there bears the name of a young settler by the name of Jones who died in 1797 at the age of 26. In 1800 the population of Oxford plantation numbered sixty-nine.

The following "Rate list" shows the names of the taxpayers in Oxford in 1802 and includes the names of such families as had become permanent settlers in town at that time.

A Rate List committed to James Russell to collect for the Plantation of Oxford, 1802, showing the amount of tax assessed on real and personal property.

HISTORICAL Page 35

An additional poll tax of $0.27 was paid by all except

Name.	Tax. Pers. Prop,	Real Est.
Abner Abbott,	$0 14	$0 26
Asa Cummings,	16	41
Jacob Chandler,	12	14
Philemon "	(poll tax only)	
Eph'm Flint,	12	16
John Foster,	02	04
Jno. Holt,	03	05
Bani Haskell,	16	17
Nehemiah Holt,	17	26
Abner Holt (2 polls)	18	26
Stephen " (3 polls)	25	44
Moses Holt,	(poll tax only)	
Uriah Holt,	06	41
Parsons Haskell,	$0 10	$0 18
J. Kettredge,	05	14
John [L]ovejoy,	27	41
Jno. Longley,	01	-
Benj. Proctor, (mills)		42
Daniel Mears, (poll tax only)		
Wm. Newel	04	05
Chas. "	05	07
Jas. Russell,	04	13
Israel Sweat (2 p)	06	08
Samuel Town,	04	05
Isaac Wardwell,	-	07

Mr. Proctor, who was a non-resident.

A list dated two years later shows that the following had become residents between 1802 and the spring of 1804: John Bell, James Flint, Joseph Holt, Enoch Holt, Paul Holt, Joel Jenkins, Jonathan Jenkins, Thomas Russell and Wm. Sweat.

ORGANIZATION AND INCORPORATION.

Oxford plantation was organized under plantation government in what must have been the only school house,

ALBANY Page 36

Sept. 27, 1802. Asa Cummings, who was the leading man in the settlement, was made moderator, and Uriah Holt plantation clerk. These two men, with Abner Holt were chosen assessors, and Stephen Holt treasurer. James Russell bid off the collection of taxes at 13%. At the second meeting, Dec. 28, 1802, it was decided to petition the General Court for incorporation, and Asa Cummings, Capt. Bani Haskell and Jonathan Holt were chosen a committee to draw up the petition; Mr. Cummings being delegated to present it to Judge Frye. There seems to have been much difficulty in choosing a name for the new town. The petition was first for the name the township had borne since 1792, and was presented in Jan. 1803; early in the year the voters decided on the name "Montgreen." In May following, another vote was taken on a name, and Albany was chosen. Had not the legislature passed the act before another meeting it is impossible to say what name this town might have borne. The final meeting of the plantation of Oxford was held May 18, 1803, the first of the town of Albany, on Sept. 19, the same year. The town became fully organized on the latter date, taking her place among the corporate bodies of the District of Maine, the 144th town in the District and one of twenty-one incorporated that year.

ACT OF INCORPORATION. COMMONWEALTH OF MASSACHUSETTS.

In the year of our Lord one thousand eight hundred and three.

An act to incorporate the Plantation of Oxford, in the County of York, into a town by the name of Albany.

Be it enacted by the Senate and the House of Representatives in General Court assembled, and by the authority of the same, that the Plantation heretofore known by the name of Oxford, in the County of York, as described within the following bounds, together with the inhabitants thereon, be and hereby are incorporated into a town by the name of Albany; beginning at a pond at the north-easterly corner of Waterford, thence north 20 degrees west six miles and one hundred rods to the south line of Bethel, thence west 20 degrees south on the Bethel line five and one half miles, thence south 20 degrees east about seven miles to Waterford line, thence northerly by said Waterford line to the bounds first mentioned-and the said town is hereby vested with all the powers, privileges and immunities which other towns do or may enjoy by the Constitution and Laws of this Commonwealth.

Signed by the governor, Caleb Strong, June 20, 1803.

TOWN OFFICIALS

CLERKS.

Zadoc Saunders, 1850-51 Geo. W. Saunders, 1852-55 John Hunt, 1856; Geo. W. Saunders, 1857-62; David F. Cummings, 1863; Hermon Cummings, 1864; Addison Love-

joy, 1865; H. D. Haskell, 1866-71; Amos G. Bean, 1872-75; Amos French, 1876; D. C. Healy, 1877; Josiah Wheeler, 1878; Dexter A. Cummings, 1879-83; Wallace B. Cum-

mings, 1884-96; Dexter A. Cummings, 1897-99; Amos G. Bean, 1900; Wallace B. Cummings, 1901-02; Herbert I. Bean, 1908-04; Roy G. Wardwell, 1905.

TREASURERS.

Hiram Bisbee, 1850; Jacob H. Lovejoy, 1851; John Hunt, 1852; J. H. Lovejoy, 1853-55; John G. Hunt, 1856; J. H. Lovejoy, 1857-90; Gilbert Wardwell, 1891-92; Wallace B. Cummings, 1893-1905.

SELECTMEN.

1850-Geo. W. Saunders, Geo. French, James Wheeler.
1851-Geo. French, Geo. W. Saunders, Geo. H. Brown.
1852-541 Geo. W. Saunders, Abenethy Grover, Reuben Libby
1855-Samuel P. Haskell, James French, Wm. Coleman.
1856-James French, Samuel Brown, Isaiah French.
1857-58-James French, Geo. French, Wm. Coleman.
Abernethy Grover, Gilbert Wardwell, Isaiah French.
1860-61-S. P. Haskell, James. French, Gilbert Wardwell
1862-Gilbert Wardwell, Thos. Morrill, Phineas P. Dresser
1863-Thos. Morrill, P. P. Dresser, E. S. Hutchinson.
1864 Gilbert Wardwell, P. P. Dresser, Elijah K. Morrill.

1865-Gilbert Wardwell, Lewis H. Sawin, John Haseltine.
1866-Geo. French, John Haseltine, Washington French.
1867-John Haseltine. L. H. Sawin, Lawson Mason.
1868-L. H. Sawin, Lawson Mason, Joseph Wheeler.
1869-Gilbert Wardwell, J. H. Lovejoy, Joseph Wheeler.
1870-72-Gilbert Wardwell, J. H. Lovejoy, John Haseltine.
1873-J. H. Lovejoy, Geo. W. Beckler, David A. Edwards.
1874-J. H. Lovejoy, Gilbert Wardwell, John Haseltine.
1875-Gilbert Wardwell, John Heselton, Wm. M. Robbins.
1876-J. Heselton, Geo. W. Beckler, D. C. Healy.
1877-78-G. Wardwell, W. M. Robbins, W. M. Brown.
1879-Geo. French, W. M. Robbins, G. W. Beckler.
1880-W. M. Robbins, G. W. Beckler, Elbridge Prince.
1881-G. Wardwell, H. C. Lawrence, F. H. Bennett.
1882-81 G. W. Beckler, F. H. Bennett, A. Hutchinson.
1885-F. H. Bennett, A. Hutchinson, J. K. Wheeler.
1886-G. W. Beckler, J. K. Wheeler, Shirley Haselton.
1887-G. W. Beckler, S. E. Haselton, L. H. Sawin.
1888-G. W. Beckler, C. H. Fernald, L. H. Sawin.
1889-L. A. Sawin. F. H. Bennett, J. W. Dresser.
1894-92-F. H. Bennett, A. Hutchinson, Chas. Flint.
1893-F. H. Bennett, A. Hutchinson, L. H. Burnham.
1894-J. Dresser, A. Hutchinson, S. E. Haselton.
1895-A. Hutchinson, S. E. Haselton, J. Flint.
1896 G. Wardwell, G. W. Beckler, H. B. McKean.

1897-F. H. Bennett, H. B. McKean, Z. K. Wheeler.

1898-H. B. McKean, J. K. Wheeler, W. E. Herrick,
1899-1900 F. H. Bennett, W. E. Herrick, A. Hutchinson.
1901-W. E. Herrick, A. Hutchinson, C. G. Beckler.
1902-W. E. Herrick, C. G. Beckler, F. G. Sloan.
1903-0-W. E. Herrick, C. G. Beckler, Geo. Cummings.
1905-W. E. Herrick, C. N. Eastman, E. W. Rolfe.

INDUSTRIES.

As an inducement to settlers, the proprietors hired Benj. Proctor to erect a saw mill, about 1793, and a grist mill the same year, or soon after. These mills, located near the southern border of the township, were owned and operated by Mr. Proctor until sold to Daniel Brown about 1830, together with about 2500 acres of valuable timber land. Soon after this they passed into the hands of Jas. Osgood, of Fryeburg, who built new mills and manufactured more extensively, sending some lumber to Portland by the canal. About 1840 Moses Petty became owner and proprietor. Ten years later he sold to John Lynch who did an extensive business in manufacturing lumber, and at one time had a match factory and a box factory. After carrying on an extensive business for many years he assigned and the mills went to Samuel Spring. In 1890 his heirs sold to Clark &

HISTORICAL Page 41

Burnham, and soon after Mr. Clark sold his interest to his partner, L. H. Burnham, the present operator.

The large spool mill at Lynchville was erected in 1871 by Elliott, Bartlett & Lynch. Soon Mr. Lynch sold his interest to Elliott & Bartlett who have continued to carry on an enterprising and succeedful business for thirty-five years. Around 700,000 feet of spool stock is consumed annually, thirty men being employed at the mill. The entire product of number 1 spools is shipped to the Coats Thread Co.

What is known as the Petty mill, in the northern part of the town, was built early in the last century, and was owned by Jeremiah Niles in 1825. John W. Dana bought it soon after, and about 1830 sold to Moses Petty, who operated it for half a century. Later owners have been A. S. Bean, Herrick & Kilburn, of Bethel, and F. L. Edwards, the present owner. N. Moore & Co.'s mill, better known as the Grover mill, was built by Francis Cummings about 1840. Later owners have been Abernethy Grover, Wilbur Bros., Eliphalet Burbank (who rebuilt it about 1880 after being burned,) Stephen S. Libby, and N. Moore & Co. A grist mill was connected with the mill until it was burned. An early mill was built on the site of Fernald's mill early in the century. Chesley Fernald and Parker N. Flint built the present mill about 1875. A saw and grist mill was built about 1814 on Swett Brook by John Lovejoy. This was washed out after which a clover mill was built. In 1871 this was destroyed in the same manner. In 1875 a steam mill was built just above this site by James McAllister; this

ALBANY Page 42

was burned a few years later.

Other mills, on sites now vacant, were: one on the outlet to Round Pond, built about 1835; one at the outlet to Kneeland Pond below the Albany Basins built by Cyrus Kneeland probably to replace the above. This was burned about 1873 and Mr. Kneeland built a steam mill on the shore of the pond, and one built by Andrew Bean in 1859 on Bean Brook. S. W. Libby built the mill in No. Albany about 1875. He later sold to Wm. Mason.

In the early days agriculture was perhaps the chief industry. The soil is fertile and productive, but uneven and generally hard to work except in the valleys. There are many good farms now under cultivation while many of those cleared one hundred years ago have been allowed to return to woodlands.

CONGREGATIONAL CHURCH.

The first and only religious organization ever effected in Albany was made very soon after the incorporation of the town. With the assistance of Rev. Joseph Strong of the N. H. Missionary Society, twenty-two residents formed an organization Sept. 15, 1803, which they named "the church of Christ in Albany." Asa Cummings was chosen clerk and first deacon. Jonathan Holt was chosen second deacon. The next year the church received from the church at

HISTORICAL Page 43

Andover the gift of a flagon, tankards, and linen for a communion service. No settled pastor was employed for several years, and the early records of the society are rather meagre. In June 1818, Asa Cummings, Jr., was dismissed to the church connected with the theological seminary, Andover, Mass. He it was who became the founder and editor of the Christian Mirror.

The church was supplied occasionally by missionaries of the N. H. Missionary Society; and also by Revs. Church of Bridgton, Ripley of Waterford, Dutton of No. Yarmouth, and Cressey of Norway. In 1823 Rev. Henry A. Merrill; in 1824 Timothy Hilliard; and in 1826 Joseph Walker were with the church. Sermons were sometimes read by the deacons or leaders when no preacher could be had, and the society continued to thrive and prosper.

On Feb. 1, 1828 the church extended a call to Rev. Thomas Ayer to become its installed pastor. This call he accepted, Mar. 4, and on the 25th a council was convened which ordained him, the first settled pastor in Albany. After a successful pastorate of five years he was dismissed by mutual consent and supplies were again had until Rev. Geo. F. Tewksbury was installed Jan. 31, 1838. The pastorate of Mr. Tewksbury lasted for fourteen years, which were fruitful years to the church. During the first three years twenty-eight were received into membership. He was also a stanch advocate of temperance. In the record of Mr. Tewksbury's ordination is found the first mention of the meeting house and its use. The early services had been held in a house which served for all public gatherings including

ALBANY Page 44

the school. A meeting house seems to have been erected about 1831 on a height of land toward North Waterford from Hunt's Corner. Mr. Tewksbury's pastorate lasted over fourteen years during which time forty-five members were added to the church. Rev. Isaac Weston preached a few months during the summer of 1853, and Rev. Lorraine Read from April 1854 to Sept. 1855. On July 6, 1856, began the pastorate of Rev. Samuel L. Gould, which also covered a period of fourteen years. Mr. Gould was beloved as a pastor and esteemed as a citizen. Since his dismissal the church has been regularly supplied during the summer months, mostly by students. Rev. E. J. Converse supplied from May 1896, to the autumn of 1897, and greatly strengthened the church. The latter year the church was incorporated for the transaction of business. The meeting house was removed to Hunt's Corner, to the present site, rebuilt, and greatly improved, at a cost of about $2000. It was rededicated in June 1878, a neat and convenient house of worship. W. J. Hail was the supply during 1902 and 03 and Carl Thompson, of Yale Divinity School, during 1904 and 05. A Sunday school is sustained, and a Young People's Society of Christian Endeavor, organized in 1889, has done much to sustain public worship.

SCHOOLS.

We have very good proof that a public school was maintained in Number Four prior to the organization of the

HISTORICAL Page 45

Plantation of Oxford, in the statement that the meeting organizing that body was called in the school house. This was located on a road (now discontinued) leading from the Haskell place to the Valley road and was a low, log structure. Later a house occupied for town purposes, schools and religious services, was erected on the right hand side of the road from Hunt's Corner to the Haskell place.

We do not find that any sum of money was raised for education by the voters in the plantation, but a vote taken in Jan., 1803, was made to petition the General Court to remit the state tax of the plantation and "to order the same to be laid out in educating the youth in said Oxford." Soon after, a school committee was chosen. At the town meeting held in November, after the incorporation, Asa Cummings, Uriah Holt, Lieut. Stephen Holt, Ephraim Flint, and James Russell were chosen a committee to rebuild the school house. The house then constructed was doubtless that afterwards used for town purposes and religious worship. We find record of much lumber being furnished for it by various citizens, in fact most if not all of the voters had a part in its construction, and it was doubtless the pride of the settlement when completed.

The first appropriation we find recorded was made in Feb., 1804, when ten dollars was voted "to be laid out in a summer school to be taught by a Mrs. & $20 for a winter school to be taught by a master." Among the early teachers were Anna Bean, Uriah Holt, Esq., Sarah Holt, Lydia Swift (of No. Norway), Eben Hutchinson, Esq., Samuel Haskell, Dea. J. H. Lovejoy and Hepsibah Gould, who became

ALBANY Page 46

the wife of Dea. Lovejoy. As the settlement increased in population and prosperity the amount of the annual appropriation was increased. In 1825, Albany had four school districts, 126 scholars between the ages of four and twenty-one, 84 of whom attended school. $120 was then raised for the cause of education; the population of the township was 307. Five years later the population was 387, and in 1840, 691.

We do not need to trace the later development of the schools. New d[i]stricts were established or discontinued as the settlement or desertion of different neighborhoods called for. The house now standing at Hunt's Corner was erected about 1840, rebuilt and remodeled in 1885.

SOLDIERS OF THE CIVIL WAR.

No section of Maine responded more readily and truly to the call for men to defend the Union than did Oxford county and the town of Albany. The following list we have compiled from the Maine Adjutant General's reports. We have endeavored to include all that were credited to the town of Albany, and if any omissions are found in the list they are unintentional.

Justis Aspenwall, Edgar D. Andrews, John L. Beckla, Andrew J. Bean, Edward Burk, Ephraim K. Buxton, Sergt. Thos. J. Bennett, Jos. H. Briggs, Parker V. , Geo. H. Brown, Jas. Brown, Frank Burnham, Walter Bisbee,

HISTORICAL Page 47

Ephraim K. Baston, John Cowie, Wm. Coffron, Ruel Cofran, Corp. Lorenzo N. Cole, Lewis F. Cummings, Danl. Cummings, Granville W. Drew, Osgood Drew, Thos. Dalton, Osias B. Edgecomb, John Flint, Jacob F. French, Geo. W. Fernald, Chas. W. Farnum, Sergt. Wm. H. Foster, Capt. Abernethy Grover, Corp. Wm. H. Henry, Chas. P. Jordan, David A. Jordan, Saml. E. Kenniston, Watson Kenniston, Geo. W. Kimball, Moses E. Kimball, Edw. R. Kneeland, Alonzo Littlefield, Vincent Mason, David Mitchell, Corp. Cyrus B. Morse, John Marshall, Bravity Marston, Thos. G. Morrill, Wm. H. Pingree, Solomon S. Pingree, Asa B. Pingree, Ai E. Searer, Seth E. Seavey, Henry C. Scribner, Oscar D. Wilber, Henry Wilber, Benj. Wilbur, Oscar D. Wilbur, Sergt. Gilbert Wardwell, Isaac W. Wardwell, Elbridge P. Wardwell, Nathan M. York, John York, Jr.

ITEMS OF INTEREST.

Round Mountain Grange, No. 162, was organized in 1875 with 27 charter members. The Grange Hall, located at Hunt's Corner, was purchased by the society in 1884. Here a store is conducted, now under the care of James Flint. The following is the list of men who have occupied the Master's chair: Herman Cummings, Geo. W. Beckler, Justis Aspinwall, James Flint, Gilbert Wardwell, James Kimball, Isaac Wardwell, Geo. Cummings, Wallace B. Cummings, Leon Kimball, Roy Wardwell and Irving Beckler. The leading officers for 1906 are Leon Kimball, master:

ALBANY Page 48

Cecil Kimball, overseer; W. B. Cummings, steward; Geo. Cummings, treas.; Ella Cummings, lect. and A. E. Cummings, sec. The society now numbers fifty-one members.

The Albany public library was organized about 1870, at Hunt's Corner. Mrs. J. H. Lovejoy became the first librarian. It has been successively under the care of Amos G. Bean, Hattie C. Wilbur, and Mrs. Nancy C. Andrews who now has it at her home. There are around 2250 volumes which number is increased as time and funds permit.

Hunt's Corner tavern was opened about 1805, to accommodate travelers over the new stage line from Portland to Bethel. It was built by one Holt who was succeeded in 1818 by Timothy Hutchinson. Two years later he was followed by Reuben Beard who remained four years when he sold to John Hunt for whom the place was named. He became well known throughout this part of the state as the genial proprietor until the year 1870 when he sold to A. S. Cole, the present owner. About 1825 a U. S. mail route was established running through the Albany Basins from North Waterford to West Bethel, on the way to Lancaster, N. H. On a certain trip the driver came near freezing and soon the old route by Hunt's Corner was resumed.

A local telephone has been recently installed in Albany by a local company incorporated as the Albany Telegraph and Telephone Company

L-> R: Chester Wheeler, Danny Wheeler, Albert Wheeler
Courtesy Addison Saunders.

Skiving Mica at Wheeler Brothers' Mica Shop Main Street, Bethel, Maine.

c1958

Courtesy Benjamin Shaub.

Albany Sign. 2008

Chapter Sixteen: Bumpus Quarry Today
By Bruce Barrett

On October 8, 2004, Dr. Lawrence Stifler and his wife, Mary McFadden purchased the Bumpus Quarry to ensure the preservation of its historical and educational significance. The Stiflers, from Brookline, MA and Albany Township, Maine and active land conservationists in both states, recognized the importance of conserving the quarry despite having no background in geology or mineralogy. Jim Mann, miner, gem cutter and owner of Mt. Mann Jewelers in Bethel, made their acquaintance when he reopened a rose quartz prospect on land they owned in Albany Township in

L to R, Jane Perham, Frank Perham, Bruce Barrett, Larry Stifler, Molly Stifler, and Woody Thompson, Bumpus Quarry, August, 2005

2003. Knowing the Stiflers' land conservation work, Mann told them about his long-standing desire to preserve the Bumpus Quarry, which he had worked in the 1960s. At a time when most pegmatite quarries no longer have any public access, Mann saw the value in preserving this important historic site for its educational and historic qualities. Mann introduced the Stiflers to Rodney Kimball, who had inherited the Bumpus Quarry from his father. After a lengthy and occasionally arduous negotiation process, the Stiflers came to terms with Kimball, and set up the Bumpus Historical Mine, L.L.C. as a not for profit venture to acquire the quarry. They also purchased an additional three acre "camp lot" along the Crooked River, across the road from the quarry, to prevent any development and provide a scenic picnic site for visitors.

Shortly after the purchase, the Stiflers invited geologist Woodrow B. "Woody" Thompson of the Maine Geological Survey and mineralogists Ray Woodman and Jim Greenlaw to visit the quarry. They discussed the issues of access and how best to open the quarry to the public. Al-

though it had not been worked in decades and the landscape was very overgrown, they expressed great excitement about various possibilities for the future of the quarry. Thompson explained that a recent effort to establish a Newry Mineral Park for the public had met with disappointment when the land was sold and closed to the public. He suggested that the proximity of the Bumpus Quarry to Route 5 and its easy access for cars and school buses would make it a suitable alternate site. While no firm plans were made at that time all parties expressed interest in the goal of historic preservation of the site.

Early in 2005, the Stiflers moved forward by hiring Steve Shearer, a local excavation contractor, to make the quarry safer and more accessible to the public. Equipped with a small Yanmar excavator, he was able to develop easy access for foot travel into the quarry by clearing the dense vegetation, which had taken over in the years since the quarry was last worked. He continues to turn over and refresh the tailings each year to expose new collecting areas for the hundreds of visitors to the quarry.

In August of 2005, the Stiflers welcomed Frank and Jane Perham along with Woody Thompson to the quarry. It was Frank's first visit since he ceased mining operations there. With his typical humility, Frank was clearly impressed by how well the tunnels had endured over the previous four decades, remaining completely intact and safe for entry. Thompson commented that it was extremely fortunate to have the wit and wisdom of Frank and Jane Perham, and that this was the first time he had ever been out in the field with both of them at the same time. The Stiflers were inspired by Jane Perham, author of Maine's Treasure Chest Gems and Minerals of Oxford County, and Frank, who had mined the Bumpus Quarry in years past. Hearing first-hand the history of the Bumpus Quarry from the Perhams was invaluable for the Stiflers.

Moving forward, Jim Mann formed a partnership with Seabury Lyon and Jeff Parsons of Bethel to give interpretive guided tours of the Bumpus Quarry to the general public for a modest fee. Building on Parson's Bethel Outdoor Adventures, they resurrected the earlier tradition of tourist operations at the quarry, calling their enterprise "Maine Mineralogy Expeditions," and creating a website, www.rocksme.biz. The Stiflers have generously allowed them to use the quarry for no charge and have supported their efforts by incurring all of the costs of site improvements and upkeep. Led primarily by Lyon with Mann's input, the Bumpus Quarry tours are very popular, and include a lecture on the local geology, mineralogy and history of the quarry, a walking tour, as well as collecting opportunities in the tailings. Numbering in the hundreds, clients of Maine Mineral Expeditions are mostly tourists visiting the Bethel area and local campgrounds, mineral clubs, and school groups from as far away as Indiana. In June of 2005, a news team from WHDH-Channel 7 in Boston produced a short segment about the quarry called "Great Escapes: Mineral Mining in Bethel, Maine." There have been many additional feature articles in a number of Maine newspapers and publications, which have further boosted interest in the quarry.

In May of 2006, Woody Thompson invited the Stiflers to attend the Maine Mineral Symposium in Augusta to give a special presentation about their goals for Bumpus Quarry. Dr. Stifler's talk included ideas for public access to the quarry and introduced the idea of a new mineralogy museum to be located in a new wing to be constructed at the Bethel Historical Society. His presentation was met with great enthusiasm by the attendees at the Symposium. Vandall T. King gave a keynote lecture about the Bumpus Quarry and other Oxford County localities, and the Bumpus Quarry hosted a site visit for over 50 of the Symposium's naturalists and mineralogists in attendance, who came from as far away as California.

The Stiflers also presented their vision of a Maine Mineral Museum to the board of the Bethel Historical Society. They quoted George Howe's "A Message to Mineral Collectors," written in the yearbook of the Oxford County Mineral and Gem Society in 1948: "I am especially interested in having located in Oxford County a museum of Oxford County minerals. It is very regrettable that a great many valuable specimens peculiar to Oxford County have been sold and allowed to leave the State of Maine, many going to foreign countries. It is something for this club and the people of Oxford County to work for. This museum should be centrally located, preferably at or near the county seat. If these specimens and the ones we are now finding could be preserved here in our own Oxford County, we would have something to be found nowhere else in the world." These

Maine Mineral Symposium Field Trip to Bumpus Quarry, May, 2006

WHDH TV – Channel 7 crew filming "Great Escapes" at the Bumpus Quarry

words were written sixty years earlier, and the time to make them a reality had finally come.

After several meetings with the staff and board of the Bethel Historical Society, the idea of a museum was formally adopted by the Society. Two committees, one chaired by Frank Perham, were appointed to study the museum's feasibility and implementation. In addition to providing a secure place for important collections of Maine minerals, the museum's goal is to return important mineral specimens to Maine for viewing by the public. A number of collectors and authorities on Maine minerals have volunteered their efforts to help make the museum a reality. The need for a museum to preserve and display these collections is great and comes at a time when museums like Harvard University's Museum of Natural History are closing their New England mineral and gem sections and important collections are being de-accessioned by other museums around the country. Given the significance of Maine's mineral localities, there is consensus that a museum centered in mineral rich Oxford County is needed to preserve important specimens, artifacts and documents.

The Bumpus Quarry will serve as an offsite adjunct to the mineral museum for field study and historical tours by the Society. Dr. Stifler's vision is to restore the Bumpus Quarry to "appear as if it is a Sunday afternoon and all the miners have gone home, but their equipment and tools remain." A campaign to establish the museum in Bethel will need to be undertaken by the Society, and initial major funding has already been committed by the Stifler Family Foundation to help make the museum a reality. The Stiflers have met with curators from many museums, and commitments to loan or gift minerals to the new museum have come from The Museum of Natural History, The Smithsonian, the Harvard University Museum and others. Perhaps most notable is the offer by the Museum of Natural History to loan one of the remaining sections of massive Bumpus beryl crystals originally acquired by the J.P. Morgan Fund and currently in storage. The return of this specimen and others to the state of Maine will be a momentous occasion and cause for great celebration by the entire mineralogy community. The Stiflers

Ava Bumpus and daughter Beth Hartford visit the Bumpus in October 2005

have also acquired a number of other private collections of important Maine mineral specimens and gems with the goal of keeping them in Maine.

To enhance the historical aspect of the quarry, a track drill from 1972 has been acquired and placed at the site along with a stiff leg derrick. In time the Bumpus Quarry will serve as a living history museum honoring the role of mining and mineralogy in Western Maine. Jim Mann has conceived the idea of a "Friends of the Bumpus Quarry" to care for the locality and further the educational and historic value of the site. In addition to the for-fee tour groups led by Maine Mineralogy Expeditions, Bumpus site manager Bruce Barrett from Albany Township has provided free access to the quarry for local school groups and mineral clubs in order to further the educational mission of the Bumpus Historic Mine L.L.C. Barrett, who has a background in geology and training at the Maine Pegmatite Workshop, has developed an age appropriate program for school groups that visit the quarry. Supported by the Stifler Family Foundation, this program has grown to serve hundreds of students at no cost to the students or participating schools. The Sti-

flers have paid for school transportation costs and the salary of the site manager to host the groups, furthering their commitment to developing the educational and historical use of the quarry.

In 2007 the Maine Conservation School in Bryant Pond began to bring groups of campers to the quarry and also began hosting teacher workshops. In 2008 the School, restructured and renamed the University of Maine 4H Center at Bryant Pond, offered weekly visits to the quarry with as many as 30 students per group.

Supported by a grant from the Stifler Family Foundation, the University of Maine 4H Center at Bryant Pond hired an Americorps Vista Volunteer to develop a geology and mineralogy curriculum at the School. In the fall of 2008 Sienna Tinsley from West Paris was hired for a period of one year. With a bachelor's degree in geology and a growing interest in Maine mineralogy, Sienna is well suited for this position. In 2009 the University of Maine 4H Center at Bryant Pond will offer mineralogy camps for students ages 8 – 18 to study and explore the important mineral resources and geology of Maine. The Bumpus Quarry will be used as a field study site for the introduction of pegmatite geology to the students from the 4H Center as well as to western Maine school groups. The quarry's easy access to school buses and awe-inspiring tunnels make it an ideal educational site for all ages.

Due to the risks inherent in all inactive quarries, the Bumpus Quarry is not open to the public on a self-guided basis. As a precaution, visitors wear hard hats and are accompanied by seasoned guides. Although there are no current plans for further mining at the Bumpus, the fascinating history of the quarry continues to engage and inspire the next generation - Maine's future geologists and rock hounds. Since the Stiflers acquired the quarry, well over 2,000 students, mineralogists, rock hounds and future geologists, have already had the delight of visiting the quarry and learning about its interesting mineralogy and history.

Jim Mann with campers from Maine Conservation School after a visit to the Bumpus Quarry in 2007.

End Notes

[1] Giant beryls, for the time period, had been discovered in Fryeburg, near Stark Hill, in 1825 and were reported by young enthusiastic mineralogist Charles Upham Shepard [1804-1886], of Amherst College, in 1830:

8. Fryeburg Beryls

We are indebted to Mr. Cook, the Preceptor of the Fryeburg Academy, for a knowledge of the interesting deposit of *Beryls* that occurs in this town. It is situated about half a mile west of the public house, upon the western declivity of a granite hill, which lies directly upon the public road. The Beryls occupy a vein a few feet in width, and ten or twelve in length. In dimensions, they vary from one, to two or three inches in diameter; but the closely aggregated manner in which they occur, is not very favorable to a high degree of finish in their form. Crystals that are tolerably complete, may however, be obtained, and occasionally, those with polished, terminal, faces; but we find them, more generally, with faces very unequally produced; as, with two lateral faces very widely extended and imparting a tabular appearance to the crystal, or with four planes so enlarged as to give a rhomboidal aspect, or finally, with the alternate faces protracted in such a manner as to form a trihedral prism. The most interesting circumstance connected with these crystals, however, is their color, which varies from a delicate bluish green to a white; those of the first mentioned shade, possessing the ordinary transparency of the species, while those of the latter are only transparent on the edges. The vein stone is quartz, slightly brown, with an occasional intermixture of imperfectly formed feldspar crystals. The longer crystals of Beryl offer the same peculiarity as respects the fractures and reunion of the laminae at right angles to their axes, as were noticed in the large Beryls of Acworth.* The deviaion [sic] from a straight line which the axis suffers, in consequence of this disturbance, amounts in some instances to 5 or 10° in; and the quartz which penetrates between the joints, is in layers of half an inch in thickness. In addition to this peculiarity, we observed here, also a slight curvature in some of the crystals, unattended by any fracture in the prism. These observations I am induced to make with the more particularly, since they appear to me important in the consideration of the much agitated question among geologists, respecting the origin of granite. With a celebrated writer upon geology, I am persuaded, 'that much light must, at some period, inevitably be thrown on the greater geological phenomena, by considering the chemical and mechanical relations existing among the smaller portions which constitute them; and that the language of Nature is often as intelligibly spoken in the minute space of an inch, as in the immensity of a mountain. ... * A large joint of the Acworth crystal, in the possession of my fellow traveler, Dr. Heermann, illustrates these fractures in a very interesting manner.

Amos Jones Cook (July 7, 1778 Templeton, MA - April 7, 1836 Fryeburg) was the long-term principal of Fryeburg Academy and was the immediate successor of the later to be famous, Daniel *"Liberty and Union, now and forever, one and inseparable!"* Webster. Webster was an unsuccessful candidate for the presidency, but was Secretary of State in 1842 when the Webster-Ashburton Treaty was negotiated with England, which treaty settled the northeast boundary between Maine and other states and the British North American Provinces (later Canada). Cook had an influence on Enoch Lincoln, teacher at Fryeburg and later lawyer and Governor of Maine, as well as boarder in the Cyrus Hamlin home in Paris Hill. Wilkinson (1942) wrote of Cook: "To add to student interest, Mr. Cook started a museum [at Fryeburg Academy] which increased rapidly. Included in this museum were letters... geological specimens, old fire-arms; and Indian relics." The Fryeburg beryls were not a record size, except for Maine and Oxford County. The world record beryls, at that time, were from what is now known as Beryl Mountain, Acworth, Sullivan County, New Hampshire. [124] It is an interesting coincidence that Charles Upham Shepard's nephew, Bertram Boltwood, was one of the scientists who laid the groundwork for the formal recognition of isotopes – a finding which a half century later was a concept utilized in manufacturing nuclear weapons and the eventual utiliza-

Fryeburg Academy. From a 1920s postcard.

Bumpus and Getchell Pharmacy, Goff Block, Auburn. 1920s Postcard.

tion of Maine's and other States' beryl.

[2] Edwin Kemble Gedney (AM Harvard University, Ph. B. Brown University, 1928, d. October 18, 1980) wrote a number of religious/paleontological articles and books attempting to reconcile the sudden appearance of fossils in the geological record with the Biblical account of creation and is regarded by modern creationists as an influential thinker contributing to their philosophy. Before his death, Gedney rose to Director of Church Planting of his faith in 1975.

Dr. Harry Berman (February 16, 1902 Boston, MA - August 27, 1944 Prestwick, Scotland). Berman attended, but did not graduate from Carnegie Institute of Technology. [111] In 1922, Berman became an assistant in the U. S, National (Smithsonian) Museum, where Auburn, Maine native, George Perkins Merrill, was the curator of mineralogy. Berman became part of the Harvard University Museum staff in 1924 and soon became part of the mineralogy department's teaching staff. In 1931, he received an Adjunct of Arts degree from Harvard University. After a post graduate scholarship experience in Europe, including a study under noted mineralogist, Victor Goldschmidt at Göttingen, Germany in 1932-1933, Berman returned to Harvard and received his M. A. in 1935 and his Ph. D. in 1936. Bermanite from Arizona was named for Berman and it is an interesting coincidence that this mineral is also known from a location less than 7 km to the east of the Bumpus Quarry at the Emmons (Uncle Tom) Quarry in Greenwood as well as the Lord Hill Quarry in Stoneham to the southeast and the Dunton Quarry in Newry to the north. Berman died from an airplane crash while on a consulting trip for the war effort. His greatest contributions were a classification of the structures of silicate minerals, used worldwide, even to this day, and his research and co-authorship of the Dana's System on Mineralogy, seventh edition.

[3] Rose quartz from Albany enjoyed notoriety, even before the Bumpus discoveries. Kunz (1892) wrote of Maine rose quartz: "At Stow, Albany, Paris, and a number of other localities in Maine, the veins of quartz shade from white – transparent and opalescent, resembling hyaline quartz, often without any imperfections – through faintly tinted pink and salmon into a rich rose color, thus forming a beautiful series

of tints for gems or for ornamental stone-work. Specimens of this rose quartz, when cut into double cabochons, or sphere-shaped objects, distinctly show the asteria effect, similar to the star sapphire, if viewed by sunlight or artificial light, a peculiarity which has been observed in specimens obtained from a number of other localities. Possibly as fine transparent opalescent rose quartz as has ever been found was obtained at Round Mountain, Albany, Me., in pieces free from all flaws and of a fine rose-red, with a beautiful, milky opalescence, measuring 4 x 5 inches in size. A sphere 2 inches in diameter, a small dish, and other objects have been cut from this material."

[4] The article was authored "Special" and we might imagine that it was a news release written by Stan Perham especially as the technical article stuck to the point and it did not seem to suffer from errors of fact commonly seen in newspaper articles about Maine mining. Although the article indicated that a new record size crystal had been exposed, the article was accompanied by a photograph of one of the 1928 beryls.

[5] The term Cold War was originally introduced to describe the political tensions between the USA and Russia, then called the Union of Soviet Socialists Republics. This Cold War lasted from 1945 through 1991. Although there are varying claims to its origin, the first published use of the phrase seems to be by noted author George Orwell, who in October 19, 1945 used it in a newspaper article for the (New York) *Tribune*. The phrase was almost immediately popularized in speeches by financier, presidential advisor, and later member of the United Nations Atomic Energy Commission, Bernard Baruch, who said: "Let us not be deceived – we are in the midst of a cold war [with Russia]." British Prime Minister, Winston Churchill, also frequently used the phrase.

During the Cold War, the USA accelerated its research on and production of nuclear weapons and there was a consequent rush for the raw minerals from which they might be constructed. Uranium was certainly the primary ingredient and the uranium rush of the late 1940s and early 1950s is well-known, but weapons manufacturing technology also depended on a wide variety of other minerals including beryl, then the only economic ore of beryllium.

[6] Lise Meitner has element 109 named for her and Otto Hahn had element 105 named for him, but a controversy over the data and other considerations regarding hahnium caused that element to be renamed dubnium by the International Union of Pure and Applied Chemistry in 1997.

[7] All unattributed locations, even those with international sounding names, such as Paris or Norway, should be understood to be for Maine locations. The following sign was erected about 1938-1939 and calls attention to the richness of internationally sounding names in Maine. Unfortunately, this sign has been a target of pranksters and the sign has been damaged or stolen a number of times.

Quotations are *verbatim* and have not been edited to correct minor errors including those of grammar.

It also must be noted that there are many contradictions in prices quoted for various materials, even from the same sources, and there must be an awareness that a strict accounting is not possible, although we might imagine that the majority of the figures quoted are reasonably close to reality. Some numbers may have had an ephemeral validity, while others may have had a vagueness, fish-story, or wishful thinking component. Various Government reports frequently announce later revised statistics, but there are frequently reports that have disinformation in them. However, there is no way to sort out those reports that have selective reporting.

[8] The gem name emerald was used as a synonym for gem green beryl for many years into the nineteenth century, but none of the Maine specimens, so called, are what would be called emerald, today.

[9] A symmetrical hexagonal prism 18 feet long and 4 feet in diameter, with an assumed density of 2.7 g/cm^3, should weigh 15.7 avoirdupois tons or 14.3 metric tons.

[10] As Mount Mica is on a low shoulder of Streaked Mountain, it is uncertain how far away the locality was. Streaked Mountain is granitic in composition and granite pegmatites abound there and into Buckfield on the eastern side.

[11] Feuchtwanger (1832) advertised one of the giant Beryl Mountain crystals for sale: "An American Beryl, of 70 lbs. weight, and 620 cubic inches. A full six sided Prism, of 27 [inches] circumference."

[12] Vincent Shainin learned to fly a small plane as a personal interest. Potter (*Boston Sunday Post*, February 20, 1949) wrote an extended article about Shainin's exploits including his study of pegmatites probably in Topsham: "Once a farmer in another part of Maine refused to let the [survey] party of geologists on his land. But Professor Shainin was determined to make the survey cover the entire area assigned to him, so he took a plane and flew over the farm, sometimes being down to only a few feet above pasture land, and he managed in this manner to get a complete report to the national and State authorities. ... He decided that the present-day geologist, roaming about wide areas, can do a much faster job if he flies. So he went up to Vermont and learned to fly."

[13] The population of Albany was, by decade, starting in 1850 ending in 1950: 747, 853, 651, 693, 645, 538, 410, 360, 309, 288, 242. With the increased building around Songo Pond, the 2000 population was 515. The value of the property did not keep pace with inflation. In 1860, the town was worth $140,847 and by 1930 the valuation was $182,268. Most of the few commercial establishments in Albany have been historically located near a small village, Lynchville, located near State routes 5 and 35 between North Waterford and East Stoneham. It is interesting to note that the junction of route 5 and 35 has the most famous signpost in the State featuring distances to local towns which are also the names of countries or famous cities. A somewhat similar sign is located in China, Maine. A road sign featuring internationally sounding Maine town names is in Norway.

[14] Harlan Bumpus was Harry and Laura's second child. 1) Thelma Bumpus [August 1, 1901 - infancy], 2) Harlan M. Bumpus [January 17, 1903 – April 27, 1976 Lewiston, ME],

3) Sybil H. Bumpus (later Staula) [May 6, 1905 – November 25, 1985], 4) Dorothy P. Bumpus (later Merriam) [February 19, 1907 – February 17, 1989 Lewiston], 5a) Margaret Lillian Bumpus [July 27. 1909 – March 20, 1969], 5b) Madeline L. Bumpus [July 27, 1909 – October 10, 1987], 7) Cora B. Bumpus [November 16, 1910 – May 19, 1980 Auburn], 8) Ruth Elizabeth Bumpus [January 4, 1914 - infancy]. Harry wanted Harlan to go to college to become a pharmacist, but Harlan, a man of particular spirit, refused and there was much disappointment by his father. Nonetheless, Harlan, who particularly enjoyed the out-of-doors, worked at the Bumpus Quarry beginning at the very first in 1927.

Harlan's children and step-children were: Arthur Wallace Hazelton [stepson, November 27, 1919 Lewiston - July 27, 2006], Edwin Charles Bumpus [May 1, 1930 – December 23, 2002], Ruth Anne Bumpus [June 16, 1931 – February 22, 2002], and Kenneth Harlan Bumpus [b. December 10, 1936].

[15] Salo (1988) noted that Harry E. Bumpus graduated from the Chicago College of Pharmacy in 1898 and that his drugstore was located at the corner of Court and Main Streets: "He was ... a member of Elm Street Universalist Church; Tranquil Lodge, F. & A. M. [Free and Accepted Masons]; the Board of Incorporators of the Central Maine Hospital (now = Central Maine Medical Center, Lewiston)." Bumpus & Getchell, located in the Goff Block, on Court and Main Streets, was classed under Apothecary Druggists in the *Maine Register* from 1904 – 1918. Beginning in 1919, the business is listed at 32 Court Street only under Getchell's name until 1927 and the following two years were similarly listed at Court and Main Streets until 1929 and after that no drug store is listed for those locations although there were competing drug stores listed on Court Street. The change in name may have suggested that the partnership was bought out by Getchell and that Bumpus continued as the pharmacist, there and elsewhere. An unidentified (Lewiston newspaper?) obituary supports this notion as it was stated: "... also had been employed at Turgeron's Pharmacy, the Perryville drug store, and most recently at Kenney's Pharmacy." He was also a member of United Commercial Travelers. Harry Erwin Bumpus was the second son of George Washington and Mary E. Albee Bumpus. George W. Bumpus's family resided in Hebron. George was also Auburn city clerk for 36 years.

[16] The Federal Census of 1920 showed an extended family living on the Cummings farm including: Abbie W. Cummings ("mother" = Abigail {A. Wiley} Jackson Cummings), age 75; Allen E. Cummings (head of household); Wallace E. Cummings (brother of head of household); Sibyl Cummings (sister of head of household); and Adelia A. Cummings (niece, age 13; [later Adelia Waterhouse]). [Adelia was the sixth child of George W. Cummings (1856-1932) and the third child in his second marriage (to Cora Wiley Cummings).] Viola Eleanor Cummings, a practical nurse, married Charlie A. Dunham of Bethel on December 21, 1904 and, later, Leon L. Kimball on October 27, 1936.

[17] Ronald E. Holden, Jr. [November 19, 1959 – August 12,

1998 Buckfield, ME].

[18] The historian and serious mineralogist are advised that the various books by Jean Blakemore, although interestingly written and sure to promote enthusiasm, have been carefully reviewed by Wintringham (2000) who found that Blakemore's works were fraught with misinformation, on almost every page, if not in many different successions of sentences. The few instances when her text has withstood the test of time have been judiciously cited. For example, the same paragraph in her book relating to Dana Douglass' collecting fees, an otherwise well-known story, related that beryl was an ore of uranium! Unfortunately, virtually all of her Maine "firsts" and "largests" in the 1976 edition are incorrect in some way, many being irreconcilable.

[19] An odd glassy substance, nicknamed "trinitite", was formed during the heat of the atomic bomb test blast at the Trinity Site. It consisted of fused sand composed mostly of quartz with traces of olivine and feldspar. [194] For many years, tiny bits of trinitite were sold as souvenirs of the blast, but, given the restrictions on entering the test site, it is uncertain how many specimens were fabricated.

[20] The search during WWII for beryl resulted in a number of spectacular discoveries, none of which were publicized for over a decade and usually with no claim for record size. As a result, Maine claimants had no basis to compare their finds.

[21] Jahns (1953) summarized USA beryl records:

"Beryl, a widespread though not particularly abundant accessory constituent of many pegmatite bodies, forms some crystals of truly exceptional size. One of these was partly exposed for several years in the Ingersoll No. 1 pegmatite, in the Black Hills region of South Dakota; a gigantic hexagonal prism [Waldschmidt, 1919], it was measured at different times by Hess [1925, p. 289], who reported a basal diameter of 45 inches, and by Connolly and O'Hara [1929, p. 254], who reported the diameter of another section as 46 inches. Although its length never was determined accurately, this prism probably weighed more than 30 tons. A mass of beryl weighing approximately 100 tons is reported from the same pegmatite by Page [personal communication, 1952], who also notes a tapering prismatic crystal, once exposed in the Bob Ingersoll No. 2 mine, that was 18 feet long and 6 feet in maximum diameter. Another immense crystal, 18 feet long, 4 feet in diameter, and weighing about 18 tons, was noted from Albany, Maine, by Gedney and Berman [1929].

Several large 'logs' of beryl were taken from partly kaolinized feldspathic pegmatite at the Herbb No. 2 mine, in Powhatan County, Virginia, during operations for sheet mica in 1944. This interesting occurrence has been described by Brown [1945, p. 265]. One of the crystals, as 'reconstructed' by the writer from huge segments on the dump, must have been about 14 feet long and 17 to 23 inches in diameter. A somewhat different but no less spectacular occurrence was encountered at the Harding mine in northern New Mexico, where a 23-ton lens of beryl was mined in 1944 by Arthur Montgomery [1951, p. 33]. The irregular mass appeared to consist of several very large anhedral crystals."

Additional giant beryls have been found in other pegmatite districts. Cobban et, al, (1997) wrote about Colorado:

"Fremont County – Texas Creek District … E. W. Heinrich (written communication, 1980) reported beryl crystals up to 4 ft in diameter; crystals as long as 20 ft have been mined. … Eight Mile Park District … rarely as large as 1 ft in diameter and 6 ft in length. … Jefferson County – Bigger mine … Light-greenish blue to green beryl occurs in huge crystals nearly 3 ft in diameter and as long as 10 ft. … Larimer County – Crystal Mountain District … Big Boulder prospect was 6 to 7 ft long and 1.8 ft in diameter… "

Roberts and Rapp (1965) reported on beryl crystals in South Dakota: "Pennington County: … A 26 ton crystal of beryl is reported to have been mined in 1945 from the Dan Patch feldspar mine one mile west of Keystone. … One mass of beryl crystals containing about 100 tons was found in the glory hole of Dike No. 11 in 1944 at the Bob Ingersoll mine two miles northwest of Keystone. … A crystal 18 feet long and 6 feet in diameter and weighing approximately 40 tons was mined from Dike No. 2; a large white specimen … from this occurrence is on display in the South Dakota School of Mines and Technology museum."

[22] Adelia A. Cummings married Clarence N. Waterhouse, also of Albany, on September 30, 1927.

[23] **Ozymandias** by Percy Bysshe Shelly [1792-1822]

*Two trunkless beryls
from an antique land, 1989.*

I met a traveler from an antique land
Who said: Two vast and trunkless legs of stone
Stand in the desert. Near them, on the sand,
Half sunk, a shattered visage lies, whose frown,

And wrinkled lip, and sneer of cold command,
Tell that its sculptor well those passions read,
Which yet survive, stamped on these lifeless things,
The hand that mocked them, and the heart that fed,
And on the pedestal these words appear:
"My name is Ozymandias, King of Kings:
Look upon my works, ye Mighty, and despair!"
Nothing beside remains. Round the decay
Of that colossal wreck, boundless and bare
The lone and level sands stretch far away.

[24] Because beryl may have considerable impurities, the BeO content is usually less than the theoretical 14.02% and, for that reason, many beryl payments had to wait for assays to be made, miners not wanting to get paid the base rate "for beryl accepted on visual inspection". Page and Larrabee (1962) reported: "The price of analyzed beryl is based on the content of BeO. Beryl containing 8.0 to 8.9 percent BeO is bought at $40 per unit, 9.0 to 9.9 percent BeO at $45 per unit, 10.0 to 1.9 percent BeO at $50 per unit, and 12 percent or more at $55 per unit. As an example, beryl having 12 percent BeO is worth $660 per ton." Neumann (1952) praised the Bumpus Quarry mineral: "The beryl is good grade, averaging 13.0 percent beryllia content; and the large crystals, having a very distinctive color, can be readily hand-sorted." Cameron et al. (1954) reported that beryl from Waisanen Quarry, Greenwood was 11.88% BeO and Bennett Quarry, Buckfield beryl was 12.55%. Non-Oxford County beryl analyses included Standish (12.4%) and Starrett Prospect, Warren (11.70%).

[25] Elijah's brother was Hannibal Hamlin, vice-president under Abraham Lincoln. Elijah's grandfather Major Eleazer Hamblen sired two U. S. vice-presidents, if you follow the "begats". Eleazer Hamblen (July 1732 Billingsgate (Wellfleet), Plymouth Colony, MA) m. June 30, 1752 to Lydia Bonney (1735 Pembroke, MA – August 12, 1769); m(2) to Sarah Lobdell. Eleazer's son Isaac Hamlin (1728 Eastham, MA - 1762 Bridgewater, MA) married Sarah Shaw (b. October 1, 1726). Their son, also Isaac (b. November 14, 1748) married Polly Beals (April 18, 1774 Bridgewater, MA - July 18, 1833 Cummington, MA). Their daughter Sarah Hamlin (~1779 Cummington, MA - January 6, 1822 Cayuga Co., NY) married Peleg Standish (b. May 3, 1774, but perhaps as late as 1776 Plymouth Co., MA - July 10, 1853 Huron Co., OH). Their daughter Sarah "Sally" Standish (April 27, 1815 Cayuga, NY - March 30, 1866 Huron Co., OH) married Reuben M. Burras (~1812 Cayuga, NY - <1850). Their son Oscar Burras (~1839 Huron, OH - ~1880) married Annette Hakes (May 19, 1841 Bronson, OH - ~1910 CA). Their daughter Adelia Burras (October 19, 1865 North Fairfield, OH - December 29, 1940 Los Angeles, CA) married William O. Cline (May 3, 1864 North Fairfield, OH - March 18, 1941 San Bernardino, CA). Their daughter Marie Pauline Cline (July 27, 1890 Norwalk, OH - October 5, 1975 Wayne Co., IL) married Robert Harwood Quayle (September 13, 1887 Clinton, IA - January 19, 1968 St. Charles, IL). Their son James Cline Quayle (May 25, 1921 Joliet, IL - July 7, 2001 Sun West City, AZ) married Martha Corinne Pullman (b.

~1925 IL). Their son James Danforth Quayle (b. 1947) was vice-president under the first George H. W. Bush. The genealogical data for the Eleazer Hamblen connection came from a variety of sources, while there are many Dan Quayle family trees: for example, *The Ancestry of Overmire Tifft Richardson Bradford Reed* by Larry Overmire, 2005, www.rootsweb.com; *Cousins of John Crossley*, by John Crossley, homepage.mac.com/jcrossley; see also www.eric-james.org; //reimert.org; etc. *Ancestry of Dan Quayle* by William Addams Reitweisner, (2005, www.wargs.com/po-litical/quayle.html) contained many typographical errors when accessed and is only partly used, although it does genealogically tie together many famous personages, entertainers, particularly seven presidents and almost as many vice-presidents of the USA to George H. W. Bush.

[26] Winfield Scott Robinson and James A. Gerry sold their mineral rights to Francis H. Cobb. [341]

[27] On September 18, 1935, the Oxford Mining and Milling Company leased the Fred E. Scribner land for mining rights [342]: "The Lessee agrees to upon [sic] up and develop during 1935, a mine on the above leased premises." Mineral royalties were $0.50/ton for Number 1 feldspar, $0.25/ton for Number 2 grade feldspar, $0.10/ton quartz, $1.00/ton for beryl, $1.00/ton for mica, and 10% of the "selling price" for all other minerals. The low figure for beryl is noteworthy, although that quarry produced very little beryl.

[28] The various sales of companies, name changes, lease reassignments, etc. appear to have been partially influenced by company reorganizations. The actions being requested or forced so that the companies could finish certain obligations before being acquired by other companies sometimes with no interest in continuing business in far away places such as Maine.

[29] Bickmore graduated from Dartmouth College with the Class of 1860. He went to Harvard University to work with Louis Agassiz in 1860-1861 and was there while Maine-born students, later internationally famed scientists, Addison E. Verrill (born in Greenwood, cousin of Charles A. Stephens, and father-in-law of Vivian Akers), Alpheus Spring Packard, Jr. (February 19, 1839 Brunswick – February 14, 1905 Providence, RI), and Edward Sylvester Morse (June 18, 1838 Portland – December 20, 1925, eventually malacologist, director Peabody Museum, editor and co-founder of the *American Naturalist*), were there.

[30] Maine reporters have always made important mineral discoveries the province of "boys". The tradition started with Mount Mica in Paris with Elijah Hamlin, then a lawyer, and Ezekiel Holmes, then a doctor, who discovered gem tourmaline east of Paris Hill, probably on or about October 18, 1821. [124] Similar claims have been made for "boys" first finding minerals at Mount Apatite in Auburn, "boys" first finding minerals at Mount Rubellite in Hebron, and a "boy" first finding minerals at Newry. When the feldspar was first found in 1927, Allen and Wallace, the two Cummings "boys", were ages 61 and 58, respectively.

[31] Gregory (1968) indicated that local folklore suggested that Harry Erwin Bumpus discovered the minerals when he was

Wayne Ross and father, West Paris miner, Reginald Ross, 1940s. Courtesy Irene Card.

hoeing "his" garden for peas or potatoes, but Harry Bumpus, a Cummings family in-law, did not live in Albany and was a pharmacist in Auburn. Of course, Harry may have been helping out with chores, but that seems unlikely.

[32] See also Rask (1983) for summary of the USA programs of the procuring beryl and working with industry to maintain beryllium metal production.

[33] Roy Perham did not stay in mining long and left Maine for many years with a career as a minor league professional baseball player. [401]

[34] Harold's second wife, Mary Slattery Perham, was named "Mother of the Year" by the State of Maine in 1961.

[35] Frank graduated from West Paris High School in 1952 and entered Bates College with the Class of 1956, but entered the Army during the Korean War. He returned to Bates and graduated with his B.A. degree in geology in 1959. After a year as a miner in the Cobalt, Ontario silver mines, he returned to Maine and pursued a career as a miner. He was also active in community service, a virtue shared by many Perhams, and was variously a Civil Defense Officer, volunteer in the West Paris Fire Department, Selectman, Assessor, and Overseer of the Poor, and board member of the Stephens Memorial Hospital. [171] Frank worked at the Tamminen Quarry (1961) and Waisanen Quarry (1962-1963), Greenwood, then at several prospects in Newry and Rumford on

Halls Ridge (1964), then at Mount Mica (1964-1965) in addition to other Oxford County locations (Perham, 1961, 1963, 1965, 2000). After the West Paris Feldspar Mill closed, Frank worked as a contract driller and blaster for the Maine Department of Transportation and other organizations and in 1972-1976 worked at the Dunton Gem Quarry, Newry and at Mount Mica, Paris for the Plumbago Mining Company. . In the 1970s, Frank worked at the Auburn Auto Repair and soon founded his own company, the Route 219 Garage in West Paris. As a hobby, Frank was a competitive race car driver at Oxford Plains Speedway and other tracks in central Maine. In the 1990s, Frank mined at the Whispering Pines Quarry, Tamminen Quarry, Noyes Mountain (Harvard) Quarry, B.B. #7 Quarry (with Tom Ryan), and into the 2000s has been working at the Tamminen Quarry and the Morgan Quarry, both Greenwood and in 2007 at the General Electric #2 Quarry in Buckfield, often with his son-in-law, Dana Morgan. In 2008, he and Barry Heath re-opened the Albany Rose Quarry in Albany.

[36] "Ever since the writer's brother, Stanley, was old enough to wear knee pants, he showed that he possessed the inner makings of a born naturalist. Every bird, every butterfly, and every moth, became objects of personal interest. As a boy he studied their habits in the field, and read their life histories in books. Soon, he was putting up Frames of Beautiful Moths and Butterflies for display purposes.

Every aspect of Out-of-door life was of added interest. He was enrolled in Scouting, and I remember that he was a First Class, Life and Star Scout with many merit badges to his credit. [George Howe was a Boy Scouts counselor in Norway and may have been the man who interviewed Stan for some of the badges' requirements.] While in Bates College he served as Pres. Of the Bates Outing Club.

Every sunrise was a thing of beauty to my brother, and every sunset was a benediction. When he walked the forest paths, he was at home. Every tree, every animal, every bird became a natural part of his heritage, and the desire to kill was not in his heart.

How early he began his activities in Mineralogy, only the Stars can tell. But, it must have been very early in youth, for his older brothers used to object to the lack of small boxes in the house, and the fact that Stanley had them all filled with beautiful specimens of rocks and strange formations.

We older brothers simply had to hunt up more boxes, and soon found out in our small village that small boxes were hard to find, and that Stanley had the 'inside track'.

Even then, my brother Stanley knew that all the ladies were the first Naturalists at heart, and he never forgot that fact to his dying day." [187]

[37] Stan's first marriage was to Gwendolyn Louise Wood [b. January 29, 1904] of Naugatuck, Connecticut on January 3, 1931. They had one child together: Frank Croydon Perham, born March 5, 1934. Stan and Gwendolyn met at Bates College where she graduated with the class of 1927. However, their marriage was short and Gwendolyn Perham died while

she was in Waterbury, Connecticut, on her wedding anniversary, January 3, 1935. Stan later married Hazel L. Scribner, November 28, 1935. Stan and Hazel adopted Jane Charlotte Perham, and together raised their two children in a style that brought forth two great mineralogists. In 1977, perhamite was named, in Frank's honor, from specimens discovered at the Bell Pit and Dunton Quarry, both Newry. It was later found at the Emmons Quarry, Greenwood and the Ski Pike Quarry, West Paris. There are now several non-Oxford County locations: Funderburk Prospect, Pike Co., Arkansas, Silver Coin mine, Humboldt County, Nevada, Utahlite Prospect, Lucin, Box Elder Co., Utah, Les Montmins Mine, Auvergne, France, Hardtkopf Mt., North Rhine-Westphalia, Germany, Laggerhof, Carinthia, Austria, Zhilandy Mine, Maikain, Northern Kazakhstan, and Moculta quarry and Penrice Quarry, Mount Lofty Ranges, and Toms Quarry, Kapunda, all South Australia, where the species has been found, to date.

[38] Long after cesium was of interest to the electronics industry, Hess et al. (1943) reported that cesium was not detectable by spectroscopic methods in Bumpus Quarry microcline. However, the sometimes associated element, rubidium, was present and qualitatively estimated at 1 weight percent, more or less.

[39] There are several Farwell Mountains, pronounced as though spelled "Farrell", within sight of each other. One is northeast of Bethel's village and the Albany Farwell Mountain straddles the Albany-Mason township boundaries, just west of the Pingree Ledge Quarry.

Farrell Mountain, Bethel.
From a 1920s postcard.

[40] In the February 14, 1995 issue of *Commerce Daily*, the Defense Logistics Agency (DLA) solicited bids relating to its beryl stockpile: "Project consists of crushing and sorting approximately 11 million pounds of Beryl Ore. Other service includes cobbing the sorted ore to obtain the gemstone rough and pre-grading the gemstone rough achieved." In the January, 1996 *Jeweler's Circular Keystone*, it was reported: "Defense Department to Sell Gem Rough. Looking for bargain gem rough? Try the U.S. Department of Defense booth at the gem and mineral shows in Tucson, Ariz. The department's Defense National Stockpile Center will set up its booth Jan. 31-Feb. 9 at the Pueblo Inn to sell mainly soapstone, beryl ore, synthetic ruby boules and synthetic sapphire

and ruby rods. The boules are described as "pear" or "carrot" shaped and the rods average 6"-8" in length, all of which makes for excellent cutting grade material. Among the beryls represented are aquamarine, heliodor, morganite and goshenite. While DNSC has invited dealers to bid on materials in the past, this sale will be open to anyone. DNSC representatives will process orders on the spot." Keller (2006) reported that the DLA sold minerals from the DNSC (Defense National Stockpile Center) at the 1996 Tucson Gem and Mineral Show in Tucson, Arizona from a display tent outside of the Sheraton Pueblo Inn hotel, although the beryl crystals observed did not have transparent areas, despite being labeled "faceting grade". The gem rough was $40 per pound. Beryl crystal sections about 7 x 7 x 4 cm were offered at $10 each, but there were no known localities for the specimens. The reporter may have seen the booth after all of the genuinely "gem grade" beryl had been sold or the initial offering may have been ineptly sorted or selected. One wonders if any Bumpus beryl made it to Tucson or if all of it was crushed into an anonymous aggregate? DLA did not repeat its attempt at direct sales to mineral naturalists and subsequently only offered bulk beryl ore lots in auctions.

[41] Thomas (1947) wrote about a bizarre incident concerning beryl, perhaps from the Bumpus Quarry, that has not been corroborated by a primary reference: "During the recent World War II tour of the world, [I] encountered one mineral attache who spoke words of wisdom off the record. ... The Germans came to Maine before World War II with tongue in cheek. They found beryllium in large amounts thrown aside at mine dumps in the form of useless beryl. This was ideal ballast for sailing vessels, it was heavy and because of the shape did not tend to shift. Of course we fell for it and Germany had one of the largest stock piles of beryl in the world. The idea of beryllium-copper alloy had just become important..." The article also made some unwarranted allusions to beryl's uses in Germany's atomic bomb experiments, etc. Germany <u>was</u> at the forefront of technical uses of beryllium-bearing alloys, but the presence of German agents in Oxford County buying beryl is a story unrecorded elsewhere. One might imagine that small quantities of Maine beryl were purchased by German industry, but the activity seems unchronicled. There does not seem to have been any exports of beryl, reported as beryl, from the USA in the 1930s. Beryl used as ballast would not have been inventoried, however. The term "sailing ship" wouldn't have meant anything except to distinguish it from an "air ship".

[42] The Portland Society of Natural History was inaugurated in 1837 with Charles T. Jackson as its first speaker. The Society's museum was destroyed by fire on July 4, 1866. After an attempted fund-raising campaign, the museum was given space by the City in 1869 and the society was revitalized, but went into decline in the twentieth century until it's dissolution in the 1960s, when its exhibits were dispersed to other organizations. The University of Maine at Orono received some of the mineral and fossil specimens, including a few type specimens of fossils. The type specimens, by which the original descriptions of some fossil organisms

were published, were soon donated to the Agassiz Museum of Comparative Zoology at Harvard University. [415]

[43] Waterhouse Mountain appears to be an obsolete local name of unknown provenance. Attwood (1977) does not list it.

[44] After several decades of neglect, the abandoned Trenton Flint and Spar Mill was renovated into apartments in the early 1970s

[45] Burnham and Morrill [B&M] is a vegetable canning company, particularly known for its baked beans: "By July 1868: A corn cannery is established at South Paris, Maine, and the "Paris" trademark is acquired for their canned corn. Their first sales of canned corn begin that same fall. B&M corn was also packed under the SACO label, as well as being distributed under many private labels. Although corn was especially difficult to can at this time, B&M soon solved the major production problems and their canned corn business quickly boomed. At one point, there were as many as nineteen corn canneries scattered throughout Maine." (www.bgfoods.com/bm/bm_history.asp)

[46] Twenty years later, Bert bequeathed his mineral collecting and related materials to Stan.

[47] The Farley family had some interesting connections with Oxford County before investing in the West Paris Mill. Charles Farley [June 14, 1791 Ipswich, MA – December 20, 1877 Portland] was a silversmith in Portland by about 1812. He was probably an apprentice before then, but his mentor is not known. By 1814, Farley was a junior partner with Eleazer Wyer until 1818, when they separated to resume independent businesses. Farley married Elijah L. Hamlin's cousin, Rebecca Faulkner Hamblen [February 25, 1805 Warterford-1878], on September 6, 1826 and early the next year, Rebecca's brother, Cyrus Hamlin, began three years of apprenticeship in the silversmithing business. This Cyrus Hamlin eventually became known as a missionary and he founded Roberts University in Istanbul, Turkey. [There are four Cyrus Hamlin's to keep track of regarding Oxford County mining, gemology, jewelry manufacturing, etc. For convenience, this one is Cyrus (IV) Hamlin. Elijah Hamlin's father is Cyrus (I). Elijah's brother is Cyrus (II) while the son of Hannibal Hamlin, the vice-president, is Cyrus (III). It is equally confusing to keep track of the various Hannibal Hamlin's.]

Farley is not known to have been a silversmith after 1830 and he may have gone into a retail business. In 1866, the Great Portland Fire of July 4th destroyed hundreds of businesses, including C. H. Farley Nautical Instruments. The original Charles Farley had no middle initial and his son, Cyrus Hamlin Farley [August 29, 1839 Harrison - April 9, 1934 Portland] was the probable owner of the company by mid-century. Portland was one of the major seaports of the USA, in the time, and one might imagine there was a reasonable business in selling and repairing such instruments in the City. C. H. Farley is listed in the *Maine Register* dealing in nautical instruments until 1893. However, C. H. Farley also became a retailer of ornamental glass, being listed as early as 1877, and briefly, 1886-1898, was an optician.

All are at least mildly related professions. By 1894, C. H. Farley was noted as a manufacturer of stained glass windows for home decoration as well as for churches until immediately after WWII, although there in no listing in 1939. In 1940, C. H. Farley & Company is listed as a division of Boston Plate and Window Glass with a specialty in stained glass, but by 1955 the company is no longer identifiable as no glass business is located at its former address on Milk Street and the names of Farley and Boston Plate are no longer listed in Portland.

[48] Edward Nelson Dingley, Jr. [February 15, 1832 Durham, ME – January 13, 1899 Washington, DC] was Governor of Maine in 1874-1876. He had succeeded Paris native, Sidney Perham, in that office. Dingley was a powerful national level politician in the U.S. Congress, as were Maine's two other politicians of note in about the same time period: James G. Blaine and Thomas Reed. Dingley was Maine representative from 1881 until 1899 and was already been chairman of the Ways and Means Committee after William McKinley's re-election.

Edith Cummings Stearns and Hugh W. Stearns
October 1, 1977
50th Weddings Anniversary

References

Many references have been identified by a number to reduce the obtrusiveness of direct citation.

Abrahamson, James L. and Carew, Paul H., 2002, *Vanguard of American Atomic Deterrence: The Sandia Pioneers, 1946-1949,* Praeger Publishers, Westport, CT, pp. 208.

Alger, Francis, 1856, *On the Beryl Formation of Grafton, NH,* **Proceedings of the Boston Society of Natural History,** v. 6, p. 22-23.

[255] **AMNH,** 1931, *[Expenses],* **Annual Report of American Museum of Natural History for 1930.**

[256] **AMNH,** 1937, *New Exhibits,* **Annual Report of American Museum of Natural History for 1936,** p. 8.

Anderson, Alfred L., 1954, *Memorial to Edson Sunderlund Bastin (1878-1953),* **Proceedings of the Geological Society of America for 1954,** p. 87-92.

[48] **Anonymous,** 2006, *Beryllium,* www.ngdir.ir/Minemineral/MineMineralChapterDetail.asp?PID=3220.

[49] **Applin, Kenneth and Hicks, Brian,** 1987, *Fibers of Dumortierite in Quartz,* **American Mineralogist,** v. 72, p. 170-172.

[51] **Attwood, Stanley Bearce,** 1977, **The Length and Breadth of Maine,** University of Maine Press, Orono, ME, pp. 279 + 1949 Supplement pp. 12 + 1953 Supplement pp. 30.

Austin, Gordon T., 1993, *Abrasive Materials,* **Minerals Yearbook 1993,** U.S. Bureau of Mines, p. 65-89.

[252] **Babitzke, Herbert R., Bagley, Edwina F., and Doyle, Robert G.,** 1978, *The Mineral Industry of Maine,* **Minerals Yearbook 1975, Volume II, Area Reports, Domestic,** US Bureau of Mines, 349-357.

[52] **Barton, W. R. and Goldsmith, C. E.,** 1968, **New England Beryllium Investigations, U. S. Bureau of Mines Report of Investigations 7070,** 177 p.

Bastin, Edson S., 1910, **Economic Geology of the Feldspar Deposits of the United States, U. S. Geological Survey, Bulletin 420,** pp. 81.

[53] **Bastin, Edson S.,** 1911, **Geology of the Pegmatites and Associated Rocks of Maine, United States Geological Survey Bulletin 445,** Government Printing Office, Washington, DC, pp. 152 + map.

[54] **[BILS] Bureau of Industrial and Labor Statistics,** 1902, *The Feldspar, Mica and Tourmaline Industries,* **Fifteenth Annual Report of the Bureau of Industrial and Labor Statistics for the State of Maine 1901,** Kennebec Journal Print, Augusta, p. 93-98.

[250] **[BILS] Bureau of Industrial and Labor Statistics,** 1904, **Seventeenth Annual Report of the Bureau of Industrial and Labor Statistics for the State of Maine 1903,** Kennebec Journal Print, Augusta, p. 239.

[BILS] Bureau of Industrial and Labor Statistics, 1908, *Maine Mining Bureau* and *Requests, Differences and Strikes,* Sentinel Publishing Company, Waterville, Maine, p. viii-ix, 403-435.

[55] **Blakemore, Jean,** 1952, **Gems and Minerals. Treasure Hunting in Maine,** Smiling Cow Shop, Boothbay Harbor, pp 117.

[56] **Blakemore, Jean,** 1976, **We Walk on Jewels. Treasure Hunting in Maine for Gems and Minerals,** Courier of Maine Books, Rockland, pp. 175.

Bowles, Oliver and Middleton, Jefferson, 1930, *Feldspar,* **Mineral Resources of the United States 1928, Part II – Non-metals, U. S. Bureau of Mines,** p. 67-80.

[57] **Bowles, Oliver and Middleton, Jefferson,** 1932, *Feldspar,* **Mineral Resources of the United States 1929, Part II – Non-metals, U. S. Bureau of Mines,** p. 83-92.

[58] **Bowles, Oliver and Stoddard, Blanche H.,** 1936, *Mica,* **Minerals Yearbook 1936, U. S. Bureau of Mines,** p. 1037-1046.

[59] **Bradshaw, John J.,** 2000, *Gemstones of Maine,* in King, Vandall T. (editor) **Mineralogy of Maine. Volume 2: Mining History, Gems, and Geology,** Maine Geological Survey, Augusta, ME, p. 283-305 + 4 color plates.

[60] **Brown, W. R.,** 1945, *Some Recent Beryl Finds in Virginia,* **Rocks and Minerals,** v. 20, p. 264-265.

[61] **Brush Wellman Company,** accessed 2006, *Brush Wellman Company History,* http://www.brushwellman.com/aboutus/co_history.asp.

[62] **Burr, Freeman F.,** 1930, *The Minerals of Maine,* in Merrill, Lucius H. and Perkins, E. H. (editors), **First Annual Report on the Geology of the State of Maine (Report of the State Geologist),** Augusta, ME, p. 25-47.

[63] **Caldwell, Dabney W. and Austin, Muriel B.,** 1957, **Maine Pegmatite Mines and Prospects and Associated Minerals, Minerals Resources Index No. 1,** Department of Industry and Commerce, Augusta, ME, pp. 43.

Cameron, Eugene N., Jahns, Richard H., McNair, Andrew H., and Page, Lincoln R., 1949, *Internal Structure of Granitic Pegmatites,* **Economic Geology Monograph #2,** p. vii + 115 + plates..

[64] **Cameron, Eugene N., Larrabee, David M., McNair, Andrew H., Page, James J., Stewart Glenn W., and Shainin, Vincent E.,** 1954, **Pegmatite Investigations 1942-45 New England, USGS Professional Paper 255,** pp. 352 + 48 plates.

[65] **CDNR (California Department of Natural Resources),** 1952, *Beryllium,* **Mineral Information Service.** California Department of Natural Resources, Division of Mines, v. 5, #5, p. 1-3.

[66] **Chadbourne, Elron R.,** 1904, *The Minerals of Maine,* **Mineral Collector,** v. 11 (5), p. 67-69.

[67] **Clark, Jack W.,** 1950, *Minor Metals,* **US Bureau of Mines Minerals Yearbook 1948,** p. 1310-1350.

[68] **Clark, Jack W.,** 1951, *Minor Metals,* **US Bureau of Mines Minerals Yearbook 1949,** pp. 1294-1324.

[69] **Clark, Jack W.,** 1953, *Minor Metals,* **US Bureau of Mines Minerals Yearbook 1950,** p. 1309-1342.

[71] Cleaveland, Parker, 1814, *Emerald,* The American Mineralogical Journal, v. 1, p. 263-264.

[72] Cleaveland, Parker, 1816, An Elementary Treatise on Mineralogy and Geology, being an introduction to the study of these sciences, and designed for the use of pupils, - for persons, attending lectures on these subjects, - and as a companion for travelers in the United States of America, Cummings and Hilliard, Boston, pp 668.

[73] Cobban, Robert R., Collins, Donley S., Foord, Eugene E., Kile, Daniel E., Modreski, Peter J., and Murphy, Jack A., 1997, Minerals of Colorado by Edwin B. Eckel, Updated and Revised, Fulcrum Publishing, Golden, CO, pp. 665.

[74] Conklin, Lawrence A., 1986, Notes and Commentaries on Letters to George F. Kunz; Correspondence from Various Sources, including Clarence S. Bement with Facsimiles, Conklin Library, New Canaan, CT, pp. 137.

[75] Connolly, J. P. and O'Harra, C. C., 1929, *The Mineral Wealth of the Black Hills,* South Dakota School of Mines Bulletin 16, 418.

[76] Corson [original surname Korsunsky changed by court order], Michael George, 1926, *The Copper-Beryllium Alloys,* Brass World, v. 22, p. 289-314.

[77] Cunningham, Larry D., 1998, *Beryllium—1997 annual review:* U.S. Geological Survey Mineral Industry Surveys, July, pp. 7.

[78] Cunningham, Larry D., accessed 2006, *Metal Prices in the United States through 1998. Beryllium,* USGS, http://minerals.usgs.gov/minerals/pubs/commodity/beryllium/.

[79] Dana, James D., 1837, A System of Mineralogy: including an Extended Treatise on Crystallography: with an Appendix, Containing the Application of Mathematics to Crystallographic Investigation, and a Mineralogical Bibliography, Durrie & Peck and Herrick and Noyes, pp. 452 + Appendixes and Index pp 199 + 1 errata + 4 plates.

[80] Dana, James D., 1844, A System of Mineralogy Comprising the Most Recent Discoveries, second edition, Wiley & Putnam, New York, pp. 633 + errata + 4 plates.

[81] Dana, James D., 1854, A System of Mineralogy Comprising the Most Recent Discoveries, fourth edition, George P. Putnam & Co., New York, pp. 533 + errata.

[82] Dunn, Pete J. and Appleman, Daniel E., 1977, *Perhamite, a New Calcium Aluminum Silico-Phosphate Mineral, and a Re-Examination of Viseite,* The Mineralogical Magazine, v. 41,p.437-442.

Dunnack, Henry E., 1920, The Maine Book, privately published, Augusta, ME, pp. 338.

[83] Ela, Robert E., 1967, *The Mineral Industry of Maine,* US Bureau of Mines Minerals Yearbook 1965, volume III, p. 391-397.

[84] Ela, Robert E., 1967a, *The Mineral Industry of Maine,* US Bureau of Mines Minerals Yearbook 1966, volume III, p. 377-383.

[85] Ela, Robert E., 1969, *The Mineral Industry of Maine,* US Bureau of Mines Minerals Yearbook 1967, volume III, p. 385-390.

[86] Ela, Robert E., 1970, *The Mineral Industry of Maine,* US Bureau of Mines Minerals Yearbook 1968, volume III, p. 353-358.

[87] Ela, Robert E., 1971, *The Mineral Industry of Maine,* US Bureau of Mines Minerals Yearbook 1969, volume III, p. 361-367.

Elwell, Edward H., 1881, Portland and Vicinity, W. S. Jones and Loring, Short, and Harmon, pp. 142.

Emery, Ruby C., 1980, *Pinhook and Lines,* privately published.

Emery, Ruby C., 1988, *History of Woodstock, Maine, 1797-1988,* privately published.

Emery, Ruby C., 2001, Hamlin's Gore, 1816-1973, Woodstock Historical Society.

[253] Evans, James G. and Moyle, Phillip R., 2006, *U. S. Garnet Production,* Contributions to Industrial-Mineral Research, Bulletin 2209-L, U. S. Geological Survey, pp. 54.

[88] Farrington, Oliver Cummings and Tillotson, E. W., Jr., 1908, *Notes on Various Minerals in the Museum Collection,* Field Columbian Museum (Fieldiana), v. 3, p. 131-163.

[89] Feuchtwanger, Lewis, 1832, *Cabinet of Minerals, &c.,* American Journal of Science, first series, v. 22, p. 180.

[90] Fisher, Irving S., 1962, *Petrology and Structure of the Bethel Area, Maine,* Geological Society of America Bulletin, v. 73, p. 1495-1419.

Foster, Angela, 2004, Vital Records of Hartford, Maine, Picton Press, Camden, Maine, p. 218.

[91] Frank, David G., Galloway, John P., Weathers, Judy, Kiilsgaard, Thor H., and Wallis, John, 2003, *Index to the United States Minerals Exploration Assistance Records from the DMA, DMEA, OME, Mineral Exploration Programs, 1950-1974,* U. S. Geological Survey Open-File Report 03-94, pp. 96.

[92] Gedney, Edwin K. and Berman, Harry, 1929, *Huge Beryl Crystals at Albany, Maine,* Rocks and Minerals, v. 4, p. 78-80.

[93] Gibson, David and Lux, Daniel R., 2006, *Plutonic Style And Tectonic Setting Of Early Devonian Plutons, Central Maine,* Geological Society of America Abstracts with Programs, v. 38 (2), p. 31.

Goldschmidt, Victor, 1916, Atlas der Krystallformen, Datolith-Feldspat Gruppe, volume III plates, C. F. Winter Verlag, Heidelberg, Germany, 247 plates.

Goldschmidt, Victor, 1918, Atlas der Krystallformen, volume V, C. F. Winter Verlag, Heidelberg, Germany, 123 plates + 199 pages text.

Gordon College Alumnus, 1974, *"Mr. Gordon." Dr. Edwin K. Gedney Retires,* Gordon College Alumnus, Spring 1974, p. front cover, 2-5, 8-9.

[94] Goreva, Julia, Ma, Chi, and Rossman, George, 2001, *Fibrous Nano-inclusions in Massive Rose Quartz: The Origin of Rose Coloration,* American Mineralogist, v.

86, p. 466-472.

[95] **Gregory, Gardiner E.**, 1968, *The Bumpus Mine, Albany, Maine,* **Rocks and Minerals,** v. 43, p. 904-907.

[96] **Griffith, Robert F.**, 1955, *Beryllium,* **Minerals Yearbook 1952. Metals and Minerals (Except Fuels), Volume 1, U. S. Bureau of Mines,** p. 203-214.

[97] **Guidotti, Charles V.**, 1965, **Geology of the Bryant Pond Quadrangle Maine, Quadrangle Mapping Series No. 3,** Maine Geological Survey, Department of Economic development, Augusta, Maine, pp. 116 + 4 maps.

[98] **Gustavson, Samuel A.**, 1949, *Minor Metals,* **Minerals Yearbook 1946, U. S. Bureau of Mines,** p. 1240-1263.

[99] **Hall, Frederick**, 1824, **Catalogue Minerals, Found in the State of Vermont and in the Adjacent States: together with the Localities, including a number of the Most Interesting Minerals, which have been Discovered in Other Parts of the United States; Arranged Alphabetically,** P. B. Goodsell, Hartford, Connecticut, pp. 44.

[100] **Hamlin, Augustus Choate**, 1873b, **The Tourmaline. Its relation as a gem; its complex nature; its wonderful physical properties, etc., etc.; with special reference to the beautiful and matchless crystals found in the state of Maine**, J. R. Osgood, Boston, pp. 107 + 4 plates.

[101] **Hamlin, Augustus Choate**, 1895, **The History of Mount Mica, U.S.A. and its wonderful deposits of matchless tourmalines,** Augustus C. Hamlin, Bangor, ME, pp. 72 + 50 plates.

[102] **Hamlin, Elijah L.**, September 30, 1824, *Sketches of the Mineralogy and Geology of Oxford County*, **Oxford Observer.**

[103] **Hess, Frank L.**, 1925, *The Natural History of the Pegmatites,* **Engineering and Mining Journal,** v. 120, p. 289-298.

[104] **Hess, Frank L., Whitney, Roscoe J., Trefethen, Joseph, and Slavin, Morris**, 1943, **The Rare Alkalis in New England, U. S. Bureau of Mines Information Circular 7232,** pp. 51.

[105] **Horton, F. W. and Stoddard, Blanche H.**, 1934, *Mica,* **Minerals Yearbook 1932-1933, U. S. Bureau of Mines,** p. 787-794.

[106] **Horton, F. W. and Stoddard, Blanche H.**, 1934a, *Mica,* **Minerals Yearbook 1934, U. S. Bureau of Mines,** p. 1057-1074.

[107] **Horton, F. W. and Stoddard, Blanche H.**, 1935, *Mica,* **Minerals Yearbook1935, U. S. Bureau of Mines,** p. 1177-1186.

[108] **Horton, F. W. and Stoddard, Blanche H.**, 1936, *Mica,* **Minerals Yearbook 1936, U. S. Bureau of Mines,** p. 1037-1046.

[109] **Hughes, H. Herbert and Middleton, Jefferson**, 1932, *Feldspar,* **Mineral Resources of the United States 1930, Part II – Non-metals, U. S. Bureau of Mines,** p. 137-149.

[110] **Hughes, H. Herbert and Middleton, Jefferson**, 1932a, *Feldspar,* **Mineral Resources of the United States 1931, Part II – Non-metals, U. S. Bureau of Mines,** p. 179-190.

[110] **Hunt, Thomas Sterry**, 1875, *On Granites and Granitic Vein-Stones,* **Chemical and Geological Essays,** James R. Osgood and Company, Boston, p. 200.

[111] **Hurlbut, Cornelius S., Jr.**, 1945, *Memorial of Harry Berman,* **American Mineralogist,** v. 30, p. 124-129.

[112] **Jackson, Charles Thomas**, 1838, **Second Annual Report on the Geology of the Public Lands, Belonging to the Two States of Massachusetts and Maine,** Luther Severance, Augusta, pp. 100 + 9 plates + xxxviii.

[113] **Jacob, Walter W.**, 2000, *Stanley Non-sparking Beryllium Copper Tools,* **Chronicle of the Early American Industries Association Inc.,** v. 53 (3), p. 105-107.

[114] **Jahns, Richard H.**, 1953, *The Genesis of Pegmatites. I. Occurrence and Origin of Giant Crystals,* **American Mineralogist,** p. 563-598.

[115] **Johnson, Bertrand L. and Cornthwaite, M. A.**, 1937, *Mica,* **Minerals Yearbook 1937, U. S. Bureau of Mines,** p. 1399-1411.

[249] **Kantz, Maurice**, 1977, *Recollections (Albert Kimball),* **Bittersweet,** v.1 #1, November, 1977, p. 16-17, 28.

[116] **Katz, Frank J.**, 1913, *Feldspar,* **Mineral Resources of the United States, Calendar Year 1912,** U. S. Geological Survey, p. 1007-1015.

[117] **Katz, Frank J.**, 1914, *Feldspar,* **Mineral Resources of the United States, Calendar Year 1913, Part II Non-metals,** U. S. Geological Survey, p. 145-151.

[118] **Kaufman, Alvin and Cleary, C. Geraldine**, 1957, *The Mineral Industry of Maine,* **US Bureau of Mines Minerals Yearbook 1954, volume III,** p. 509-515.

[119] **Keller, Bob**, accessed 2006, *The La Quinta Inn, Back on the Strip, The Executive Inn,* **Snapshots from the 1996 Tucson Show,** http://www.rockhounds.com/tucson-show/reports/tucson96/snapsh10.shtml.

[120] **Kendall, James**, 1929, **Smith's College Chemistry, revised edition,** The Century Company, New York, pp. 759.

[121] **Kennedy, J. S. and O'Meara, R. G.**, 1948, **Flotation of Beryllium Ores, Report of Investigations 4166,** USBM, pp. 18.

[122] **King, Vandall T.**, 1980, **Distribution of Alkali and Alkaline-Earth Elements in a Newry, Maine Pegmatite,** M.S. thesis, State University of New York at Buffalo, privately published, Rochester, NY, pp. 131.

[123] **King, Vandall T.**, 2000, **Mineralogy of Maine, Mining History, Gems, and Geology,** Maine Geological Survey, Augusta, Maine, pp. 524 + 25 plates.

[124] **King, Vandall T.**, 2006, *Primary Sources and a Partial Analysis of a Revered History Book –This [sic] History of Mount Mica,* **Journal of the Geo-Literary Society,** v. 21 (3), p. 5-23.

[125] **King, Vandall T.**, unpublished a, **History of Mining Activity at Newry, Rumford, the Goldfields, and Nearby Localities, Oxford County, Maine** (tentative title).

[126] **King, Vandall T.**, unpublished b, **Mount Mica, Paris,**

District (tentative title)..

[127] **King, Vandall T. and Foord, Eugene E.**, 1994, **Mineralogy of Maine. Volume 1: Descriptive Mineralogy,** Maine Geological Survey, Augusta, ME, pp. 418 + 18 color plates + 80 black and white plates.

[128] **King, Vandall T. and Foord, Eugene E.**, 2000, *Mineralogy of Maine – Addenda to Volume 1,* in King, Vandall T. (editor) **Mineralogy of Maine. Volume 2: Mining History, Gems, and Geology,** Maine Geological Survey, Augusta, ME, p. 427-512.

[129] **King, Vandall T. and Sanderson, David**, 2005, **George Howe of Norway, Maine, 1860-1950, A Brief Biography,** Norway Summer Festival Committee, pp. 8.

[130] **King, Vandall T., Shaub, Benjamin M., and Shaub, Mary C.**, unpublished, **George Howe, Shavey Noyes, the Town of Norway, and the Stoneham Mining District.**

[131] **Knapp, Wallace**, 1952, *{Bumpus Minerals Collected], Maine - World News On Mineral Occurrences,* **Rocks and Minerals,** v. 27, p. 488.

[132] **Krichich, Joseph**, 1964, *The Mineral Industry of Maine,* **US Bureau of Mines Minerals Yearbook 1963, volume III,** p. 519-526.

[133] **Krichich, Joseph and Otte, Mary E.**, 1963, *The Mineral Industry of Maine,* **US Bureau of Mines Minerals Yearbook 1962, volume III,** p. 517-524.

Kunz, George Frederick, 1885, *Precious Stone,* **Mineral Resources of the United States, Calendar Years 1883 and 1884,** U. S. Geological Survey, p. 723-782.

Kunz, George Frederick, 1887, *Precious Stone,* **Mineral Resources of the United States, Calendar Year 1886,** p. 596-605.

[134] **Kunz, George Frederick**, 1892, **Gems and Precious Stones of North America, second edition,** Scientific Publishing Company, pp. 367 + 8 plates.

[135] **Kunz, George Frederick**, 1930, *Precious and Semi-precious Stones,* **Mineral Industry During 1929,** v. 38, p. 530-554.

[136] **Kusler, David J.**, 1965, *The Mineral Industry of Maine,* **US Bureau of Mines Minerals Yearbook 1964, volume III,** p. 483-491.

Landers, Helen, 1955, *My Machine and I,* **Oxford County Mineral and Gem Association Yearbook 8,** p. 10-11.

Landes, Kenneth Knight, 1925, *The Paragenesis of the Granite Pegmatites of Central Maine,* **The American Mineralogist,** v. 10: 355-411.

Landes, Kenneth Knight, 1925a, **A Study of the Paragenesis of the Pegmatites of Central Maine,** Ph.D. Dissertation.

Landes, Kenneth Knight, 1933, *Origin and Classification of Pegmatites,* **The American Mineralogist,** 18(2): 33-56.

[137] **Lasmanis, Raymond**, 1958, *My Mineral Activities in 1957,* **Rocks and Minerals,** v. 33, p. 500-503.

Lapham, William Berry, 1882, The History of Woodstock, Maine: with family sketches and an appendix, S. Berry, Portland Me., pp. 315.

Lucas, Frederic A., 1936, **Guide to the Exhibition Halls, 16th edition,** American Museum of Natural History, NY, p. 141.

[138] **Ma, Chi, Goreva, Julia and Rossman, George**, 2001, *Colored Varieties of Quartz Arising from Inclusions,* **Eleventh Annual V. M. Goldschmidt Conference,** #3676.

[139] **Ma, Chi, Goreva, Julia and Rossman, George**, 2002, *Fibrous Nano-inclusions in Massive Rose Quartz: HRTEM and AEM Investigations,* **American Mineralogist,** v. 87, p 269-276.

[140] **Maillot, E. E., Boos, Margaret Fuller, Mosler, McHenry**, 1949, **Investigations of the Black Mountain Beryl Deposit, Oxford County, Maine, USBM Report of Investigations 4412,** pp. 10.

[141] **Matthews, Allan F.**, 1943, *Minor Metals,* **Minerals Yearbook 1941, U. S. Bureau of Mines,** p. 793-810.

[142] **Matthews, Allan F.**, 1943a, *Minor Metals,* **Minerals Yearbook 1942, U. S. Bureau of Mines,** p. 819-830.

[143] **Matthews, Allan F.**, 1945, *Minor Metals,* **Minerals Yearbook 1943, U. S. Bureau of Mines,** p. 816-831.

[144] **Matthews, Allan F.**, 1948, *Minor Metals,* **Minerals Yearbook 1946, U. S. Bureau of Mines,** p. 1266-1284.

[145] **Merrill, Lucius H. and Perkins, Edward H.**, 1930, *Report on Feldspar Quarries Visited Aug. 21-24, 1929,* **First Annual Report on the Geology of the State of Maine,** Augusta, p. 18-23.

[146] **Metcalf, Robert W.**, 1935, *Feldspar,* **Minerals Yearbook 1935, U. S. Bureau of Mines,** p. 1107-1113.

[147] **Metcalf, Robert W.**, 1936, *Feldspar,* **Minerals Yearbook 1936, U. S. Bureau of Mines,** p. 981-987.

[148] **Metcalf, Robert W.**, 1937, *Feldspar,* **Minerals Yearbook 1937, U. S. Bureau of Mines,** p. 1353-1362.

[149] **Metcalf, Robert W.**, 1938, *Feldspar,* **Minerals Yearbook 1938, U. S. Bureau of Mines,** p. 1211-1220.

[150] **Metcalf, Robert W.**, 1939, *Feldspar,* **Minerals Yearbook 1939, U. S. Bureau of Mines,** p. 1297-1307.

[151] **Metcalf, Robert W.**, 1940, *Feldspar,* **Minerals Yearbook 1940, U. S. Bureau of Mines,** p. 1353-1362.

[152] **Metcalf, Robert W.**, 1943, *Feldspar,* **Minerals Yearbook 1941, U. S. Bureau of Mines,** p. 1417-1426.

[153] **Metcalf, Robert W.**, 1958, *The Mineral Industry of Maine,* **US Bureau of Mines Minerals Yearbook 1955, volume III,** p. 507-515.

[154] **Metcalf, Robert W.**, 1958a, *The Mineral Industry of Maine,* **US Bureau of Mines Minerals Yearbook 1956, volume III,** p. 545-554

[155] **Metcalf, Robert W. and Otte, Mary E.**, 1958, *The Mineral Industry of Maine,* **US Bureau of Mines Minerals Yearbook 1957, volume III,** p. 519-527.

[156] **Metcalf, Robert W. and Otte, Mary E.**, 1959, *The Mineral Industry of Maine,* **US Bureau of Mines Minerals Yearbook 1958, volume III,** p. 439-446

[157] **Metcalf, Robert W. and Otte, Mary E.**, 1960, *The Mineral Industry of Maine,* **US Bureau of Mines Minerals Yearbook 1959, volume III,** p. 465-473.

[158] **Metcalf, Robert W. and Otte, Mary E.**, 1961, *The Min-*

[158] *eral Industry of Maine,* **US Bureau of Mines Minerals Yearbook 1960, volume III,** p. 475-484.

[159] **Metcalf, Robert W. and Otte, Mary E.,** 1962, *The Mineral Industry of Maine,* **US Bureau of Mines Minerals Yearbook 1961, volume III,** p. 493-502.

[160] **Middleton, Jefferson,** 1927, *Feldspar,* **Mineral Resources of the United States 1924, Part II – Non-metals, U. S. Bureau of Mines,** p. 19-25.

[161] **Middleton, Jefferson,** 1928, *Feldspar,* **Mineral Resources of the United States 1925, Part II – Non-metals, U. S. Bureau of Mines,** p. 39-46

[162] **Middleton, Jefferson,** 1929, *Feldspar,* **Mineral Resources of the United States 1926, Part II – Non-metals, U. S. Bureau of Mines,** p. 109-117.

[163] **Middleton, Jefferson,** 1930, *Feldspar,* **Mineral Resources of the United States 1927, Part II – Non-metals, U. S. Bureau of Mines,** p. 57-65.

[164] **Middleton, Jefferson,** 1931, *Feldspar,* **Mineral Resources of the United States 1928, Part II – Non-metals, U. S. Bureau of Mines,** p. 67-80.

[165] **Mikhailov, Viktor N.,** 1996, **USSR Nuclear Weapons Tests and Peaceful Nuclear Explosions: 1949 through 1990,** Ministry of the Russian Federation for Atomic Energy, and Ministry of Defense of the Russian Federation, (not seen).

[166] **Miller, Robert and Wing, Lawrence,** 1945, *Appendix A,* in Trefethen, Joseph M., **Report of the State Geologist 1943-1944,** Maine Development Commission, Augusta, ME, pp. 29-64.

Mitchell, Harry Edward and Davis, B. V., 1906, **The Town Register: Waterford, Albany, Greenwood, East Stoneham,** H. E. Mitchell Company, Brunswick, ME, p. 31-48.

Mitchell, Daggett, and Weston, 1904, The Topsham and Bowdoinham Register for 1905, H. E. Mitchell Company, Brunswick, Maine, p. 29.

[167] **Montgomery, Arthur,** 1951, *The Harding Pegmatite – Remarkable Storehouse of Massive White Beryl,* **Mining World,** v. 13, p. 32-35.

[168] **Morrill, Philip,** 1939, *The Maine Pegmatite Belt,* **Rocks and Minerals,** v. 14, p. 272-274.

Morrill, P., and a Lot of Other People, 1958, **Maine Mines and Minerals. Western Maine, volume 1,** Dillingham Natural History Museum, pp. 80.

[169] **Morrill, Philip,** 1955, **Maine Mines and Mineral Locations,** privately published, pp. 45.

[170] **Morrill, Philip,** 1966, **Tales of a Homemade Naturalist. The Maine Diaries of Herbert M. W. Haven,** Park View Press, Harrisonburg, VA, pp. 153.

[171] **Morrill, Philip and a Lot of Other People,** 1958, **Maine Mines and Minerals, volume 1, Western Maine,** Winthrop Mineral Shop, Winthrop, ME, pp. 80.

[172] **Mote, Richard H.,** 1954, *The Mineral Industry of Maine,* **US Bureau of Mines Minerals Yearbook 1953, volume III,** p. 475-484.

[173] **Myers, W. M. and Stoddard, Blanche H.,** 1928, *Mica,* **Mineral Resources of the United States 1925, Part II** – Non-metals, U. S. Bureau of Mines, p. 181-193.

[174] **National Jeweler,** 1992, *Morganite: It's Pink Emerald Now. Gemstone News,* **National Jeweler,** v. 36 (#12), p. 22.

[175] **Needleman, Stanley,** 1954, *Beryllium,* **Minerals Yearbook 1951, U. S. Bureau of Mines,** p. 209-218.

[176] **Neumann, G. L.,** 1952, **Bumpus Pegmatite Deposit, Oxford County, Maine, U. S. Bureau of Mines Report of Investigations 4862,** pp. 15.

[177] **Nighman, C. E.,** 1946, *Minor Metals,* **Minerals Yearbook 1944, U. S. Bureau of Mines,** p. 807-822.

[178] **Nighman, C. E.,** 1947, *Minor Metals,* **Minerals Yearbook 1945, U. S. Bureau of Mines,** p. 811-832.

[179] **Ordway, Richard J.,** 1951, *Radioactivity of Some Maine Pegmatites,* **Report of the State Geologist 1949-1950,** Maine Development Commission, Augusta, ME, p. 91-106.

OCMGAY Oxford County Mineral and Gem Association Yearbook Editor, 1962, *Sifting More Drift,* **Oxford County Mineral and Gem Association Yearbook 15,** p. 13, 25.

[180] **Page, James J. and Larrabee, David M.,** 1962, **Beryl Resources of New Hampshire, U. S. Geological Survey Professional Paper 353,** pp. 49 + 16 plates.

Palache, Charles, 1924, *The Chrysoberyl Pegmatite of Hartford, Maine,* **American Mineralogist,** v. 9, p. 217-221.

[181] **Palache, Charles,** 1932, *The Largest Crystal,* **American Mineralogist,** v. 17, p. 362-363.

Penfield, Samuel Lewis, 1894, *On the Crystallization of Herderite,* **American Journal of Science, third series,** v. 47, p. 329-339.

[182] **Perham, Frank C.,** 1961, *A Find of Rare Pseudo-cubic Quartz Crystals in Maine,* **Rocks and Minerals,** v. 36, p. 240-241.

[183] **Perham, Frank C.,** 1963, *Mining Notes of 1963,* **Oxford County Mineral and Gem Association Yearbook,** v. 16, p. 24.

[184] **Perham, Frank C.,** 1965, *Waisanen Mine Operation – Summer 1963,* **Rocks and Minerals,** v. 39, p. 341-347.

[185] **Perham, Frank C.,** 1965, *Newry Hill Area Workings - 1964,* **Oxford County Mineral and Gem Association Yearbook,** v. 17, p. 16.

[186] **Perham, Frank C.,** 2000, *Mining Experiences and Observations at Mount Mica in 1964 and 1965,* in King, Vandall T. (editor) **Mineralogy of Maine. Volume 2: Mining History, Gems, and Geology,** Maine Geological Survey, Augusta, ME, p. 271-276.

[187] **Perham, Harold C.,** 1975, **Book No. 4 – Maine Families Perham-Pratt-Fogg-Nelson-Williams Families,** privately published.

[188] **Perham, Jane C.,** 1987, **Maine's Treasure Chest. Gems and Minerals of Oxford County, second edition,** Quicksilver Publications, West Paris, ME, pp. 269.

Perham, Stanley I., 1955, *Notes on Maine Gems and Minerals,* **Oxford County Mineral and Gem Association Yearbook 8,** p. 32.

Perham, Stanley I., 1959, *Maine Meteorites,* **Oxford County Mineral and Gem Association Yearbook 8,** p. 27-28.

[189] **Perham, Stanley I.,** 1971, *1970 Mineral Notes,* **Oxford County Mineral and Gem Association Yearbook, Volume 21,** p. 24-26.

[190] **Perham Stevens, Jane C.,** 1972, **Maine's Treasure Chest. Gems and Minerals of Oxford County,** Perham's Maine Mineral Store, West Paris, ME, pp. 216.

[191] **Petkof, Benjamin,** 1985, *Beryllium,* **Mineral Facts and Problems: U.S. Bureau of Mines Bulletin 675,** p. 75-82.

[192] **Pink, Louis H. and Delmadge, Rutherford E.,** 1957, **Candle in the Wilderness. A Centennial History of the St. Lawrence University,** Appleton-Century-Crofts, Inc., New York, pp. 304.

[193] **Rask, William C.,** 1983, *Beryllium,* Office of Strategic Planning & Analysis, Albuquerque, NM, p. 1-17 (full report not released).

[194] **Ratkevich, Ron,** 1981, *Trinitite: The Origin of a Rare Atomic Mineral,* **Lapidary Journal,** v. 34, p. 2276-2278.

[195] **Rickwood, P. C.,** 1981, *The Largest Crystal,* **American Mineralogist,** v. 66, p. 885-907.

[196] **Roberts, Willard Lincoln and Rapp, George, Jr.,** 1965, **Mineralogy of the Black Hills, South Dakota School of Mines Bulletin 18,** Rapid City, SD, pp. 268.

[197] **Robinson, Samuel,** 1825, **A Catalogue of American Minerals, with their Localities; including all which are known to exist in the United States and British Provinces, and having the Towns, Counties, and Districts in Each State and Province Arranged Alphabetically. With an Appendix, containing Additional Localities and a Tabular View,** Cummings, Hilliard, & Co., Boston, pp. 316.

[198] **Rogers, H. O. and Galiher, Claude,** 1934, *Feldspar,* **Minerals Yearbook 1933, U. S. Bureau of Mines,** p. 735-742.

[199] **Rogers, H. O. and Metcalf, Robert W.,** 1934, *Feldspar,* **Minerals Yearbook 1934, U. S. Bureau of Mines,** p. 999-1007.

[200] **Salo, Ruth Brown and Berndt, Julia Bumpus,** 1988, **Some Bumpus Families of Hebron, Maine,** privately published.

[201] **Sampter, E. Lawrence,** 1945, *Our Report for the Summer of 1945,* **Rocks and Minerals,** v. 20, p. 594-595..

[202] **Sampter, E. Lawrence,** 1946, *Our Report for the Summer of 1946,* **Rocks and Minerals,** v. 21, p. 743-745.

[203] **Sampter, E. Lawrence,** 1947, *Our Report for the Summer of 1947,* **Rocks and Minerals,** v. 22, p. 1103-1106.

[204] **Sampter, E. Lawrence,** 1953, *Oxford County, Maine – 1952,* **Rocks and Minerals,** v. 28, p. 236.

[205] **Schaller, Waldemar T.,** 1916, *Mineralogical Notes, Series 3,* **U. S. Geological Survey Bulletin 509,** pp. 115.

[251] **Schlenker, Jon A., Wetherington, Norman A., and Wilkins, Austin H.,** 1988, **In the Public Interest. The Civilian Conservation Corps in Maine,** University of Maine at Augusta Press, Augusta, pp. 164.

Scoville, O. J., 1937, *Liquidating Town Government in Decadent Rural Areas of Maine,* **Journal of Land and Public Utility Economics,** v. 13 (3), pp. 285-291.

SDFWP, 1938, **A South Dakota Guide,** Federal Writer's Project, HE Publishing Company, Pierre, SD, pp. 416 + plates.

[206] **Shainin, Vincent E.,** 1948, **Economic Geology of Some Pegmatites in Topsham, Maine, Maine Geological Survey Bulletin 5,** Augusta, ME, pp. 32.

[207] **Shainin, Vincent E.,** 1949, Preliminary Report of the Pegmatites on Red Hill, Rumford, Maine, **Report of the State Geologist 1947-1948,** Maine Development Commission, Augusta, ME, p. 90-102.

[208] **Shaub, Benjamin Martin,** 1960, *Notes on the Dunton Gem Quarry, Newry Hill,* **Oxford County Mineral and Gem Association Yearbook,** v. 13, p. 27-32.

Shaub, Benjamin Martin, 1961, *Seeing Things in Rocks and Minerals? Why?,* **Rocks and Minerals,** v. 36, p. 233—234.

[209] **Shepard, Charles U.,** 1830, *Mineralogical Journey in the Northern Parts of New England,* **American Journal of Science, first series,** v. 17, p. 353-360, v. 18, p. 126-136, 289-303.

[210] **Sinkankas, John,** 1959, **Gemstone of North America,** D. Van Nostrand Company, Inc., Princeton, NJ, pp. 675 + 7 color plates.

[211] **Sinkankas, John,** 1981, **Emerald and Other Beryls,** Chilton Book Company, Radnor, PA, pp. 665 + 12 color plates.

[212] **Sinkankas, John,** 1997, **Gemstones of North America, volume III,** Geoscience Press, Inc., Tucson, AZ, pp. 526.

[213] **Smith, George Otis,** 1904, *Quartz Veins in Maine and Vermont,* **Contributions to Economic Geology, 1903, U. S. Geological Survey Bulletin 225,** p. 81-88.

[214] **Spence, Hugh Swaine,** 1929, *[Beryl Crystals 18 Ft. Long],* **Engineering and Mining Journal,** v. 128, p. 858.

Sterrett, Douglas B., 1907, *Precious Stones,* **Mineral Resources of the United States Calendar Year 1906,** U. S. Geological Survey, p. 1213-1252.

[215] **Sterrett, Douglas B.,** 1911, *Gems and Precious Stones,* **U. S. Geological Survey, Mineral Resources of the United States, Calendar Year 1909, Part II Nonmetallic Products,** p. 739-808.

[216] **Sterrett, Douglas B.,** 1923, **Mica Deposits of the United States Geological Survey Bulletin 740,** pp. 341.

[217] **Stewart, Glenn W.,** 1952, *Logs of Diamond Drill Holes at Bumpus Quarry, Oxford County, Maine,* in Neumann, G. L., 1952, **Bumpus Pegmatite Deposit, Oxford County, Maine, U. S. Bureau of Mines Report of Investigations 4862,** p. 7-15.

[218] **Stoddard, Blanche H.,** 1930, *Mica,* **Mineral Resources of the United States 1927, Part II – Nonmetals, U. S. Bureau of Mines,** p. 187-198.

[219] **Stoddard, Blanche H.,** 1931, *Mica,* **Mineral Re-**

sources of the United States 1928, Part II – Non-metals, U. S. Bureau of Mines, p. 607-614.

[220] **Stoddard, Blanche H.**, 1932, *Mica,* **Mineral Resources of the United States 1929, Part II – Non-metals, U. S. Bureau of Mines,** p. 373-388.

[221] **Stoddard, Blanche H.**, 1933, *Mica,* **Mineral Resources of the United States 1930, Part II – Non-metals, U. S. Bureau of Mines,** p. 387-395.

[222] **Stoddard, Blanche H.**, 1933a, *Mica,* **Mineral Resources of the United States 1931, Part II – Non-metals, U. S. Bureau of Mines,** p. 279-287.

[223] **Sturtevant, Lawrence M.**, 1948, Ezekiel Holmes and his Influence, 1801-1865, MA Thesis, University of Maine, Orono, pp. 87.

[224] **Swan, Ralph F.**, 1969, *Field Trip to the Bumpus Mine,* **Oxford County Mineral and Gem Association Yearbook, Volume 20,** p. 6.

[225] **Templeton, Douglas M.**, 2004, *Mechanisms of Immunosensitization to Metals,* **Pure and Applied Chemistry,** v. 76, p. 1255-1268.

[226] **Thomas, William B. S.**, 1947, *A National Stockpile of Strategic Minerals,* **Rocks and Minerals,** v. 22, p. 315.

[227] **Thompson, Woodrow B., Wintringham, Neil A., and King, Vandall T.**, 2000, *Maine Mineral Locality Index,* in King, V. T. (editor), **Mineralogy of Maine, volume 2, Mining History, Gems, and Geology,** Maine Geological Survey, Augusta, p. 355-426.

[228] **Trefethen, Joseph M.**, 1945, **Report of the State Geologist 1943-1944,** Maine Development Commission, Augusta, ME, pp. 64 + 2 maps.

[229] **True, Nathaniel Tuckerman**, 1861, *Geology of Bethel,* **Sixth Annual Report of the Secretary of the Maine Board of Agriculture 1861,** Stevens & Sayward, Augusta, ME, p. 221-225 (reprinted from *Bethel Courier).*

[230] **True, Nathaniel Tuckerman**, 1869, *New Localities of Minerals in Maine,* **Proceedings of the Portland Society of Natural History,** v. 1, p. 163-165.

Twitty, Eric, 2001, **Blown to Bits in the Mine. A History of Mining & Explosives in the United States,** Western Reflection Publishing Company, Ouray, Colorado, pp. 208.

[231] **Tyler, Paul M.** , 1937, *Minor Metals,* **Minerals Yearbook 1937, U. S. Bureau of Mines,** p. 759-785.

[232] **Tyler, Paul M.**, 1938, *Minor Metals,* **Minerals Yearbook 1938, U. S. Bureau of Mines,** p. 671-685.

[233] **Tyler, Paul M.**, 1939, *Minor Metals,* **Minerals Yearbook 1939, U. S. Bureau of Mines,** p. 747-761.

[234] **Tyler, Paul M.**, 1940, *Minor Metals,* **Minerals Yearbook 1940, U. S. Bureau of Mines,** p. 759-722.

[235] **Tyler, Paul M. and Petar, A. V.**, 1934, *Minor Metals: Beryllium, Bismuth, Cadmium, Cobalt, Selenium, Tantalum, Tellurium, Titanium, and Zirconium,* **Minerals Yearbook 1932-1933, U. S. Bureau of Mines,** p. 347-367.

[236] **Tyler, Paul M. and Petar, A. V.**, 1934a, *Minor Metals: Beryllium, Bismuth, Cadmium, Cobalt, Selenium and Tellurium, Tantalum and Columbium, Titanium, and Zirconium,* **Minerals Yearbook 1934, U. S. Bureau of Mines,** p. 517-541.

[237] **Tyler, Paul M. and Petar, A. V.**, 1935, *Minor Metals: Beryllium, Bismuth, Cadmium, Cobalt, Selenium, Tantalum, Tellurium, Titanium, and Zirconium,* **Minerals Yearbook 1935, U. S. Bureau of Mines,** p. 581-61.

[238] **Tyler, Paul M. and van Siclen, A. P.**, 1936, *Beryllium, Bismuth, Cadmium, Cobalt, Selenium, Tantalum and Columbium, Tellurium, Titanium, and Zirconium,* **Minerals Yearbook 1936, U. S. Bureau of Mines,** p. 525-541.

[239] **Tyler, Paul M. and Warner, K. G.**, 1939, *Mica,* **Minerals Yearbook 1939, U. S. Bureau of Mines,** p. 1343-1353.

[240] **Tyler, Paul M. and Warner, K. G.**, 1940, *Mica,* **Minerals Yearbook 1940, U. S. Bureau of Mines,** p. 1403-1418.

[241] **Verrill, Addison Emery**, 1861, *[Report on Maine Minerals],* **Proceedings of the Boston Society of Natural History,** v. 7, p. 423-424.

[242] **Waldschmidt, W. A.**, 1919, *The Largest Known Beryl Crystal,* **Pahasapa Quarterly,** v. 9, p. 11-16 (not seen).

[243] **Watts, Arthur S.**, 1916, **The Feldspars of the New England and Northern Appalachian States, U.S. Bureau of Mines Bulletin 92,** pp. 187.

[254] **Watts, Arthur S.**, 1922, *Feldspar,* **The Mineral Industry. Its Statistics, Technology, and Trade During 1921, Volume XXX, McGraw-Hill Book Company, Inc., New York,** p. 226-228.

[244] **Wentworth, Bertram F.**, 1977, **Waterford, Maine. 1875-1976,** Waterford Historical Society, Waterford, ME, pp. 308 + 2 plates.

Wheeler, George Augustus and Wheeler, Henry Warren, 1878, **History of Brunswick, Topsham, and Harpswell, Maine including the Ancient Territory known as Pejepscot,** Alfred Mudge & Son, Boston, pp. 959 + plate.

[245] **Wilhelm, Sandy**, 1976, *Former Maine Guide Tracks Bees,* **Bethel Citizen,** September 23, 1976.

[246] **Wilkinson, V. June**, 1942, *The Middle Years,* **Fryeburg Academy 1792-1942 Sesquicentennial,** p. 36-39.

Williams, Albert, J., 1888, **Useful Minerals of the United States,** U. S. Geological Survey, Government Printing Office, Washington, DC, pp. 812.

[247] **Wintringham, Neil A.**, 1955, **A Week with Maine Minerals, second edition,** Holiday Hill Press, Mountainside, NJ, pp. 174.

[248] **Wintringham, Neil A.**, 2000, *Maine Mineral Checklists and Guidebooks – The Collector's Literature: A Review and Personal Evaluation,* in King, V. T. (editor), **Mineralogy of Maine. Volume 2: Mining History, Gems, and Geology,** Maine Geological Survey, Augusta, ME, p. 341-354.

Woodbury, David Oakes, 1957a, *That Gem Man at Trap Corner,* **Down East,** v. 3 (6): p. 14-15, 32.

Woodbury, David Oakes, 1957b, *Gem Man at Trap Corner,*

Reader's Digest, May 1957, p. 167-169, 171-172.
[249] **Zodac, Peter,** 1929, *A Real Rock Crystal,* **Rocks and Minerals,** v. 4, p. 119.

Oxford County Registry of Deeds

[300] March 16, 1967, OCRD v. 659, p. 154.
[301] February 26, 1927, OCRD, v. 376, p. 346-347.
[302] May 18, 1927, OCRD v. 376, p. 398-399.
[303] March 1, 1926, OCRD v. 376, p. 118-119.
[304] September 18, 1925, OCRD v. 366, p. 542-543.
[305] October 4, 1927, OCRD v. 376, p. 522-523.
[306] September 21, 1927, OCRD v. 376, p. 510-511.
[307] December 31, 1927, OCRD v. 376, p. 589-590.
[308] September 13, 1951, OCRD v. 508, p. 429-431.
[309] October 7, 1930, OCRD v. 389, p. 152-153.
[310] September 11, 1930, OCRD v. 416, p. 372-373.
[311] November 13, 1930, OCRD v. 389, p. 198-201.
[312] November 21, 1930, OCRD v. 394, p. 305; also v. 394 p. 166.
[313] July 14, 1931, OCRD v. 389, p. 365.
[314] August 28, 1931, OCRD v. 389, p. 391-392.
[315] May 19, 1931, OCRD v. 389, p. 320-322.
[316] August 28, 1931, OCRD v. 389, p. 392-393.
[317] September 21, 1931 OCRD v. 403, p. 16-18.
[318] February 10, 1933, OCRD v. 403, p. 175-176.
[319] August 17, 1933, OCRD v. 403, p. 254-255.
[320] April 24, 1933, OCRD v. 401, p. 539.
[321] April 9, 1935, OCRD v. 408, p. 171.
[322] January 27, 1936, OCRD v. 422, p. 5.
[323] September 7, 1937, OCRD v. 409, p. 287-288.
[324] September 6, 1938, OCRD v. 417, p. 404.
[325] February 6, 1939, OCRD v. 417, p. 616-617.
[326] September 11, 1943, OCRD v. 444, p. 408-409.
[327] October 22, 1949 OCRD v. 496, p. 590-591.
[328] December 15, 1949, OCRD v. 497, p. 79-81.
[329] December 26, 1952, OCRD v. 463, p. 568-577.
[330] June 25, 1953, OCRD v. 530, p. 285-286.
[331] October 17, 1953, OCRD v. 530, p. 502.
[332] July 10, 1956 OCRD v. 559, p. 54-56.
[333] December 29, 1963, OCRD v. 530, p. 600.
[334] March 31, 1965, OCRD v. 654, p. 579-580.
[335] March 16, 1967, OCRD v. 659, p. 154.
[336] March 9, 1971, OCRD v. 700, p. 17-120.
[337] July 13, 1943, OCRD v. 409, p. 551.
[338] July 24, 1943, OCRD v. 409, p. 566-568.
[339] July 20, 1943, OCRD v. 409, p. 559-562.
[340] September 25, 1930, OCRD v. 389, p. 166.
[341] January 16, 1907, OCRD.
[342] October 10, 1935, OCRD v. 413, p. 448-449.
[343] February, 1925, OCRD v. 366, p. 423-424.
[344] July 29, 1943, OCRD v. 409, p. 559-562.
[345] December 26, 1952, OCRD v. 463, p. 568-577.
[346] July 24, 1943, OCRD v. 409, p. 566-568.
[347] July 24, 1943, OCRD v. 409, p. 564-566.
[348] July 13, 1943, OCRD v. 409, p. 551-553.
[349] July 13, 1959, OCRD v. 580, p. 207.

[350] June 6, 1937, OCRD v. 346, p. 326.
[351] January 10, 1940, OCRD v. 423, p. 418-421.
[352] August 19, 1925, OCRD v. 366, p. 528-529.
[353] February 17, 1927, OCRD v. 376, p. 340-342.
[354] August 27, 1927, OCRD v. 376, p. 496.
[355] September 23, 1927, OCRD v. 376, p. 511.
[356] OCRD v389, p155-156. (not seen)
[357] OCRD v389 p130-131 (not seen)
[358] Siddall, Cecil J., 1950, *United Feldspar & Minerals Corporation vs .Harry E. Bumpus et al.,* **Maine Reports 142, Cases Argued and Determined in the Supreme Judicial Court of Maine January 5, 1946 to June 16, 1947,** *United Feldspar & Minerals Corporation v. Bumpus et al.,* Anthoensen Press, Portland, p. 230-234.; **Cases Argued and Determined, Atlantic Reporter, second series,** 1947, v. 49, A. sd, p. 473-475.
[359] Office of Maine Secretary of State, volume 103, p.135-137.
[360] Office of Maine Secretary of State, volume 16, p. 359.
[361] Office of Maine Secretary of State, volume 7, p. 322.

Personal Communications

[400] Ben Conant, 2006
[401] Frank Perham, 2006
[402] Ava Bumpus, 2006
[403] Charles Guidotti, 1991
[404] Taisto Koskela, 1991
[405] Dana C. Douglass, Jr., 1995
[406] Richard V. Gaines, 1982
[407] Ben Shaub, 1992
[408] Jim Mann, 2006
[409] Jan Brownstein, 2006
[410] Addison Saunders, 2006
[411] John Bradshaw, 2006
[412] Bob Hinckley, 1999
[413] Barbara Inman, 2006
[414] Ober Kimball to Ben Shaub, 1958
[415] Bradford Hall, 1970
[416] Linc Page, 1995
[417] Tom Lewis, 2001
[418] Winsor Rippon, April 24, 2008. It has been widely reported that Harold Rippon was an teacher at Lynn High School in Massachusetts. Harold was born in Lynn and did teach in Lynn for the first few years of his career, but almost all of his career, including the time he developed the Mercropon faceting lap, was spent at Rice Grammar School, Boston.

Index

Brewster, Ralph Owen - 129
Bricks - 231, 251
Bridgton - 58, 163, 220, 256
Briggs, Joseph H. - 257
Briggs, Donald "Stone Rock" - 165
Brimstone Corner - 62
Brock's Service Station - 25
Brodie, G. H. - 249
Brown Quarry (Oxford County) - 6, 223, 238
Brown Quarry (Sagadahoc County) - 235
Brown, Daniel - 256
Brown, David D. - 31
Brown, Emma - 31
Brown, George H. - 255, 257
Brown, H. G. - 62
Brown, James - 257
Brown, Parker V. - 257
Brown, Randall - 174
Brown, Robert - 8
Brown, Samuel - 255
Brown, Steven - 174
Brown, W. M. - 255
Brown-Thurston Quarry - 249
Brownfield - 130, 251
Brownville - 28
Brunswick - 31, 40, 42, 73, 141, 212, 232, 235, 238, 249, 250, 266
Brush Beryllium Company - 245
Brush Wellman Company - 245
Brush, Charles, Jr. - 245
Bryant Pond - 49, 98, 189, 204, 205, 206, 207, 225, 251
Bryant, Priscilla (See Chavaree)
Bryant, Stearns J. - 124
Buchanan, Brenda - 213
Buchanan, Dave - 162
Buchanan, Frank - 213
Buchanan, Gaylene - 213
Buck & Baker - 235
Buck, Russell "Rouse" - 162
Buckfield - 5, 6, 8, 32, 34, 35, 44, 45, 47, 55, 57, 58, 59, 60, 64, 68, 75, 83, 84, 88, 102, 110, 116, 124, 130, 141, 142, 161, 166, 168, 180, 181, 184, 196, 236, 246, 247, 250, 251, 264, 265, 266, 267
Bumpus and Getchell Pharmacy - 263, 264
Bumpus Historical Mine LLC - 186, 259, 260, 261
Bumpus Mining District - 6, 7, 8, 98, 164, 193, 196, 224
Bumpus Quarry - 2,3, 5-8, 15, 17, 22, 23, 25, 26, 49, 51, 56, 57, 59, 63-70, 74, 77, 79-88, 92-99, 102, 104, 105, 107-111, 115, 117, 122-125, 127-140, 142, 144, 150-162, 165, 166, 168-172, 174-179, 183-186, 196, 198, 200, 201, 209, 211, 212, 219, 220, 221, 231, 232, 236, 237, 239, 240, 241, 244, 245, 249, 250, 260, 261, 263, 265, 274, 259, 260, 261
Bumpus, Anne Elizabeth Cummings Hazelton - 69
Bumpus, Ava M. Hutchinson - 8, 23, 53, 63, 64, 65, 66, 70, 87, 90, 129, 131, 137, 138, 139, 150, 151, 152, 153, 154, 260
Bumpus, Cora B. - 264
Bumpus, Daniel - 54, 74, 115
Bumpus, Dorothy P. (See Merriam)
Bumpus, Dulcina Rebecca (See Hibbs)
Bumpus, Edwin Charles - 23, 264
Bumpus, George Washington - 116, 261
Bumpus, Harlan Maurice - 23, 52, 53, 54, 64, 65, 66, 69, 70, 76, 84, 86, 87, 92, 95, 98, 127, 131, 132, 133, 137, 139, 151, 152, 264
Bumpus, Harold Perham - 116
Bumpus, Harry Erwin - 51, 53, 54, 63, 64, 65, 67, 69, 77, 82, 84, 88, 98, 104, 106, 108, 109, 112, 115, 116, 132, 136, 137, 138, 139, 142, 143, 150, 184, 264, 267
Bumpus, Hermon Carey - 88
Bumpus, Kenneth Harlan - 137, 139, 264
Bumpus, Laura Josephine Cummings - 3, 51, 52, 53, 54, 63, 64, 65, 67, 82, 84, 98, 104, 105, 106, 108, 109, 112, 115, 132, 136, 137, 138, 139, 142, 143, 150, 184, 264
Bumpus, Mabel Loverna Perham - 116
Bumpus, Madeline L. - 264
Bumpus, Margaret Lillian - 264
Bumpus, Mary E. Albee - 264
Bumpus, Morris - 114
Bumpus, Raleigh Martin - 116
Bumpus, Ruth Anne - 23, 264
Bumpus, Ruth Elizabeth - 264
Bumpus, Sybil H. (See Staula)
Bumpus, Thelma - 264
Burbank, Benjamin B. - 73

Mining scenes in West Paris, Oxford County, Maine

1925-1926

Photographs by Vivian Akers

Courtesy Sid Gordon